Lecture Notes in Physics

The Editorial Policy for Proceedings

The series Lecture Notes in Physics reports new developments in physical research and teaching – quickly, informally, and at a high level. The proceedings to be considered for publication in this series should be limited to only a few areas of research, and these should be closely related to each other. The contributions should be of a high standard and should avoid lengthy redraftings of papers already published or about to be published elsewhere. As a whole, the proceedings should aim for a balanced presentation of the theme of the conference including a description of the techniques used and enough motivation for a broad readership. It should not be assumed that the published proceedings must reflect the conference in its entirety. (A listing or abstracts of papers presented at the meeting but not included in the proceedings could be added as an appendix.)

When applying for publication in the series Lecture Notes in Physics the volume's editor(s) should submit sufficient material to enable the series editors and their referees to make a fairly accurate evaluation (e.g. a complete list of speakers and titles of papers to be presented and abstracts). If, based on this information, the proceedings are (tentatively) accepted, the volume's editor(s), whose name(s) will appear on the title pages, should select the papers suitable for publication and have them refereed (as for a journal) when appropriate. As a rule discussions will not be accepted. The series editors and Springer-Verlag will normally not interfere with the detailed editing except in fairly obvious cases or on technical matters.

Final acceptance is expressed by the series editor in charge, in consultation with Springer-Verlag only after receiving the complete manuscript. It might help to send a copy of the authors' manuscripts in advance to the editor in charge to discuss possible revisions with him. As a general rule, the series editor will confirm his tentative acceptance if the final manuscript corresponds to the original concept discussed, if the quality of the contribution meets the requirements of the series, and if the final size of the manuscript does not greatly exceed the number of pages originally agreed upon.

The manuscript should be forwarded to Springer-Verlag shortly after the meeting. In cases of extreme delay (more than six months after the conference) the series editors will check once more the timeliness of the papers. Therefore, the volume's editor(s) should establish strict deadlines, or collect the articles during the conference and have them revised on the spot. If a delay is unavoidable, one should encourage the authors to update their contributions if appropriate. The editors of proceedings are strongly advised to inform contributors about these points at an early stage.

The final manuscript should contain a table of contents and an informative introduction accessible also to readers not particularly familiar with the topic of the conference. The contributions should be in English. The volume's editor(s) should check the contributions for the correct use of language. At Springer-Verlag only the prefaces will be checked by a copy-editor for language and style. Grave linguistic or technical shortcomings may lead to the rejection of contributions by the series editors.

A conference report should not exceed a total of 500 pages. Keeping the size within this bound should be achieved by a stricter selection of articles and not by imposing an upper limit to the length of the individual papers.

Editors receive jointly 30 complimentary copies of their book. They are entitled to purchase further copies of their book at a reduced rate. As a rule no reprints of individual contributions can be supplied. No royalty is paid on Lecture Notes in Physics volumes. Commitment to publish is made by letter of interest rather than by signing a formal contract. Springer-Verlag secures the copyright for each volume.

The Production Process

The books are hardbound, and the publisher will select quality paper appropriate to the needs of the author(s). Publication time is about ten weeks. More than twenty years of experience guarantee authors the best possible service. To reach the goal of rapid publication at a low price the technique of photographic reproduction from a camera-ready manuscript was chosen. This process shifts the main responsibility for the technical quality considerably from the publisher to the authors. We therefore urge all authors and editors of proceedings to observe very carefully the essentials for the preparation of camera-ready manuscripts, which we will supply on request. This applies especially to the quality of figures and halftones submitted for publication. In addition, it might be useful to look at some of the volumes already published. As a special service, we offer free of charge LATEX and TEX macro packages to format the text according to Springer-Verlag's quality requirements. We strongly recommend that you make use of this offer, since the result will be a book of considerably improved technical quality. To avoid mistakes and time-consuming correspondence during the production period the conference editors should request special instructions from the publisher well before the beginning of the conference. Manuscripts not meeting the technical standard of the series will have to be returned for improvement.

For further information please contact Springer-Verlag, Physics Editorial Department V, Tiergarten-strasse 17, W-6900 Heidelberg, FRG

Aa. Sandqvist   T. P. Ray (Eds.)

Central Activity in Galaxies

From Observational Data to Astrophysical Diagnostics

Lectures Held at the
Predoctoral Astrophysics School III
Organized by the European Astrophysics Doctoral Network
(EADN) in Dublin, Ireland, 10-22 September 1990

Springer-Verlag
Berlin Heidelberg New York
London Paris Tokyo
Hong Kong Barcelona
Budapest

Editors

Aage Sandqvist
Stockholm Observatory
S-133 36 Saltsjöbaden, Sweden

Thomas P. Ray
Dublin Institute for Advanced Studies
School of Cosmic Physics
5 Merrion Square, Dublin 2, Ireland

ISBN 3-540-56371-7 Springer-Verlag Berlin Heidelberg New York
ISBN 0-387-56371-7 Springer-Verlag New York Berlin Heidelberg

Typesetting: Camera ready by author/editor using the TEX macropackage from
Springer-Verlag
Printing and binding: Druckhaus Beltz, Hemsbach/Bergstr.
58/3140-543210 - Printed on acid-free paper

Preface

The European Astrophysics Doctoral Network (EADN) is an affiliation with representatives from approximately twenty universities throughout western Europe. It started in 1988 and its aim is to stimulate the mobility of graduate students who are preparing their doctoral theses within Europe. By providing mobility grants, within the EC ERASMUS scheme, it promotes international collaboration and the exchange of ideas.

In addition, each year the EADN organizes a summer school aimed at graduate astrophysics students in either the first or second year of their doctoral studies. The first such school was held in Les Houches (France) in September 1988 and a subsequent school was held in Ponte de Lima (Portugal). Dublin (Ireland) was the venue for the third school which took place from 10 to 22 September 1990. As a rule, each school proposes two closely related themes, one being astrophysical and the other more methodological, i.e. the field of technology or numerical studies. For example, at Ponte de Lima, these two themes were "Late Stages of Stellar Evolution" and "Numerical Methods of Hydrodynamics for Astrophysics". At Dublin, they were "Central Activity in Galaxies" and "From Observational Data to Astrophysical Diagnostics".

The enigma of the nuclei of galaxies with their central "monster" driving the vast range of activity observed in quasars, radio galaxies, Seyferts, starburst galaxies and even our own Galaxy are explored in this volume. Rapid development in these fields is likely to continue into the future with the availability of new large-scale facilities to European astronomers. Some of these are already in full operation, such as the William Herschel Telescope, the ESO New Technology Telescope (NTT), the Swedish ESO Submillimetre Telescope (SEST) and the James Clerk Maxwell Telescope (JCMT). More ground-based and space-borne facilities will be available in the years to come such as the ESO Very Large Telescope (VLT) and the Infrared Space Observatory (ISO), to name but two. Extracting useful information from observational data requires sophisticated physical analysis coupled to powerful computing tools such as the MIDAS and AIPS packages. In this volume, we also examine such tools and consider the growing field of multivariate statistical analysis. The only field covered at the Dublin School and not presented in this volume is Modelling and Interpretation of Nebular Spectra.

Eight lecturers from various parts of Europe delivered approximately seventy-five hours of lectures over the ten working days of the School: five lecturers concentrated on central activity in galaxies and three on data analysis. The contents of the lectures, though advanced, are aimed at a broad audience. They should, in general, be understandable by students who are still beginners in their field, and the large lists of references to the appropriate literature could function as guides to deeper studies. Although the Dublin School took place in 1990, most authors submitted their written manuscripts in 1992 with the appropriate updating of results and reference lists.

An important aspect of the Dublin School was the student lecture programme. This enabled the students to present their own research programmes to their fellow students without being intimidated by the presence of the senior scientists. However, the interaction with the lecturers *was* considered to be very important by the students, not only during the lectures but also in the free time.

As with previous summer schools, the participants also enjoyed a lively social programme (not exclusively in the famous Dublin pubs). This began with a welcoming reception by the Dublin Institute for Advanced Studies and continued with an outing the following weekend to Newgrange, possibly the oldest astronomically oriented megalithic site in the world. A visit was also made to the ancient abbey at Monasterboice. The closing of the Dublin School was marked with a banquet held in the main dining hall of Trinity College.

The organizers would like to thank the Physics Department of Trinity College in Dublin for hosting the School and the Dublin Institute for Advanced Studies for its assistance with the organization. Thanks are also due to the School secretaries, M. Callanan and C. McAuley, and, of course, to the students for all their work!

Saltsjöbaden   Tom Ray
7 October 1992   Aage Sandqvist

Contents

PARTICIPANTS

Lecturers:
Peter Biermann, MPI für Radioastronomie, Bonn
Suzy Collin-Souffrin, Institut d'Astrophysique, Paris
Reinhard Genzel, MPI für Extraterrestrische Physik, Garching b. München
Teije de Jong, Sterrekundig Instituut 'Anton Pannekoek', Amsterdam
Steven Jörsäter, Stockholm Observatory, Saltsjöbaden
Fionn Murtagh, Space Telescope ECF, ESO, Garching b. München
Daniel Péquignot, Observatoire de Paris, Meudon
Judith Perry, Institute of Astronomy, Cambridge

Local Organizer:
Tom Ray, Dublin Institute for Advanced Studies, Dublin

Scientific Director:
Aage Sandqvist, Stockholm Observatory, Saltsjöbaden

Students:
Yiannis Andredakis, University of Crete
Klaus Anton, Landessternwarte, Heidelberg
Nicole van der Bliek, Sterewacht, Leiden
Francesca R. Boffi, Observatorio Astronomico, Bologna
Torsten Böhm, Observatoire de Paris, Meudon
Marco Bondi, Instituto di Radioastromia, Bologna
Victor H. da Rosa Bonifácio, Centro de Astrofisica, Porto
Jonathan Braine, Observatoire de Paris, Meudon
Xavier Calbet, Instituto de Astrofisica de Canarias
Xavier Luri-Carrascoso, Universitat de Barcelona
David Corcoran, Dublin Institute for Advanced Studies
Rosa Gonzalez Delgado, Instituto de Astrofisica de Canarias
Christopher A. Dickson, College of Cardiff, Wales
José A. de Diego, Instituto de Astrofisica de Canarias
Jonathan Ferreira, Observatoire de Grenoble
Paul A. Foulsham, University of Leicester
Luis P. de la Fuente, Instituto de Astrofisica de Canarias
Antonio Di Giacomo, Observatorio Astrofisica di Arcetri
Paul Goudfrooij, Sterrekundig Instituut 'Anton Pannekoek', Amsterdam
Reynold B. L. Greenlaw, Leeds University
Andreas Hajek, Institut für Astronomie, Vienna
Maja Hjelm, Stockholm Observatory
Ronald Hes, European Southern Observatory
Wolf-Dietrich Kunze, Hamburg Observatory
Ariane Lancom, École Normale Supérieure, Paris
Dirk Laurent, Sterrenkundig Observatorium, Belgium
Per Lindblad, Stockholm Observatory
Mats Löfdahl, Stockholm Observatory

Dean Longley, NRAL, Jodrell Bank
Terence Mahoney, Instituto de Astrofisica de Canarias
Karl Mannheim, MPI für Radioastronomie, Bonn
Filippo Mannucci, Observatorio Astrofisica di Arcetri
Andrea Mason, SISSA, Trieste
Magnus Näslund, Stockholm Observatory
Koryo Okumura, Observatoire de Paris, Meudon
Antonio Pedrosa, Centro de Astrofisica, Porto
Josep Marti Ribas, Universitat de Barcelona
Evlabia Rokaki, I.A.P., Paris
Jacques Sebag, Observatoire de Paris
Miquel Serra, Instituto de Astrofisica de Canarias
Nikolaos Solomos, University of Patras, Greece
George Sourlantzis, University of Crete
Paul Stein, Astronomisches Institut der Universität Basel
Jose Eduardo Telles, University of Cambridge
Bob van den Hoek, Sterrenkundig Instituut 'Anton Pannekoek', Amsterdam
Antonio M. Varela, Instituto de Astrofisica de Canarias
Baltasar Vila-Vilaró, Instituto de Astrofisica de Canarias
Rosendo Vílchez, Universitat de Barcelona
Ignaz Wanders, Astronomiska Observatoriet, Uppsala
Robin Williams, Institute of Astronomy, Cambridge

The Galactic Centre

Aage Sandqvist [1], Reinhard Genzel [2]

[1]Stockholm Observatory, S-133 36 Saltsjöbaden, Sweden
[2]Max-Planck-Institut für Extraterrestrische Physik, D-8046 Garching bei
München, Germany

Abstract: This lecture series discusses the impact of recent measurements in the infrared
and radio region on our knowledge about the Nucleus of our Galaxy (the central few hundred
parsecs). The emphasis is on the determination of energetics, physical conditions and dynamics
in the Nucleus through spectroscopic and broad-band observations of interstellar gas and dust.
An overview of the phenomena found in the Galactic Nucleus (1) is followed by a discussion of
the basic interpretation of line radiation (2). As examples, discussions follow of observations of
the central interstellar clouds at scales of several hundred parsecs, and of the giant cloud and
star formation region Sgr B2 (3), of the Galactic Centre itself with the Sgr A complex (4), and
of the central 10 parsec of the Galaxy including the Circumnuclear Disk, Sgr A West and Sgr
A*/IRS16 (5).

1 Overview of the Phenomena

Four general review articles on the topic of the Galactic Centre have been published
in *Annual Review of Astronomy and Astrophysics*: Burke (1965) on "Radio Radiation
from the Galactic Nuclear Region", Oort (1977) on "The Galactic Centre", Brown and
Liszt (1984) on "Sagittarius A and Its Environment", and Genzel and Townes (1987) on
"Physical Conditions, Dynamics, and Mass Distribution in the Centre of the Galaxy".
In addition there are three symposia volumes on the theme of the Galactic Centre which
contain numerous review articles and specific scientific contributions. These volumes are
edited by Riegler and Blandford (1982), Backer (1987) and Morris (1989).

The Galactic Nucleus, hidden behind about 30 magnitudes of visual extinction by
cool interstellar dust along the line of sight, cannot be studied at visible, ultraviolet or
soft X-ray wavelengths. The available information about this nearest centre of a galaxy
comes from measurements at γ-ray, hard X-ray, infrared, (sub-)millimetre and radio
wavelengths. The radio continuum emission from the inner $1°$ of the Galaxy, observed at
a wavelength of 2.8 cm by Seiradakis et al. (1989) and presented in Fig. 1, shows the two
major radio sources Sgr A and Sgr B2. The actual centre of the Galaxy is contained in
Sgr A near $\alpha = 17^h42^m29^s, \delta = -28°59'$, the source in the upper left of the figure is the
giant star formation region Sgr B2.

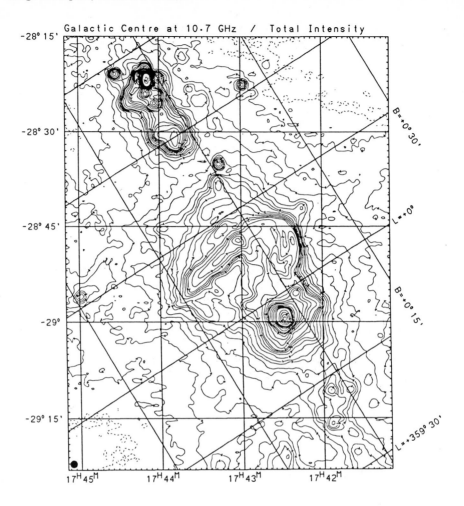

Fig. 1. Radio continuum emission map of the Galactic Centre region at a wavelength of 2.8 cm (Seiradakis et al. 1989)

Measurements of near-infrared (1 to 10 μm) continuum radiation have given detailed information about the distribution, velocity field and character (e.g. Allen, Hyland and Jones 1983; Lebofsky and Rieke 1987; Sellgren 1989) of cool stars (i.e. those with surface temperatures of a few 10^3 K) and stars with large dust shells. Observations at 1 - 2 μm have shown the presence of an extensive stellar cluster with a density distribution scaling approximately as $R^{-1.8}$ from about 1° (150 pc for an Earth-Galactic Centre distance of $R_0 = 8.5$ kpc) to a few arcseconds from the peak. The stellar light is peaked within a few arcsec of a complex of infrared sources IRS16 which is, therefore, sometimes interpreted to be the centre of the Galaxy. Lunar occultation observations at 2 μm by Adams et al. (1988), Simon et al. (1990) and Simons, Hodapp and Becklin (1990), and speckle imaging by Eckart et al. (1992) with the NTT, indicate that IRS16 is composed of a number of stellar diameter sources. These sources also appear to be intrinsically hot (T_{eff} >10 000 K). Hall, Kleinmann and Scoville (1982), Geballe et al. (1984) and Krabbe et al. (1991) have found high velocity ($\Delta v \geq 1000$ km s^{-1}) hydrogen and helium gas centered on IRS16 which they interpret as caused by mass outflow. The sources in IRS16 may thus be members of a compact cluster of hot massive stars (Allen et al. 1989, 1990), perhaps so-called luminous blue variables (Krabbe et al. 1991), that may emit a significant fraction of the Centre's 3 to 5 $\times 10^6$ L$_\odot$ of ultraviolet radiation (Becklin, Gatley and Werner 1982).

The brightness distribution even at 2 μm is heavily influenced by patchy foreground extinction which can in part be identified with known molecular cloud complexes associated with the Galactic Centre (Sandqvist 1974; Lebofsky 1979; Glass, Catchpole and Whitelock 1987). These molecular clouds have been extensively investigated in the last decade with millimetre and submillimetre techniques (e.g. Güsten and Downes 1980; Brown and Liszt 1984; Bally et al. 1987, 1988; Güsten 1989; Sandqvist 1989) and high resolution interferometry (Sandqvist et al. 1987; Okumura et al. 1989). They are massive ($> 10^5$ M$_\odot$), dense ($n(\text{H}_2) \approx 10^4$ cm^{-3}) and, in comparison with the Galactic disk Giant Molecular Clouds (GMC), unusually warm (40 to 100 K). Although some of them, such as clouds with LSR radial velocities near +20 km s^{-1} and +50 km s^{-1} lie at a projected distance of only a few arcminutes (5 to 10 pc) from IRS16, their relationship to the nuclear region has been uncertain. While initially suggested to surround and contain Sgr A (Sandqvist 1974), they were subsequently postulated by Güsten and Downes (1980) to be at distances of about 100 pc from the Centre for reasons of tidal stability. More recently, measurements seem to again favour distances closer to their projected distances (Mezger et al. 1989; Sandqvist 1989; Genzel et al. 1990; Zylka, Mezger and Wink 1990). The velocity field of a number of these clouds deviates strongly from that of general Galactic rotation, indicating that their motion must have a strong non-circular component (Liszt and Burton 1980; Bally et al. 1987, 1988).

The Galactic Centre is an intense source of far-infrared and radio continuum radiation, although its total luminosity ($\approx 10^7$ L$_\odot$) is actually not very spectacular when compared with the nuclei of other spiral galaxies (Rieke 1982, 1989). The far-infrared emission is usually interpreted to come from thermally radiating dust grains that are heated by short wavelength (UV and visible) radiation. That the Galactic Centre is a strong source of radio continuum radiation is a fact well known since the original discovery of extra-solar radio radiation by Jansky in the 1930's. The Centre's radio emission is a mixture of synchrotron and free-free radiation from HII regions. The strongest radio source, Sgr A, is identical with the peak of stellar radiation. Lunar occultations in 1968 showed that Sgr A consists of at least two components now known as Sgr A East and Sgr A West (Sandqvist 1974). Recent investigations of the radio continuum emission with interferometers (especially the Very Large Array (VLA), e.g. Ekers et al. 1983; Lo and Claussen 1983; Yusef-Zadeh, Morris and Chance 1984; Yusef-Zadeh 1989; Pedlar et al. 1989; Anantharamaiah et al. 1991) have provided beautiful arcsec resolution pictures of the radio nucleus with a plethora of fascinating structures (Fig. 2): thin arcs and threads of non-thermal radiation indicative of a large-scale magnetic field (in the region often referred to as the "Radio Arc"), compact HII regions marking the birth sites of recently formed OB stars (e.g. near the core of the +50 km s^{-1} cloud), a shell of non-thermal emission resembling a supernova remnant (Sgr A East), high-velocity streamers of hot thermally emitting plasma streamers associated with the central parsec (referred to as the Sgr A West "mini-spiral") and, finally, a remarkable compact radio source within an arcsecond or so of IRS16. This compact source, usually referred to as Sgr A*, has stellar dimensions ($\approx 10^{14}$ cm), resembles in some aspects the powerful central compact radio sources in active galactic nuclei, and may thus be our best candidate for a possible central engine or "monster" in our own Galaxy (Lo 1989).

PHENOMENA IN THE GALACTIC CENTER

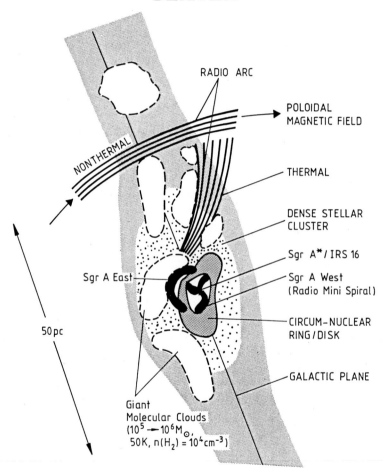

Fig. 2. Schematic diagram of the phenomena in the central 50 to 100 parsec of the Galaxy

The Galactic Centre region is also associated with a number of point-like X-ray sources as well as with diffuse X-ray emission (Watson et al. 1981). Perhaps in part because of the large intervening extinction, the Sgr A complex itself is only a weak X-ray source in the 1 to 20 keV band (Skinner 1989). The most remarkable discovery has been the detection of a strong and time variable source of 511 keV $e^+ - e^-$ annihilation line radiation within a few degrees of the Galactic Centre. The line radiation was strong in the 1970's but decreased or "turned off" rather abruptly at the beginning of the 1980's, to return to about the same strength in 1988 (Leventhal et al. 1990). A comparative review of the different results accumulated until 1988/89 can be found in Lingenfelter and Ramaty (1989). Recently, Sunyaev et al. (1991) found with the SIGMA telescope and the Russian GRANAT spacecraft that the hard X-/soft γ-ray continuum emission in the Galactic Centre comes from several variable compact sources within $\pm 5°$ of Sgr A*. The strongest source, identified with the Einstein source 1740.7-2942, at a distance of 48' from Sgr A*, exhibited in late 1990 a "hard state" with a prominent broad emission feature centered near 500 keV that could be due to $e^+ - e^-$ annihilation (Bouchet et al. 1991, Sunyaev et al. 1991). It is not yet clear how this result relates to the also variable, but *narrow* 511 keV line seen earlier by the HEAO-A (Riegler et al. 1981). One possible explanation is that the variable 511 keV line emission in the Galactic Centre originates from one (or several) compact source(s) near to, but not coincident with, Sgr A*. These sources may not even be *in* the Galactic Centre and could be stellar black holes, perhaps similar to Cyg X-1 (Ozernoy 1989, Sunyaev et al. 1991).

2 Interpretation of Infrared, Submm and Radio Radiation

The infrared to radio wavelength region contains a large number of spectral lines of abundant molecules and atoms in different ionization stages, as well as thermal and non-thermal continuum emission. Objects as diverse as planets, regions of active star formation and quasars emit the bulk of their radiation in this region, often as thermal emission of warm ($T \approx 10$ to 60 K) dust grains which absorb the intrinsic, ultraviolet and visible stellar radiation. The interpretation of thermal infrared dust emission - giving unique information about bolometric luminosity, dust temperature, total mass and physical properties of the grains - will not be discussed here further. We refer the reader to reviews by Fazio (1976) and Hildebrand (1983). Thermal (Brems-strahlung) radio continuum emission from a hot plasma and non-thermal, synchrotron radio emission are discussed in the book of Rybicki and Lightman (1979). Infrared, submillimetre and millimetre line emission is often produced by collisional excitation. As such, these lines are excellent temperature and density probes for neutral and ionized regions. The $\lambda \geq 10$ μm emission is also very much less affected by extinction than optical and near-infrared radiation. A brief discussion of atomic and molecular lines, especially of their role as probes of the physical conditions in interstellar and circumstellar matter, follows (for details, see Genzel 1992).

Consider a simple two level, atomic or molecular system with populations in the upper level n_u ([cm^{-3}], statistical weight g_u), and in the lower level n_l (g_l), with energy separation between levels $E_{ul} = h\nu$, where ν [Hz] is the transition frequency with an Einstein coefficient for spontaneous emission A say. Assume further that collisions with a collision partner at kinetic temperature T (e.g. with molecular or atomic hydrogen, or with electrons) lead from lower to upper level and vice versa. The rate of downward collisions per sec per molecule is $C_{ul} = \gamma_{ul}n$, where $\gamma_{ul}(T)$ [cm^3 s^{-1}] is the rate coefficient of that type of collision, n [cm^{-3}] is the volume density of collision partners. Detailed balance then gives $C_{lu} = C_{ul}(g_u/g_l)\exp(E_{ul}/kT)$. Including only collisions and spontaneous emission, the ratio of upper and lower level populations can then be calculated from detailed balance (upward rates = downward rates) to be

$$\frac{n_u}{n_l} = \frac{(g_u/g_l)\exp(-E_{ul}/kT)}{1 + n_{\text{crit}}/n} \tag{1}$$

$n_{\text{crit}} = A/\gamma_{ul}$ thus is an important parameter specifying what density is required to populate a given level close to thermal equilibrium (LTE: $n \to \infty$)

The Einstein A coefficient is proportional to $\nu^3\mu_{ul}^2$ (μ_{ul} is the transition matrix element). Hence, transitions at high frequencies and large transition moments require large critical densities. For typical values of $\mu \approx 1$, $D = 10^{-18}$ cgs (e.g. electric dipole transitions) and $\gamma \approx 5 \times 10^{-11}$ cm^3 s^{-1} (collisions with molecular or atomic hydrogen), the critical density is

$$n_{\text{crit}} \approx 5 \times 10^6 \mu_{1D}^2 \nu_{300\text{GHz}}^3 \quad [\text{cm}^{-3}] \tag{2}$$

For $n \approx n_{\text{crit}}$, the population in the upper level is half of that in thermal equilibrium. For infrared transitions ($E_{ul} \geq kT$), therefore, densities close to or exceeding the critical density are sufficient for "near equilibrium" population. In contrast, microwave transitions typically require $n \gg n_{\text{crit}}$ for "near equilibrium" population since $E_{ul} \ll kT$.

Since atomic fine structure lines are magnetic dipole transitions ($\mu_{\text{magdip}} \approx 1/200\mu_{\text{eldip}}$ and collisional excitation by electrons in ionized regions is $\approx 10^2$ times more effective than by neutral hydrogen, the critical densities are 10^4 to a few 10^7 times lower than given for the default parameters in (2). Far-infrared and submillimetre fine structure lines are convenient probes of a wide range of density and excitation (e.g. ionization stage) in neutral atomic and ionized gas.

For molecular transitions, however, the simple model of a two- or few-level system is usually not appropriate, especially for rotationally or vibrationally excited levels. Because of the existence of many downward (and upward) channels from an excited level to lower levels, the true critical density (in the sense of $n_u/n_l = 1/2(n_u/n_l)_{\text{LTE}}$ at $n = n_{\text{crit}}$) has to be calculated from a solution of rate equations involving many levels. It is typically an order of magnitude lower than that derived from (1). Figure 3 shows how the different infrared, submillimetre and microwave lines of a variety of abundant molecules probe a wide range of the astrophysically important parameter space in hydrogen density and temperature. For example, rotational transitions of CO in the 100 μm to 1 mm range probe warm to hot gas (100 to 1000 K) of moderately high density ($n_{\text{H}} \approx 10^4$ to 10^6 cm^{-3}) and rotational transitions of polar heavy top molecules (CS, HCN, HC$_3$N etc.) probe warm gas of very high density ($n_{\text{H}} \approx 10^7$ to 10^9 cm^{-3}).

For a realistic interpretation of line emission the effects of radiative transport (trapping of line photons in the cloud) and excitation by continuum background radiation have to be considered as well.

An important and sensitive method of deriving the physical conditions of a line emitting cloud is the analysis of a number of rotational or ro-vibrational lines with a model which includes collisional excitation (cross sections typically taken from theoretical calculations), radiative pumping and radiative transport. Another promising method is the comparison of lines in very different wavelength ranges. For example, comparison of

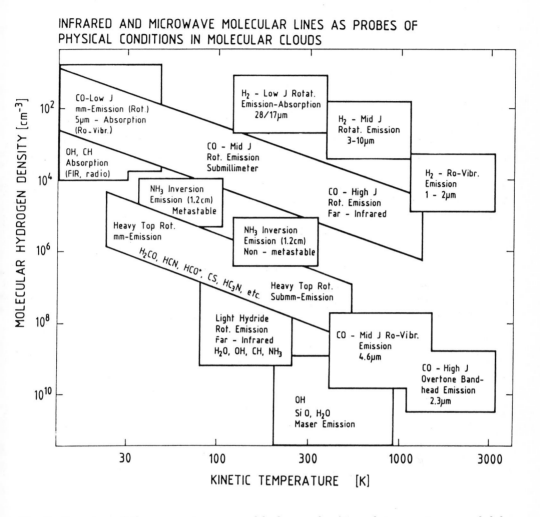

Fig. 3. Overview of the parameter space of hydrogen density and temperature sampled by different molecular lines in the infrared, submillimetre, millimetre and microwave range

far-infrared/submillimetre with optical/near-infrared lines can be used to accurately determine the extinction. In molecules which have radio and far-infrared transitions (e.g. OH, CH, NH$_3$), excitation temperature and molecular column density can be determined accurately in (low density) clouds in front of a bright continuum source where far-infrared and radio transitions occur in absorption (or in stimulated emission).

3 The Interstellar Medium in the Galactic Nucleus on Large ($\geq$ 100 pc) Scales

Surveys of the 115 GHz ^{12}CO $J = 1 - 0$ line in the 1970's (Bania 1977, Liszt and Burton 1978) established the existence of a large concentration of molecular material in the central 500 pc of the Galaxy. The density of massive ($M \approx 10^6$ M$_\odot$, diameter $\approx$ 50 pc) giant molecular clouds (GMC) is about 100 times larger in the Galactic Centre than in the disk. About 10% of the mass in the central few hundred parsec is in form of interstellar clouds. Well sampled surveys of the optically thin (or close to optically thin) ^{13}CO $J = 1 - 0$ line (see Fig. 4) and the high density trace lines of CS have been carried out with the Bell Laboratories 7 m antenna (Bally et al. 1987, 1988). Massive cloud complexes (diameter $\approx$ 50 pc) near Sgr A ($l \approx 0°$) and Sgr B2 ($l \approx 0.6°$) are apparent and are closely confined to the Galactic disk. While the radial velocities of most of these "disk clouds" are "allowed" in terms of Galactic rotation, their radial velocities as a function of distance from the Centre require that the clouds have significant non-circular motions (Bania 1977, Liszt and Burton 1978, Bally et al. 1988). The internal velocity widths ($\Delta v_{\text{FWHM}} \approx 20$ to 50 km s^{-1}) are 5 to 10 times larger than typical for clouds at distances $\geq$ 1 kpc from the centre, indicating large random gas motions within the clouds. The scale height of the disk clouds perpendicular to the Galactic plane is consistent with the magnitude of these random internal motions (Bally et al. 1988). In addition to the massive clouds concentrated near the Galactic disk, there are many clouds (such as the "molecular ring": Scoville and Solomon 1973, Bania 1977, Liszt and Burton 1978) with velocities forbidden in terms of Galactic rotation far above and below the Galactic plane. Bally et al. (1988) conclude that these clouds must be in highly elliptical orbits.

The physical conditions of the gas in the Galactic Centre clouds are also very different from those in the disk at a few kpc from the Centre. Tracers of high density, such as NH$_3$ inversion lines (Güsten, Walmsley and Pauls 1981) and CS rotational lines (Bally et al. 1987) indicate average hydrogen volume densities $\geq 10^4$ cm^{-3}. Gas temperatures derived from NH$_3$ and CH$_3$CN observations are $\geq$ 60 K, that is, significantly higher than dust temperatures (Güsten et al. 1985).

The Sgr B2 region is the most prominent and massive concentration of molecular gas and star formation activity in the vicinity of the Galactic Centre. The gas mass is about 3×10^6 M$_\odot$ over an approximately 40 pc diameter region (Bally et al. 1988). A ring of molecular gas surrounding the Sgr B2 region may be evidence for interstellar material accreting onto the complex (Bally et al. 1988). The Sgr B2 cloud has a complex chemistry and structure, with gas densities ranging from $n_{\text{H}_2} \geq 10^2$ cm^{-3} in the 40 pc "envelope" (sampled by the mm transitions of CO) to $\geq 10^7$ cm^{-3} in small (0.1 pc) condensations associated with newly-formed stars and H$_2$O masers near the cloud cores, Sgr B2(N) and Sgr B2(M). The central 5 pc diameter core of Sgr B2 is an active region of star formation with a total luminosity of $\geq 10^7$ L$_\odot$. Within this region, Benson and Johnston (1984) find about a dozen HII regions with exciting stars earlier than O6. Goldsmith et al. (1987) conclude that observed differences in spatial distributions of various chemical species in the Sgr B2 complex cannot be attributed to excitation or radiative transport effects but must be a consequence of spatial variations in chemical abundances.

Especially remarkable is a system of molecular clumps associated with the "arched" ionized filaments in the Radio Arc (Fig. 5a). Serabyn and Güsten (1987), Bally et al.

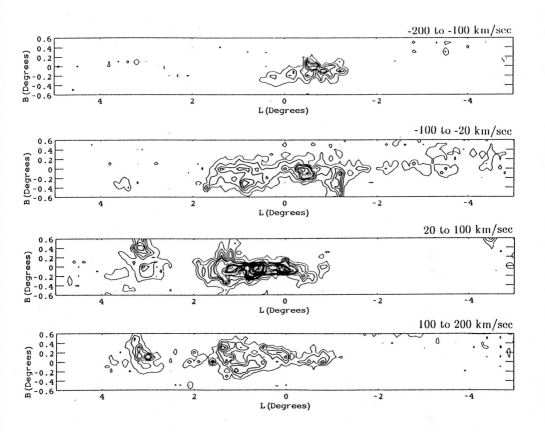

Fig. 4. ^{13}CO $J = 1 - 0$ maps in broad velocity intervals from the Bell Labs Galactic Centre survey (Bally et al. 1988). A 40 km s^{-1} interval centred on $V_{LSR} = 0$ km s^{-1} has been excluded to avoid contamination by Galactic disk emission

(1987) and Tsuboi (1988) have shown that the thermal filaments in the "bridge" and the "pistol" (see Yusef-Zadeh 1986) are associated with a series of dense molecular clouds along the ionized filaments, with a mass of two orders of magnitude larger than that in the ionized gas. From balloon-borne and airborne observations, Okuda et al. (1989) and Genzel et al. (1990) have shown that the Radio Arc is also associated with intense 158 μm [CII] fine structure line emission (Fig. 5b). The ionized emission follows the thermal filaments of the Arc and probably arises at the interfaces between ionized and molecular gas. The picture emerging from these observations is thus one of a filamentary system of dense and rather massive (a few 10^5 M$_\odot$ in total) molecular clouds which have ionized and partially ionized gas on their surfaces. As the origin of this large scale ionization

Genzel et al. (1990) propose OB stars to be present throughout this region, while Morris and Yusef-Zadeh (1990) favour collisional ionization by a "collision" of fast moving clouds with strong magnetic fields. A bright cluster of infrared sources near the Radio Arc, called the "quintuplet" (cf. Okuda et al. 1990) has recently been shown to contain infrared HeI/HI emission line stars that probably are luminous O stars (Moneti, Glass and Moorwood 1991; Krenz, Harris and Genzel 1992). If OB stars are indeed responsible for the ionization of the Arc, massive star formation must have been proceeding recently ($< 10^7$ y) throughout the central 100 pc.

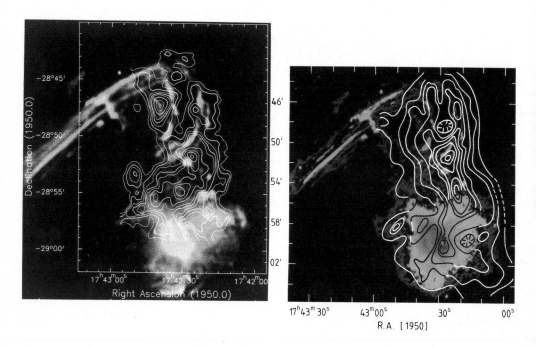

Fig. 5. Overlays of (a) CS emission in the velocity range -55 to 5 km s^{-1} from Serabyn and Güsten (1987) and (b) 158 μm [CII] line flux in the velocity range -60 to 0 km s^{-1} from Poglitsch et al. (1991) superposed on a 20 cm radiograph of Sgr A and the Arc from Yusef-Zadeh, Morris, and Chance (1984)

4 The Central 25 pc of the Galaxy (Sgr A Complex)

Evidence is accumulating that the non-thermal radio shell source Sgr A East, the thermal mini-spiral of Sgr A West and the nearby $+20$ km s^{-1} (M-0.13-0.08) and $+50$ km s^{-1} (M-0.02-0.07) molecular clouds are all in the central 10 to 20 parsec and are in part dynamically interacting with each other and possibly even the Circumnuclear Disk (CND) surrounding Sgr A West (Sandqvist 1974, 1989; Mezger et al. 1989; Pedlar et al. 1989; Genzel et al. 1990; Zylka, Mezger and Wink 1990; Anantharamaiah et al. 1991; Ho et al. 1991; Okumura et al. 1991). Figure 6 shows the molecular belt containing the $+20$ and $+50$ km s^{-1} clouds and illustrates the close projected proximity of the molecular regions to the radio continuum emission from Sgr A. The core of the $+50$ km s^{-1} cloud, as delineated by the 1.0 K contour, is elongated in a direction parallel to the Galactic plane, and is enveloped by the 20-cm continuum emission from the northwest around to the south. Running south from the core ridge is the string of thermal radio sources, A, B, C and D, found by Ekers et al. (1983). This gives the impression that the molecular core is sufficiently dense to arrest the advancing shock from Sgr A East and to keep out the Sgr A halo component. However, in regions just south of the core, the shock front from Sgr A East has sufficiently compressed the gas to produce the string of four HII regions. The fact that Sandqvist et al. (1987) did not see any OH absorption towards any of these H II regions implies that they lie on the sunward side of the molecular belt (see the $+51$ km s^{-1} frame in Fig. 7). The southern part of the Sgr A East shell seems to be plowing directly into the northern edge of the $+20$ km s^{-1} cloud (as delineated by the 0.6 K contour). The continuum and molecular components again fit well against each other. Just inside of the $+50$ km s^{-1} cloud/Sgr A East interface, Genzel et al. (1990) find blue-shifted high velocity C^{18}O gas ($v_{LSR} \approx -100$ to 0 km s^{-1}) that may be molecular material, originally part of the $+50$ km s^{-1} cloud, which has been accelerated by the expanding Sgr A East shell.

There is a gradual velocity gradient across the $+20$ km s^{-1} cloud to the $+50$ km s^{-1} cloud in the same sense as that of Galactic rotation, which can also be seen in Fig. 6. The overall positive sign of the radial velocities implies a Galactic non-circular velocity component as well. Moreover, it is apparent from Fig. 7 that the molecular gas at $+25$ and $+50$ km s^{-1} is seen clearly in absorption against the shell structure of Sgr A East but not against the mini-spiral structure of Sgr A West. (Notice that the "local" Galactic molecular gas, as exemplified by the -1 km s^{-1} frame in Fig. 7, *is* seen in absorption against at least the Northern arm of the Sgr A West mini-spiral.) This may imply that the molecular belt lies between the two continuum components, behind Sgr A West and in front of Sgr A East, and therefore that Sgr A West lies in front of Sgr A East. The mini-spiral of Sgr A West can also be seen in absorption against Sgr A East in the radio continuum at long wavelengths (Pedlar et al. 1989). Sgr A East, therefore, has to be placed behind Sgr A West (and Sgr A*) and most likely is directly associated with it.

Another argument for a location of Sgr A East and the $+50$ km s^{-1} cloud within the central 10 parsec of the Galaxy is the observation by Genzel et al. (1990) and Poglitsch et al. (1991) that the surface of the $+50$ km s^{-1} cloud that faces Sgr A* is very bright in [CII] line emission, as if this part of the cloud is directly photo-dissociated by UV radiation from the Centre itself. In contrast, [CII] emission from the slightly more distant $+20$ km s^{-1} cloud is much fainter. The relatively large velocity gradient in the gas bridging

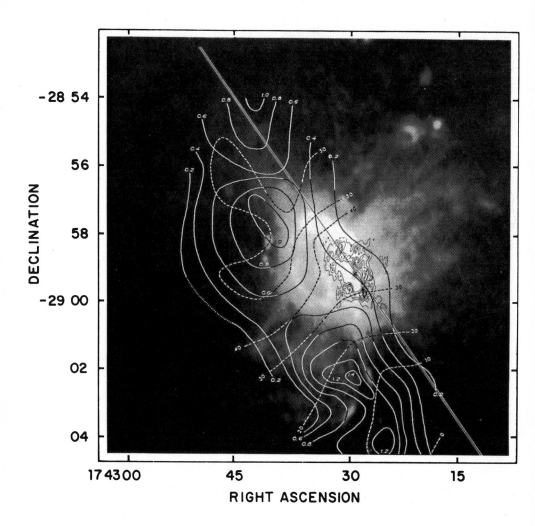

Fig. 6. The Galactic Centre molecular belt with the +20 and +50 km s^{-1} clouds. The 2-mm H$_2$CO T^*_A-distribution (solid lines) with isovelocity contours (broken lines) superimposed upon the 20-cm continuum radiograph of Yusef-Zadeh and Morris (1987) and the Circum-Nuclear Disk (thin lines) seen in the HCN observations of Güsten et al. (1987b). Sgr A* is marked by a cross. The straight diagonal line runs parallel to the Galactic plane. (From Sandqvist 1989)

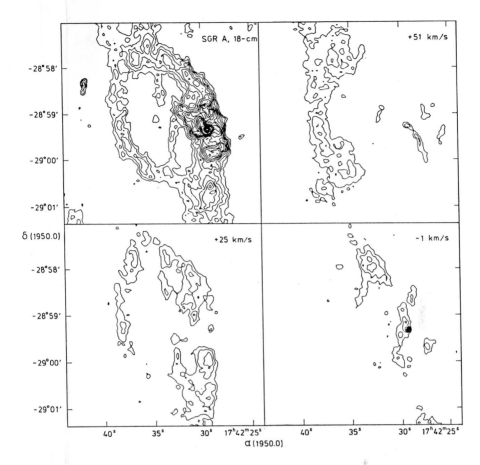

Fig. 7. Upper left frame: The 18-cm continuum emission from Sgr A, observed with the VLA at a resolution of 4 arcsec. Note the shell structure of Sgr A East, the mini-spiral of Sgr A West and the unresolved point source Sgr A*, as well as the HII regions A, B, C and D east of the Sgr A East shell. Upper right and the two lower frames: 1667-MHz OH spectral line absorption maps at radial velocities of +51, +25 and -1 km s^{-1}. The outermost contour level in the spectral line absorption maps is -20 mJy/beam and the contour interval is -20 mJy/beam. For the continuum map, these values are 5 and 10 mJy/beam, respectively. (From Sandqvist et al. 1987)

the centers of the +20 and +50 km s^{-1} clouds ($\approx$4 km s^{-1}/pc) is also consistent with a location of these two clouds in the central 20 pc of the Galaxy. The +20 km s^{-1} cloud in particular is clearly seen in absorption against the 2 μm radiation of the central stellar cluster. It should therefore be in front of the Centre, but why is it then not seen in absorption against Sgr A West in the OH maps in Fig. 7? Zylka, Mezger and Wink (1990) propose a detailed model in which the molecular belt containing the +20 and +50 km s^{-1} clouds is part of a gas streamer that is falling from a position $\approx$25 pc in front of the Centre into the inner 10 pc. A somewhat more complex model is presented by Sandqvist (1989) which tries to also take account of the confusing OH absorption results.

What is the nature of Sgr A East? On the basis of the density and amount of molecular gas in the +50 km s^{-1} cloud that is seen to be interacting with the expanding radio shell,

Mezger et al. (1989) and Genzel et al. (1990) conclude that 10^{52} ergs or more of kinetic energy are required in any explosion model of Sgr A East. This seems to support earlier proposals of a violent explosion at or near the Centre of our Galaxy within the last 10^5 years or less (e.g. Yusef-Zadeh and Morris 1987, Güsten et al. 1987a). However, an explanation of Sgr A East in terms of several consecutive supernova explosions or a combination of a supernova explosion with a high velocity wind is also possible (Mezger et al. 1989).

5 The Central 10 pc of the Galaxy (CND, Sgr A West, IRS16 and Sgr A*)

Figure 8 gives an overview of the region around the center of the stellar cluster. The spatial distribution of the thermal radio emission in Sgr A West has the morphology of a mini-spiral, consisting of several long filaments or arcs with embedded knots (Ekers et al. 1983, Lo and Claussen 1983). Velocity measurements from neutral and ionized gas indicate that these arcs (e.g. the "western arc" and "northern arm" in Fig. 8) are indeed physical streamers (Serabyn and Lacy 1985; Schwarz, Bregman and van Gorkom 1989). Near the centre of the system of streamers and the infrared source IRS16 lies a compact, non-thermal radio source (Sgr A*: Lo et al. 1985, Fig. 7). Its linear size is ≤ 20 AU and its radio luminosity is about 2×10^{34} erg s^{-1}, making it a unique object in the Galaxy (Lo et al. 1985). Sgr A* is probably the primary candidate for a possible massive black hole, in analogy to compact extragalactic radio sources (Lynden-Bell and Rees 1971).

The millimetre interferometric observations of Güsten et al. (1987a,b) have demonstrated the existence of an almost complete, clumpy ring-like structure of dense molecular material surrounding most of Sgr A West and a 1.5 pc radius inner "cavity" without much molecular gas (see Fig. 6). The ring is approximately centered on Sgr A*. Its dominant large scale motion is rotation, (≈ 100 km s^{-1}) although there are clearly clouds with large local random motions. The ring forms the inner edge of a thin disk-like structure (CND) extending to at least 8 pc. The CND is inclined by about 60° to 70° with respect to the line of sight, and its major axis is tilted between 5 and 20° relative to the Galactic plane. The western arc of ionized gas in Sgr A West almost certainly represents the photoionized inner edge of the CND. The northern and eastern arms may be streamers of ionized gas falling inward from the CND toward the Centre (Lo and Clausen 1983, Serabyn and Lacy 1985, Güsten et al. 1987a). Lacy, Achtermann and Serabyn (1991) propose an alternative interpretation in which the streamers are part of a one-armed spiral in a Keplerian disk. These streamers have been shown to also be associated with neutral atomic gas, as shown by 63 μm [OI] mapping (Jackson et al. 1992). An OH streamer, discovered by Sandqvist, Karlsson and Whiteoak (1989), shows that even molecular gas can exist *inside* the CND. This OH streamer reaches from the CND's southwestern region inwards through the "empty" cavity to the compact non-thermal radio source Sgr A* (see Fig. 9). The streamers all have different velocity fields and some may have different inclinations from the CND. They are probably temporary features with lifetimes less than a few rotation periods (a few 10^5 years) and may supply a current mass inflow rate into the central core of about 10^{-2} M$_\odot$/year.

The mid- and far-infrared continuum emission from warm dust grains is also centered on Sgr A West. The 5 to 30 μm emission follows closely the radio continuum distribution

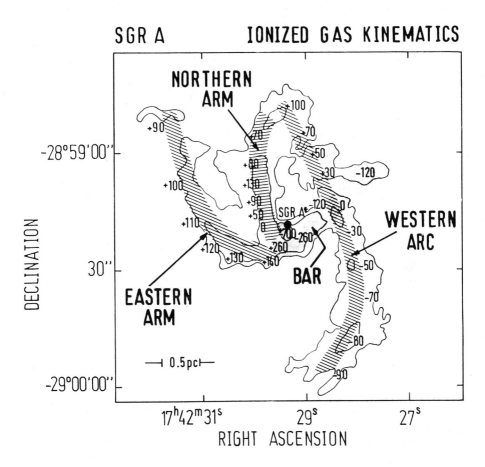

Fig. 8. Overview of the ionized velocities superposed on the Lo and Claussen (1983) 6 cm continuum map. The measurements in the western arc, bar, northern and eastern arms are from 12.8 μm [NeII] spectroscopy by Lacy et al. (1980) and Serabyn and Lacy (1985), and from H76α HI recombination line measurements by Schwarz, Bregman and van Gorkom (1989). The ±700 km s^{-1} broad line emission region is from 2 μm HeI/HI spectroscopy by Hall, Kleinmann and Scoville (1982) and Geballe et al. (1984,1987)

and probably comes from dust within the ionized streamers (Becklin et al. 1978, Gezari et al. 1985). The $\lambda \geq 50$ μm emission, however, forms a double lobed structure with two emission peaks ≈ 2 pc on either side of Sgr A*/IRS16 and is aligned approximately along the Galactic plane. Becklin, Gatley and Werner (1982) concluded from these observations that centered on the nucleus is a dust ring of ≈ 2 pc radius whose plane is close to the line of sight. This is the dust signature of the CND. The region inside of 2 pc is an ionized "cavity" of low mean gas and dust density.

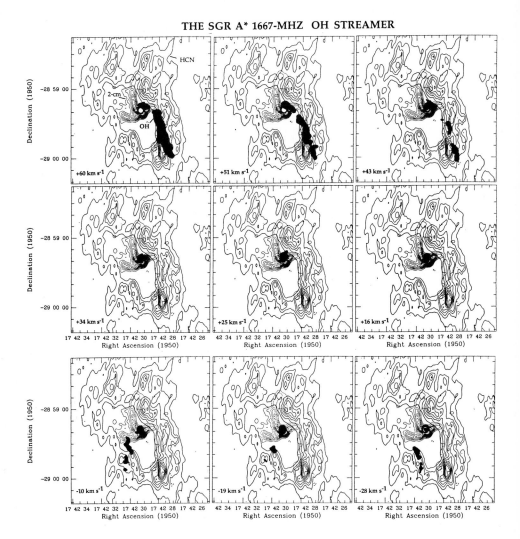

Fig. 9. Velocity maps of the Sgr A* 1667-MHz OH Streamer superimposed upon the HCN map of the CND by Güsten et al. (1987b) and the 2-cm continuum map of Sgr A West and Sgr A* by Ekers et al. (1983). (From Sandqvist, Karlsson and Whiteoak 1989)

Several important parameters can be derived from the overall characteristics of the infrared and microwave, continuum and line radiation of the Galactic Centre. First, the total ultraviolet and visible luminosity of the Galactic Centre, as derived from the integrated far-infrared radiation, is about $10^7 \, L_\odot$ (Becklin, Gatley and Werner 1982). The intensities of infrared fine-structure lines in high ionization stages (SIV, OIII, ArIII) are weak, indicating that the ultraviolet radiation field has a fairly low effective temperature ($T_{\text{eff}} \approx 35000$ K, Lacy et al. 1980, Watson et al. 1980, Serabyn and Lacy 1985). A low effective temperature is also consistent with the ratio of total luminosity to Lyman-continuum luminosity (derived from the radio flux), or to Lyman-α luminosity (derived from the mid-IR flux).

5.1 Origin of the Luminosity

There are two competing interpretations of these facts. In the "star formation" scenario the UV radiation could come from a (compact) cluster of OB stars in the vicinity of IRS16 (cf. Rieke 1982, Geballe et al. 1984, Adams et al. 1988, Simon et al. 1990, Simons, Hodapp and Becklin 1990, Eckart et al. 1992). In order to be consistent with the low effective temperature, late O and B stars must dominate the ionization, possibly because the early O stars, formed in a burst of star formation $\leq 10^7$ y ago, have by now left the main sequence (Rieke 1982). This first interpretation has recently found strong support in the discovery of a cluster of HeI emission line stars in the central 20" (Allen, Hyland and Hillier 1990; Krabbe et al. 1991). The HeI stars are probably massive (≈ 40 to $100\,M_\odot$), luminous (10^5 to $10^6\,L_\odot$) blue supergiants with large mass loss (a few 10^{-5} to $10^{-4}\,M_\odot$ y^{-1}). Their spectral characteristics are similar to the so-called "luminous blue variable" stars, like η Car and P Cyg (Krabbe et al. 1991; Allen, Hyland and Hillier 1990). If this interpretation is correct, the central HeI star cluster could contribute a significant fraction of the total and Lyman continuum luminosity of the central parsec. Because of the short lifetime of these massive stars, star formation then must have occurred in the Galactic Centre no longer than a few 10^6 years ago, in agreement with Rieke (1982).

Alternatively and/or additionally, the UV radiation could come from an accretion disk around a black hole coincident with or near Sgr A*. If that black hole would accrete at the Eddington rate, a moderate mass ($M_{hole} \approx 10^2$ to $10^3\,M_\odot$) may be sufficient to explain the ultraviolet luminosity (Ozernoy 1989). If a massive black hole ($M \approx 10^6\,M_\odot$) is present at the center, its present accretion must be much below the Eddington limit (cf. Rees 1987). Melia (1992) has recently presented a model for Sgr A* with a $\geq 10^6\,M_\odot$ black hole that accretes spherically from the gas in its vicinity. With the gas density determined from the combined mass loss rate of the IRS16 cluster, Melia (1992) finds a reasonable agreement with the observed spectrum of Sgr A*.

5.2 Circumnuclear Disk (CND)

Considering now the line emission in Sgr A, a remarkable finding is the high absolute intensity and excitation of the neutral atomic (OI, CII) and molecular (CO, OH, H_2) gas. About 1% of the total luminosity emerges in high excitation infrared and submm line emission which originate in the 2 pc radius circum-nuclear gas and dust ring. Harris et al. (1985) and Lugten et al. (1986) find that about $10^4\,M_\odot$ of warm ($T \approx 200$ K) and dense ($n_{H_2} \approx 10^4$ to 10^6 cm^{-3}) molecular gas in the ring is required to explain the excitation and intensity of the far-IR, submm and mm CO lines in Sgr A. The warm atomic gas, sampled by the OI and C$^+$ far-infrared fine structure lines, is coexistent with the warm CO and has similar physical parameters (Genzel et al. 1985, Jackson et al. 1992). Even hotter gas ($T \approx 2000$ K) at the inner edge of the ring has been found by Gatley et al. (1984) from 2 μm observations of vibrationally excited H_2, presumably indicating the presence of shock-excited molecular gas. Cloud-cloud shocks driven by the large random local motions deduced from the line widths and heating by the ultraviolet radiation via the photoelectric effect may be the dominant excitation mechanisms of the warm molecular and atomic gas (Genzel et al. 1985, Tielens and Hollenbach 1985).

5.3 Gas and Stellar Dynamics, Mass Distribution

One important question has always been to what extent the dynamics of the CND gas is influenced by magnetic forces. Evidence for a coherent, large scale magnetic field configuration within the CND has come from polarization measurements of the 100 μm dust emission (Werner et al. 1988, Hildebrand et al. 1990). Recent Zeeman-splitting measurements of HI and OH lines have given a level of 2 mG for the detection of a magnetic field in the southern part of the CND and a marginal detection in the northern part (Schwarz and Lasenby 1990; Killeen, Lo and Crutcher 1992). The energy density of the magnetic field in the CND is thus no more than about 10^{-7} erg cm^{-3}. For comparison, the energy density of the local turbulent motions in the disk is about 10^{-6} erg cm^{-3}, and that of the large scale rotational and non-circular motions is 10^{-5} erg cm^{-3} or more. Hence, the assumption that the gas dynamics is dominated by gravitation, and not by magnetic forces seems to be well justified. The gas motions may thus be a good indicator of the mass distribution in the Centre.

The velocities of the neutral and ionized gas as a function of distance from Sgr A*/IRS16, the distribution of 2 μm stellar radiation and most recently, the velocities of stars have all been used to determine the mass distribution in the Galactic Centre (see discussion in Genzel and Townes 1987). The velocities of interstellar gas clouds can probe the mass distribution if the gas rotates about the Centre (neutral CND and western arc: Güsten et al. 1987a, Serabyn and Lacy 1985), if the velocities can be interpreted in terms of other well defined orbits in the Centre's gravitational field (northern and eastern arm: Serabyn and Lacy 1985; Serabyn et al. 1988; Lacy, Achtermann and Serabyn 1991), or if the virial theorem can be applied in a statistical sense (Lacy et al. 1980, Crawford et al. 1985). The stellar surface brightness distribution can be converted to a mass distribution if a constant mass to luminosity ratio can be adopted. Finally, the mass distribution can be determined from statistics of stellar velocity through the virial theorem provided that the stellar distribution has a relatively simple symmetry (Sellgren et al. 1987; Winnberg et al. 1985; Lindqvist, Habing and Winnberg 1992).

Figure 10 is a composite of the mass distributions derived from different tracers (cf. Genzel and Townes 1987, McGinn et al. 1989). The different estimators appear to give very consistent determination of the mass distribution in the central few hundred pc. The enclosed mass decreases approximately linearly from a few hundred to a few pc, representing a stellar cluster with mass to luminosity ratio of $\approx$ 1 M$_\odot$/L$_\odot$. Inside of 2 pc, the enclosed mass decreases more slowly and may approach a constant value. In contrast, the stellar light at 2 μm is consistent with a core radius of between 0.1 and 1 pc (the two dashed curves in Fig. 10, cf. Allen 1987). The measurements are best consistent with a central point mass of 1 to 3×10^6 M$_\odot$. The central "dark" mass could be a cluster of stellar remnants (neutron stars or stellar black holes) or, more probably, a central massive black hole.

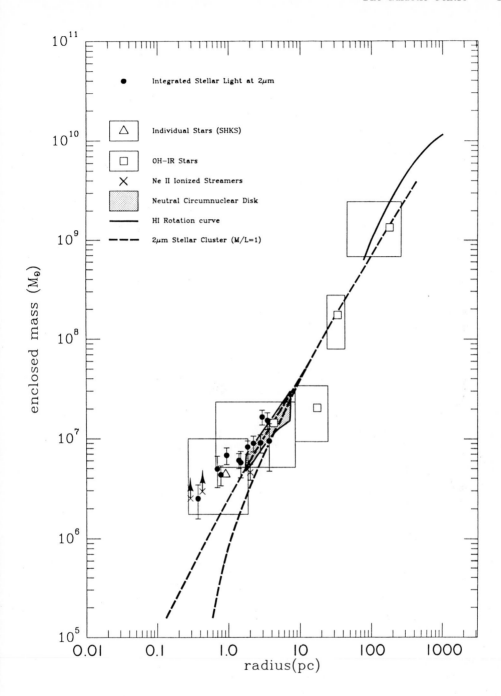

Fig. 10. The composite mass distribution in the Galactic Centre, summarizing the data from a number of available sources. (Adapted from McGinn et al. 1989)

5.4 Mass Inflow into the Central Region

A number of observations discussed in this lecture suggest that interstellar gas is streaming into the central region of the Galaxy, analogously to what is proposed in many external galaxies, but at a much smaller rate (10^{-4} to $10^{-2}\,M_\odot\,y^{-1}$). A schematic model of the proposed inflow scenario is shown in Fig. 11. Several of the massive Galactic Centre clouds at distances between 10 and 50 pc have large non-circular motions and may be feeding gas into the CND at radii between 1.5 and 2 pc. Clump-clump collisions may lead to further loss of angular momentum and thus to inflow of gas clouds in the neutral CND toward the very central region. Within about 1.5 pc from IRS16/Sgr A*, the ultraviolet radiation from a compact cluster of hot stars associated with IRS16 and possible other central source(s) rapidly ionizes first the surfaces and then the bulk of the infalling cloudlets.

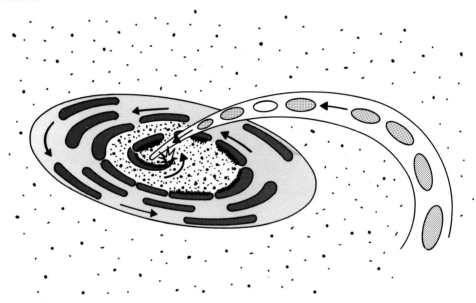

Fig. 11. Schematic model of the clumpy Circumnuclear Disk (CND), infalling ionized and neutral gas streamers, and the central stellar cluster. Most of the inner infalling clouds and the exposed inner edge of the CND are ionized by UV radiation from the nucleus

The first measurements of the magnetic field strength in the central few parsecs are fully consistent with the idea that the gas motions are mainly dominated by gravitational field and hence, that the increase of gas velocities toward IRS16/Sgr A* in fact signals the presence of a mass concentration that cannot be solely explained by the stars visible in the near-infrared. The central cavity and the Sgr A East shell may indicate that a violent explosion has occurred at the Centre within the last 10^5 years.

References

Adams D.J., Becklin E.E., Jameson R.F., Longmore A.J., Sandqvist Aa., Valentijn E. (1988): *Astrophys. J. (Lett.)*, **327**, L65

Allen D.A. (1987): in AIP Conf. Proc. 155, *The Galactic Center*, ed. Backer D.C., American Institute of Physics, New York p. 1

Allen D.A., Hyland A.R., Hillier D.J. (1990): *Mon. Not. R. astr. Soc.*, **244**, 706

Allen D.A., Hyland A.R., Hillier D.J., Bailey J.A. (1989): in Proc. IAU Symp. 136, *The Center of the Galaxy*, ed. Morris M., Kluwer, Dordrecht, p. 513

Allen D.A., Hyland A.R., Jones T.J. (1983): *Mon. Not. R. astr. Soc.*, **294**, 1145

Anantharamaiah K.R., Pedlar A., Ekers R.D., Goss W.M. (1991): *Mon. Not. R. astr. Soc.*, **249**, 262

Backer D.C. (1987): (ed.) AIP Conf. Proc. 155, *The Galactic Center*. American Institute of Physics, New York

Bally J., Stark A.A., Wilson R.W., Henkel C. (1987): *Astrophys. J. Suppl.*, **65**, 13

Bally J., Stark A.A., Wilson R.W., Henkel C. (1988): *Astrophys. J.*, **324**, 223

Bania T.M. (1977): *Astrophys. J.*, **216**, 381

Becklin E.E., Gatley I., Werner M.W. (1982): *Astrophys. J.*, **258**, 134

Becklin E.E., Matthews K., Neugebauer G., Willner S.P. (1978): *Astrophys. J.*, **220**, 831

Benson J.M., Johnston K.J. (1984): *Astrophys. J.*, **277**, 181

Bouchet L. et al. (1991): *Astrophys. J. (Lett.)*, **383**, L45

Brown R.L., Liszt H.S. (1984): *Ann. Rev. Astron. Astrophys.*, **22**, 223

Burke B.F. (1965): *Ann. Rev. Astron. Astrophys.*, **3**, 275

Crawford M.K., Genzel R., Harris A.I., Jaffe D.T., Lacy J.H., Lugten J.B., Serabyn E., Townes C.H. (1985): *Nature*, **315**, 467

Eckart A., Genzel R., Krabbe A., Hofmann R., van der Werf P.P., Drapatz S. (1992): *Nature*, **355**, 526

Ekers R.D., van Gorkom J.H., Schwarz U.J., Goss W.M. (1983): *Astron. Astrophys.*, **122**, 143

Fazio G.G. (1976): in *Frontiers of Astrophysics*, ed. Avrett E.H. Harvard, Cambridge, p. 203

Gatley I., Jones T.J., Hyland A.R., Beattie D.H., Lee T.J. (1984), *Mon. Not. R. astr. Soc.*, **210**, 565

Geballe T.R., Krisciunas K.L., Lee T.J., Gatley I., Wade R., Duncan W.D., Garden R., Becklin E.E. (1984): *Astrophys. J.*, **284**, 118

Geballe T.R., Wade R., Krisciunas K., Gatley I., Bird M.C. (1987): *Astrophys. J.* **320**, 562

Genzel R. (1992): in *The Galactic Interstellar Medium*, eds. Burton W.B., Elmegreen B.G., Genzel R., Springer, Berlin (in press)

Genzel R., Stacey G.J., Harris A.I., Townes C.H., Geis N., Graf U.U., Poglitsch A., Stutzki J. (1990): *Astrophys. J.*, **356**, 160

Genzel R., Townes C.H. (1987): *Ann. Rev. Astron. Astrophys.*, **25**, 377

Genzel R., Watson D.M., Crawford M.K., Townes C.H. (1985): *Astrophys. J.*, **297**, 766

Gezari D.Y., Tresch-Fienberg R., Fazio G.G., Hoffmann W.F., Gatley I., Lamb G., Shu P., McCreight C. (1985): *Astrophys. J.*, **299**, 1007

Glass I.S., Catchpole R.M., Whitelock P.A. (1987): *Mon. Not. R. astr. Soc.*, **227**, 373

Goldsmith P.F., Snell R.L., Hasegawa T., Ukita N. (1987): *Astrophys. J.*, **314**, 525

Güsten R. (1989): in Proc. IAU Symp. 136, *The Center of the Galaxy*, ed. Morris M., Kluwer, Dordrecht, p. 89

Güsten R., Downes D. (1980): *Astron. Astrophys.*, **87**, 6

Güsten R., Genzel R., Wright M.C.H., Jaffe D.T., Stutzki J., Harris A. (1987a): *Astrophys. J.*, **318**, 124

Güsten R., Genzel R., Wright M.C.H., Jaffe D.T., Stutzki J., Harris A. (1987b): in AIP Conf. Proc. 155, *The Galactic Center*, ed. Backer D.C., American Institute of Physics, New York, p. 103

Güsten R., Walmsley C.M., Pauls T. (1981): *Astron. Astrophys.*, **103**, 197

Güsten R., Walmsley C.M., Ungerechts H., Churchwell E. (1985): *Astron. Astrophys.*, **142**, 381

Hall D.N.B., Kleinmann S.G., Scoville N.Z. (1982): *Astrophys. J. (Lett.)*, **262**, L53

Harris A.I., Jaffe D.T., Silber M., Genzel R. (1985): *Astrophys. J. (Lett.)*, **294**, L93

Hildebrand R.H. (1983): *Quart. J. R. astr. Soc.*, **24**, 267

Hildebrand R.H., Gonatas D.P., Platt S.R., Wu X.D., Davidson J.A., Werner M.W., Novak G., Morris M. (1990): *Astrophys. J.*, **362**, 114

Ho P.T.P., Ho L.C., Szczepanski J.C., Jackson J.M., Armstrong, T., Barrett, A. (1991): *Nature*, **350**, 309

Jackson J.M., Geis N., Genzel R., Harris A.I., Madden S.C., Poglitsch A., Stacey G.C., Townes C.H. (1992): *Astrophys. J.*, (in press)

Killeen N.E.B., Lo K.Y., Crutcher R. (1992): *Astrophys. J.*, **385**, 585

Krabbe A., Genzel R., Drapatz S., Rotaciuc V. (1991): *Astrophys. J. (Lett.)*, **382**, L19

Krenz T., Harris A.I., Genzel R. (1992): (in preparation)

Lacy J.H., Achtermann J.M., Serabyn E.E. (1991): *Astrophys. J. (Lett.)*, **380**, L71

Lacy J.H., Townes C.H., Geballe T.R., Hollenbach D.J. (1980): *Astrophys. J.*, **241**, 132

Lebofsky M.J. (1979): *Astron. J.*, **84**, 324

Lebofsky M.J., Rieke G.H. (1987): in AIP Conf. Proc. 155, *The Galactic Center*, ed. Backer D.C., American Institute of Physics, New York, p. 79

Leventhal M., MacCallum C.J., Barthelmy S.D., Gehrels N., Teergarden B.J., Tueller J. (1989): *Nature*, **339**, 36

Lindqvist M., Habing H.J., Winnberg A. (1992): *Astron. Astrophys.*, **259**, 118

Lingenfelter, R.E., Ramaty R. (1989): *Astrophys. J.*, **343**, 686

Liszt H.S., Burton W.B. (1978): *Astrophys. J.*, **226**, 790

Liszt H.S., Burton W.B. (1980): *Astrophys. J.*, **236**, 779

Lo K.Y. (1989): in Proc. IAU Symp. 136, *The Center of the Galaxy*, ed. Morris M., Kluwer, Dordrecht, p. 527

Lo K.Y., Backer D.C., Ekers R.D., Kellermann K.I., Reid M.J., Moran J.M. (1985): *Nature*, **315**, 124

Lo K.Y., Claussen M. (1983): *Nature*, **306**, 647

Lugten J.B., Genzel R., Crawford M.K., Townes C.H. (1986): *Astrophys. J.*, **306**, 691

Lynden-Bell D., Rees M.J. (1971): *Mon. Not. R. astr. Soc.*, **152**, 461

McGinn M.T., Sellgren K., Becklin E.E., Hall D.N.B. (1989): *Astrophys. J.*, **338**, 824

Melia F. (1992): *Astrophys. J. (Lett.)*, **387**, L25

Mezger P.G., Zylka R., Salter C.J., Wink J.E., Chini R., Kreysa E., Tuffs R. (1989): *Astron. Astrophys.*, **209**, 337

Moneti A., Glass I.S., Moorwood A.F.M. (1991): *Mem. Astr. Soc. It.* , **62**, 755

Morris M. (1989): (ed.) Proc. IAU Symp. 136, *The Center of the Galaxy*, Kluwer, Dordrecht

Okuda H. et al. (1989): in Proc. IAU Symp. 136, *The Center of the Galaxy*, ed. Morris M., Kluwer, Dordrecht, p. 145

Okuda H. et al. (1990): *Astrophys. J.*, **351**, 89

Okumura S.K. et al. (1989): *Astrophys. J.*, **347**, 240

Okumura S.K. et al. (1991): *Astrophys. J.*, **378**, 127

Oort J.H. (1977): *Ann. Rev. Astron. Astrophys.*, **15**, 295

Ozernoy L.M. (1989): in Proc. IAU Symp. 136, *The Center of the Galaxy*, ed. Morris M., Kluwer, Dordrecht, p. 555

Pedlar A., Anantharamaiah K.R., Ekers R.D., Goss W.M., van Gorkom J.H., Schwarz U.J., Zhao J. (1989): *Astrophys. J.*, **342**, 769

Poglitsch A., Stacey G.J., Geis N., Haggarty M., Jackson J., Rumitz M., Genzel R., Townes C.H. (1991): *Astrophys. J. (Lett.)*, **374**, L33

Rees M. (1987): in AIP Conf. Proc. 155, *The Galactic Center*, ed. Backer D.C., American Institute of Physics, New York, p. 71

Riegler G.R., Blandford R.D. (1982): (eds.) AIP Conf. Proc. 83, *The Galactic Center*, American Institute of Physics, New York

Riegler G.R., Ling J.C., Mahoney W.A., Wheaton W.A., Willett J.B., Jacobson A.S., Prince T.A. (1981): *Astrophys. J. (Lett.)*, **248**, L13

Rieke G.H. (1982): in AIP Conf. Proc. 83, *The Galactic Center*, eds. Riegler G.R., Blandford R.D., American Institute of Physics, New York, p. 194

Rieke G.H. (1989): In Proc. IAU Symp. 136, *The Center of the Galaxy*, ed. Morris M., Kluwer, Dordrecht, p. 21

Rybicki G.B., Lightman A.P. (1979): *Radiative Processes in Astrophysics*, J. Wiley. New York

Sandqvist Aa. (1974): *Astron. Astrophys.*, **33**, 413

Sandqvist Aa. (1989): *Astron. Astrophys.*, **223**, 293

Sandqvist Aa., Karlsson R., Whiteoak J.B. (1989): in Proc. IAU Symp. 136, *The Center of the Galaxy*, ed. Morris M., Kluwer, Dordrecht, p. 421

Sandqvist Aa., Karlsson R., Whiteoak J.B., Gardner F.F. (1987): in AIP Conf. Proc. 155, *The Galactic Center*, ed. Backer D.C., American Institute of Physics, New York, p. 95

Schwarz U.J., Bregman J.D., van Gorkom J.H. (1989): *Astron. Astrophys.*, **215**, 33

Schwarz U.J., Lasenby J. (1990): in Proc. IAU Symp. 140, *Galactic and Intergalactic Magnetic Fields*, eds. Beck R., Kronberg P.P., Wielebinski R., Kluwer, Dordrecht, p. 383

Scoville N.Z., Solomon P.M. (1973): *Astrophys. J.*, **180**, 55

Seiradakis J.H., Reich W., Wielebinski R., Lasenby A.N., Yusef-Zadeh F. (1989): *Astron. Astrophys. Suppl. Ser.*, **81**, 291

Sellgren K. (1989): in Proc. IAU Symp. 136, *The Center of the Galaxy*, ed. Morris M., Kluwer, Dordrecht, p. 477

Sellgren K., Hall D.N.B., Kleinmann S.G., Scoville N.Z. (1987): *Astrophys. J.***317**, 881

Serabyn E., Güsten R. (1987): *Astron. Astrophys.*, **184**, 133

Serabyn E., Lacy J. (1985): *Astrophys. J.*, **293**, 445

Serabyn E., Lacy J.H., Townes C.H., Bharat R. (1988): *Astrophys. J.*, **326**, 171

Simon M., Chen W.P., Forrest W.J., Garnett J.D., Longmore A.J., Gauer T., Dixon R.I. (1990): *Astrophys. J.*, **360**, 95

Simons D.A., Hodapp K.-W., Becklin E.E. (1990): *Astrophys. J.*, **360**, 106

Skinner G.K. (1989): in Proc. IAU Symp. 136, *The Center of the Galaxy*, ed. Morris M., Kluwer, Dordrecht, p. 567

Sunyaev R. et al. (1991): *Astrophys. J. (Lett.)*, **383**, L49

Tielens A.G.G.M., Hollenbach D. (1985): *Astrophys. J.*, **291**, 772

Tsuboi M. (1988): Ph.D. Thesis, Univ. of Tokyo

Watson D.M., Storey J.W.V., Townes C.H., Haller E.E. (1980): *Astrophys. J. (Lett.)*, **241**, L43

Watson M.G., Willingale R., Grindlay J.E., Hertz P. (1981): *Astrophys. J.*, **250**, 142

Werner M.W., Davidson J.A., Morris M., Novak G., Platt S.R., Hildebrand R.H. (1988): *Astrophys. J.*, **333**, 729

Winnberg A., Baud B., Matthews H.E., Habing H.J., Olnon F.M., **1985**, *Astrophys. J. (Lett.)*, 291, L45

Yusef-Zadeh F. (1986): Ph.D. Thesis, Columbia Univ.

Yusef-Zadeh F. (1989): in Proc. IAU Symp. 136, *The Center of the Galaxy*, ed. Morris M., Kluwer, Dordrecht, p. 243

Yusef-Zadeh F., Morris, M. (1987): *Astrophys. J.*, **320**, 545

Yusef-Zadeh F., Morris M., Chance D. (1984): *Nature*, **310**, 557

Zylka R., Mezger P.G., Wink J.E. (1990): *Astron. Astrophys.*, **234**, 133

Activity in Galactic Nuclei

Judith J. Perry

Institute of Astronomy, Madingley Road, Cambridge CB3 0HA, U.K.

'To every Form of being is assigned'
Thus calmly spoke the venerable Sage,
'An *active* Principle'.

William Wordsworth

Abstract: These lectures review briefly the spectra and classification of active galactic nuclei (AGN), including quasi-stellar objects (QSOs), and the properties of their 'host' galaxies; starbursts (SB) and the possible association between galaxy interactions, SB, and AGN. I also review their cosmological distribution and evolution before discussing possible explanations for the observations. In particular, emphasis is laid on the rôle of stars, and the hydrodynamics of the Interstellar Medium (ISM) of active galaxies.

1 Introduction

Despite the attention of many sages in astrophysics, some venerable, the *active* principle underlying the multifaceted phenomena of Active Galactic Nuclei remains elusive. After a third of a century of observation most astrophysicists would agree that, although theoretical understanding of their formation, structure and evolution is incomplete (some would claim primitive), *operational* definitions exist by which observed objects can be classified. The very term 'active' (according to the Oxford English Dictionary "...5. abounding in action; energetic") is ill-defined: originally synonymous with 'non-stellar' or 'non-thermal' it has evolved to mean 'an output of energy unsustainable over the Hubble time'. Currently, objects once considered to be quite different are becoming seen as members of an extended family. Because the purpose of these lectures is to review and discuss possible foundations for their understanding, I shall begin with a review of the classification scheme used currently to encompass that broad class of objects called "active". Firstly, however, a note of caution. The Hertzsprung-Russell diagram underlies all modern astrophysics: because most stars occupy only a small band in luminosity-surface temperature space, it has been possible to classify stars, and to determine distances both within and external to our Galaxy. There is no AGN equivalent of the HR diagram – were such an *empirical* classification scheme found, which, like the HR diagram, had a physical interpretation, more progress could be made in understanding the phenomenon of *Activity*.

Observational Classification.

In all cases (except blazars) an object must possess *prominent emission lines* in its optical spectrum for inclusion in the broad class of *Active* Galaxies.

Quasi-stellar Objects (QSO) are stellar-appearing (unresolved on survey plates) objects, optically more luminous than first ranked cluster elliptical galaxies ($M \sim -22.5$ to -23.5, for $H_o = 50\,\mathrm{km\,s^{-1}\,Mpc^{-1}}$). Their spectra have broad permitted lines, narrow high-excitation forbidden lines, and a strong *very broad* infra-red, optical, ultraviolet to X-ray (IR-Opt-UVX) continuum. They are often variable, and usually have low polarization. Within this broad class the distinction is sometimes made between **Quasars** and QSOs, with the former satisfying all the criteria for QSOs but in addition producing a radio luminosity comparable to the optical.

Seyfert Galaxies are galaxies with a prominent nucleus whose optical luminosity is roughly equal to or less than a first-ranked elliptical. A **Seyfert 1** has a spectrum characterised by broad permitted lines, narrow high-excitation forbidden lines, a strong, *possibly* non-stellar IR-Opt-UVX continuum and weak radio emission. The continuum and lines are highly variable and although usually of low polarization, moderate polarization is sometimes observed. Seyfert 1s are spectrally like low-luminosity QSOs. **Seyfert 2s** do not have broad emission lines; they have narrow high-excitation forbidden and permitted lines only, and their IR-Opt-UVX continuum is weak. Intermediate classes exist.

Radio Galaxies (RG) have radio luminosities equal to or greater than their optical luminosity. **Broad Line RGs** have prominent broad permitted lines in their nuclear optical spectra, whereas **Narrow Line RGs** show only narrow high-excitation permitted lines. RGs are also divided into Powerful and Weak (**PRG** and **WRG**), accordingly as they are more or less luminous (for $H_o = 50\,\mathrm{km\,s^{-1}\,Mpc^{-1}}$) than $\sim 10^{25}\mathrm{W\,Hz^{-1}}$ at 1.4 GHz; and into steep or flat spectrum sources accordingly as their spectral index, α, at ~ 1 GHz (for a flux, $F(\nu) \propto \nu^{-\alpha}$) is greater or less than ~ 0.4.

N Galaxies is a classification based on appearance which is falling into disuse as the sensitivity and resolution of observations increase: it describes a faint nebulous image dominated by a stellar-appearing nucleus.

Blazars have a strong, highly polarized, rapidly variable non-stellar continuum, are strong radio sources, but emission lines are weak or absent. They are divided into **BL Lac** objects (named after their prototype) and **OVVs** (Optically Violently Variable quasars). OVVs show strong emission lines when the continuum is weak; many, but not all, BL Lacs also show weak emission lines in their low state. OVVs are significantly brighter than BL Lacs and are often the most luminous objects found at their redshift.

A **Starburst** denotes star formation on a time scale significantly shorter than the Hubble time, at a rate which could not be sustained by the system for the Hubble time (a phenomenological rather than observational definition.) There are *localized* regions within 'normal' galaxies, such as Orion in the Milky Way, which are clearly undergoing bursts of star formation. However, in these lectures I follow the usual extra-galactic convention to mean a burst of star formation involving a substantial fraction of the mass of the galaxy, usually located in its bulge or nucleus. Starbursts are usually strong IR sources.

LINERs or Low Ionization Nuclear Emission-Line Regions in galaxies are characterized by narrow low-excitation permitted and forbidden lines, sometimes accompanied by weak broad permitted lines.

Necessarily, these lectures must be limited in their scope. Therefore, my emphasis will be on QSOs (used henceforth to include quasars unless otherwise stated), Seyferts and Starbursts and the links between them. I shall concentrate on the *galactic* aspects, rather than the details of the physics of black holes. One of the central problems in AGN research is to explain the generation of the extra-ordinary output of energy from the nuclear regions of galaxies. There are very compelling arguments in favour of black holes in the centres of galaxies; however, no observational proof of their existence exists to date. Several excellent reviews of the detailed physics near black holes exist; particularly useful are Blandford (1990) and Rees (1984). I cannot claim to be complete in my references, but hope that the references cited will provide a useful lead-in to the literature. I hope none of my colleagues feels slighted if her or his work is not cited.

After a brief overview of the main spectral characteristics of *activity* in galaxies, I turn to the exciting new observations on the host galaxies: their morphology, environment, and the evidence pointing to the importance of interaction as a possible cause or trigger of activity. I then discuss the spectral observations and their implications for the structure of active galaxies, working my way in from the largest scales toward the nucleus. Unless otherwise stated, $H_0 = 50 \, \mathrm{km \, s^{-1} \, Mpc^{-1}}$, $q_0 = 0.5$ and $\Lambda = 0$.

2 Spectral Characteristics

The gross characteristics of the rest-frame continua of AGN, shared by almost all members of the class, is that the energy radiated, (unlike stellar black-body spectra) spans more than 10 decades in frequency and is remarkably uniform from the IR to the X-ray (illustrated in Figs. 1 and 2). The spectral energy distribution, $F(\nu)$, is represented usually by a 'power-law':

$$F(\nu) \propto \nu^{-\alpha}. \tag{1}$$

This has proved to be a convenient observational representation because, as is obvious from Figs. 1 and 2, a broadly 'flat' AGN spectrum approximates very well to a 'power-law' over the limited spectral range of most observations. It has been suggested, however, that logarithmic Gaussians are a better fit to the multi-wavelength data (Landau et al. 1986; Perry, Ward and Jones 1987). It is often more convenient to use $\nu F(\nu)$ rather than $F(\nu)$ to represent the spectra because $\nu F(\nu)$ is a measure of the energy emitted per unit $\log\nu$. [The luminosity over a band of frequency $\equiv \int F(\nu)d\nu = 2.3 \int \nu F(\nu)d(\log\nu)$.] Because of the usual problems of scale, however, it is normal to plot $\log(\nu F_\nu)$ vs. $\log(\nu)$, as in Fig. 1. In these plots it is easy to see at which frequencies the energy output peaks. 3C 273 is both near ($z = 0.158$) and intrinsically very luminous $L_{\mathrm{Bol}} \gtrsim 10^{47} \, \mathrm{erg \, s^{-1}}$; this has resulted in it having the most complete observational spectral coverage of any QSO. It exhibits the important spectral features of AGN although it is not 'typical' of most quasars as it is both a strong radio and gamma ray source.

There are substantial variations in many of the rest-frame UV spectral characteristics, which are outlined below.

The prominent emission lines which define active galaxies are a rich source of detailed physical information about the galactic regions emitting the lines. It is the *redshift* of these lines, which, using standard cosmological models, defines the distance to the sources. This, in turn, permits the conversion of apparent to absolute magnitude. The redshifts of active galaxies range from almost zero to well over 4. An ongoing search for ever higher redshift QSOs is being pursued with enthusiasm in several observatories around the world. At the time of writing the largest known redshift is 4.9 (Schneider, Schmidt and Gunn 1991a); over 20 QSOs are now known with $z > 4$ (Irwin, McMahon and Hazard, 1991; Schneider, Schmidt and Gunn 1991b). The importance of discovering high redshift objects is that under the conventional cosmological interpretation $z > 4$ represents a look-back time of more than 75% of the age of the Universe. Therefore it is believed that observations of QSOs over the full range of redshift will enable us to study the evolution of structure in the Universe. The distribution of sources in redshift space is not uniform, and will be discussed later in Sect. 5. There remain a minority of astronomers who question the cosmological interpretation of the redshift on the basis of data suggesting an apparent association between QSOs and nearby bright galaxies of significantly lower redshift (Burbidge et al. 1990 and references therein; see also Burbidge 1992). I will not discuss their doubts here, leaving the interested reader to read their arguments as they present them. Like most of my colleagues I believe that the evidence favouring the cosmological interpretation is substantial and I shall use that interpretation here.

The apparent magnitudes of QSOs and AGNs, when corrected for distance and assuming isotropic emission, imply that the integrated luminosities are $\gtrsim 10^{42} - 10^{48} \, \mathrm{erg \, s^{-1}}$. These luminosities (equivalent to $\gtrsim 10^8 - 10^{14} \, \mathrm{L_\odot}$) are inferred to be emitted from nuclear regions whose characteristic linear size is typically smaller than 10^{-3} galactic radii. There are short time scale variations in the continuum emission; for low luminosity objects the X-ray fluxes in some sources vary substantially within hours, and sometimes even minutes. These rapid variations are conventionally used to set upper limits on the size of the continuum emitting regions from light travel time arguments: a coherent source cannot be larger than $c\tau_{\mathrm{var}}$, where τ_{var} is the time scale of variation. *If* the variability is due to a single central object, then that source must, in some objects at least, be smaller than a light-hour, or $\lesssim 10^{14}$ cm. The variability appears to be correlated to luminosity, with the lowest luminosity objects varying the most rapidly; the highest luminosity objects have time scales of the order of 10^6 to 10^7 s. Thus the largest inferred sizes are $\lesssim 10^{17}$ cm. The relationship of variability to luminosity has yet to be quantified. It is this small volume, combined with the prodigious energy output, which has led to the assumption that the central source must be an accreting supermassive black hole (BH) (Zel'dovich and Novikov 1964; Lynden-Bell 1969; Rees 1984) since stars cannot generate a sufficient energy density. However, a competing proposal is that the variability arises in compact supernova remnant 'flares' in a starburst (Terlevich et al. 1992; Tenorio-Tagle et al. 1992; Filippenko 1992). Variability studies provide essential constraints on models; coordinated monitoring in many different wavebands is required to map the various continuum and line emitting regions. Several international campaigns are now in progress, designed to provide the continuous broad wavelength coverage required (Peterson 1988; Peterson et al. 1991; O'Brien and Harries 1991).

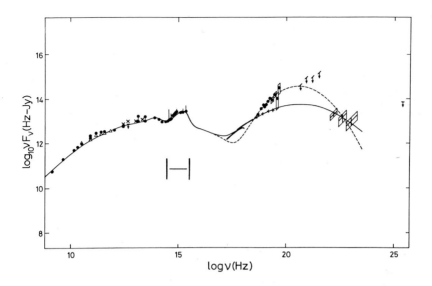

Fig. 1. The overall spectrum of 3C 273, a flat spectrum radio quasar (Perry, Ward and Jones, 1987). The solid line is the 'low-state' fit and the dashed line the 'high-state' fit to the data, as described by Perry et al. The range of the optical observations discussed in 2.1.1 is indicated

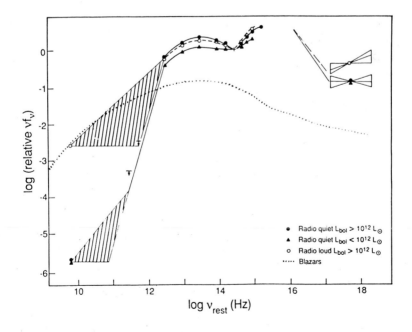

Fig. 2. "Average" spectra for radio loud and radio quiet PG QSOs. The hatched region between 10^{10} and 10^{12} Hz represents the range of spectral indices observed. (Sanders et al. 1989)

2.1 The Optical/UV Continua and Emission Lines

Although the optical, not surprisingly, is the best-studied wavelength region, it is only recently that systematic comparisons of large samples of optical AGN spectra have become available. These allow statistical comparison of the properties of different objects. For example, in a comprehensive study of 718 high luminosity objects (they define $M_{B_J}(\text{QSO}) \leq -21.5$; $-21.5 \leq M_{B_J}(\text{AGN}) \leq -20.5$) from the Large Bright Quasar Survey (LBQS: Foltz et al. 1987, 1989; Hewitt et al. 1991), Francis et al. (1991) constructed "composite" optical spectra which allowed them to identify numerous weak emission features and to define the underlying continuum. Several of their figures are reproduced in Fig. 3, and the spectral range of their study is shown on Fig. 1.

Some of the difficulties in AGN research are clearly brought out by their work. In order to construct models of the emission line regions and derive densities, temperatures and excitation conditions – discussed in detail in Prof. Collin-Souffrin's lectures (this volume) – emission lines from many ions must be simultaneously measured. Yet in the spectral range available for any one object too few lines are observed; therefore throughout the history of modeling AGN it has been necessary to construct "composite" spectra assuming that there is a continuity between objects of different z and absolute luminosity, so their spectra can be pieced together. This is illustrated in Figs. 3b and 3c, where the number and mean luminosity of the objects contributing to the composite in different frequency ranges is shown. It is clear that if there are systematic differences between high and low luminosity objects, they will be difficult to determine based on analysis of composite spectra. The development of sensitive IR arrays and the deployment of the Space Telescope may allow many more lines in bright QSOs to be studied without being forced to use composite spectra.

Francis et al. find that there is a significant scatter about the mean variations - of an order of magnitude - in the strength of the primary emission line fluxes relative to the underlying continuum. Furthermore, they find that "the continuum is not well represented by a power-law on any but the smallest of scales" and that on those scales the apparent continuum slopes have a range of $\alpha = -0.5$ to 1.5, with commonly $\alpha = -0.3$ to 1.0. A subset of quasars possesses Broad Absorption Lines (BAL) and, in extreme cases, BAL QSOs possess spectra with virtually no emission features evident and with extensive absorption troughs as prominent as emission features in many other quasars.

The spectroscopic properties of the first 10 quasars discovered with $z > 4$ were examined by Schneider, Schmidt and Gunn (1989). They found that these quasars have spectral properties quite similar to typical low redshift quasars. This result argues against *strong* spectral evolution of the population and will need to be quantified.

A prominent feature in the optical/UV which extends probably into the soft X-ray region (through the observational 'gap') is the 'big blue bump' usually attributed to the thermal emission from an accretion disc (Shields 1978; Malkan and Sargent 1982; Czerny and Elvis 1987; Rokaki, Boisson and Collin-Souffrin 1992). Detailed models of the emission from accretion discs around black holes will be discussed in Collin-Souffrin's lectures. Not all QSOs have 'big blue bumps' however (McDowell et al. 1989), and there are problems with the stability of the 'classic thin' accretion discs used to model the spectra. Amongst the alternative suggestions to explain the big blue bump in the context of BH models are a population of very small, very dense blobs in a hot accretion flow (Guilbert and Rees 1988; Lightman and White 1988). For a critical review of these possibilities see

Collin-Souffrin (1991). Alternatively, Terlevich and his collaborators propose that this feature is due to young massive stars plus compact supernova remnants (CSNR).

Despite all these caveats, a number of very important general features have been established from analysis of the observations, and these will be discussed in detail in Collin-Souffrin's lectures. The main points that are relevant to the physical models I will be discussing are the following: The line spectra are well explained assuming photo-ionization by the broad-band "underlying" continuum. There are two main physically distinct components to the emission line spectra: the low ionization lines (LIL) and high ionization lines (HIL), consisting respectively of primarily the Balmer series and lines from singly ionized species (e.g. CII, SII, FeII) and of $Ly\alpha$ and multiply ionized species e.g. CIV, NV, OVI (Wills, Netzer and Wills, 1985; Collin-Souffrin et al. 1986). The HIL is systematically blueshifted with respect to the LIL (Gaskell 1982; Espey et al. 1989; Tytler and Fan 1992) which is generally interpreted as meaning that the HIL are formed in gas having a systematic component of bulk radial motion (probably outflow, possibly inflow), and that there is some material obscuring redshifted emission. The full width at half maximum (FWHM) of the emission lines appears to be correlated with luminosity (Baldwin, Wampler and Gaskell 1989), which implies that the velocity of the emission line gas is larger in more energetic objects. The equivalent width of the CIV emission lines is anti-correlated to the continuum strength - the Baldwin Effect (Baldwin 1977; Wampler et al. 1984). The lines vary, apparently in response to variations in the ionizing continuum, on time scales ranging from several days in low luminosity objects to more than several years in high luminosity QSOs.

The 'classic' narrow emission lines (NLR) observed in most AGN spectra have widths less than 1000 $km\,s^{-1}$, although their wings may be extensive, ranging up to 2000 $km\,s^{-1}$ (van Groningen and de Bruyn 1989). From the intensity ratios of the forbidden lines, electron densities of $n_e \approx 10^3$ to $10^4\,cm^{-3}$ are deduced. If photoionized by the central continuum, they must lie at distances of the order of 10^3 times as far from the nucleus as the broad line emitting gas. They thus 'map' the outer regions of the galactic centres. In addition to the classic NLR two kpc-scale regions of narrow line emission have been discovered in high-resolution imaging of Seyfert galaxies. These various narrow line regions are discussed in Sect. 6.

2.2 Radio Structure and Properties

Only $\sim$ 10% of active galaxies are radio-loud. One reason that this small percentage of active galaxies receives so much attention is that the spatial resolution of radio interferometers considerably surpasses that of the much shorter wavelength optical telescopes. Therefore, there is a wealth of fine detail available about the morphology of radio emission which is unavailable at other wavelengths. It is not yet clear if the difference in the radio/optical properties of radio loud (RL) and radio quiet (RQ) objects is purely a reflection of the host galaxy structure (see below), or if the processes which create the large scale structure of the host galaxy simultaneously influence activity in the nucleus.

Radio activity manifests itself on two dramatically different physical scales: *compact cores* assumed to be coincident with the active nucleus, and *extended structures*, usually "jet-like" or double-lobed, with linear sizes ranging from roughly 3 kpc to 5 Mpc. Compact sources tend to be variable and have flat spectra (high frequency compact sources have $\alpha \sim 0$). New components appear often. Their optical activity is loosely correlated

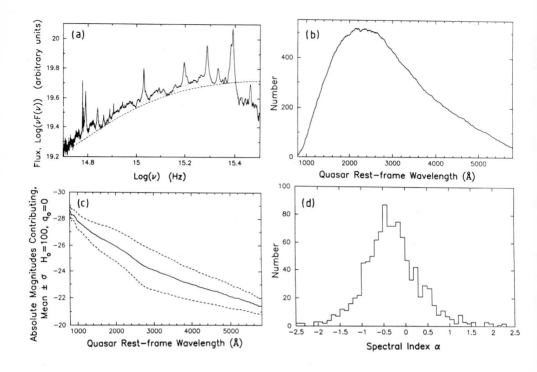

Fig. 3. The composite of 718 optical spectra (Francis et al. 1991). (a) The composite spectrum, $\log(\nu F_\nu)$ vs. rest-frame ν. (b) The number of QSOs and AGN contributing to the composite as a function of rest-frame ν. (c) The mean absolute magnitude (*solid line*) and $\pm 1\sigma$ (*dashed lines*) of the objects contributing to the composite as a function of rest-frame ν. (d) Histogram of continuum spectral indices, α, where $F(\nu) \propto \nu^\alpha$, for all objects contributing to the composite spectrum

to their radio power. Compact sources resolved by Very Long Baseline Interferometry (VLBI) usually show jet-like features with unresolved bright spots at one end: "core-jet" structure.

The morphology of extended sources is also loosely correlated to radio power: sources with $L_R \gtrsim 10^{42}\,\mathrm{erg\,s^{-1}}$ typically are between 100–300 kpc in size and are "edge-brightened", whereas weaker sources are generally more complex, "edge-darkened", and can be larger than ~ 5 Mpc. Extended sources are usually optically thin and have a steep spectrum ($\alpha \sim 0.5$–1). They often have small "hot spots": regions of intense radio emission (with $\alpha \lesssim 0.5$), accompanied by optical emission in roughly 10% of cases. Because the lifetime of relativistic electrons responsible for the emission is rather short compared to the travel time from the nucleus, it is generally agreed that some form of *in situ* acceleration is necessary. The most probable source of the required energy appears to be a jet from the nucleus.

Jets are observed in both Radio Galaxies and Quasars. They are very common in Weak Sources (WS) (WS are commonly defined as those with a radio power at 1.4 GHz of less than $10^{23.3}$ W Hz^{-1}) – jets are found in up to 80% of WS – where they are usually double. In Seyferts jets are faint, twisted and extend only a few kpc. In Intermediate Sources (IS) jets are typically appear to be one- sided, or there is a high contrast between the jet and counter jet. In most sources this is usually interpreted as being due to relativistic and orientation effects, rather than due intrinsic one-sidedness. They are straighter and better collimated in these sources than in WS. In Strong Sources (SS) the situation is mixed: jets are hard to find in powerful RGs but one-sided jets are quite common in quasars (they are found in roughly 40–70% of low-z quasars).

Many compact, flat spectrum sources display apparent superluminal motion, i.e. they have radio features whose observed angular velocities of separation, when converted into kinematic speeds through standard cosmology, imply projected separation velocities faster than the speed of light. There is still much controversy surrounding the detailed interpretation of superluminal motion (Rees 1984), however it is widely agreed that relativistic fluid motions, most probably in a deep gravitational potential well, are required.

Clear evidence of the interaction of jets with the surrounding ambient material exists. Low surface brightness "cocoons" surround a high proportion of jets in nearby RGs (this diffuse emission often has a very steep spectrum, $\alpha > 0.8$–2). Many jets show marked curvature and kinks, apparently as a result of having been 'bent' by their interaction with the interstellar or intergalactic medium through which they propagate. Most dramatic of all are the so-called "head-tail" sources which are found almost exclusively in rich clusters, and often show clear evidence of a strong bow shock at the leading "head".

Polarization mapping of jets has revealed an important morphological difference between the orientation of the magnetic fields, B, in the jets of powerful and weak radio sources. In the former, B is parallel to the jet axis. In the latter, B is generally perpendicular to the axis, but in a thin skin it is parallel to the walls, and near the nucleus it is parallel to the direction of the jet. In general the fields always seem to be *ordered* on scales comparable with those of the jet.

Although rare, the occurrence of jets has profound implications for models of the central source. The fact that radio sources 'remember' for $\gtrsim 10^6$ yr their direction of ejection of radio plasma, in the series of spatially separated blobs or hot spots, and the linear coincidence of the smallest and largest scale jets in most sources, argues forceably the necessity of a central 'gyroscope'. Since it is hard to construct a plausible model of a gyroscope which does not involve the angular momentum of a single coherent body – generally argued to be a supermassive black hole – jets are taken to be one of the strongest pieces of physical evidence for a central monster. These observations, and generally most radio loud features, are very difficult to explain in models, like the starburst (SB) models of Terlevich and his coworkers, which do not incorporate a single coherent body. In addition, the morphology of extended radio sources (non-superluminal) can be used to estimate minimum lifetimes: since double-lobed sources are observed whose nuclei are "active", it appears reasonable to argue that the activity has lasted at least as long as the light travel time from the nucleus to the radio lobes. At least a few percent of radio sources have sizes $\gtrsim 1.5$ Mpc. For these sources, at least, $\tau_{\text{life}} \gtrsim (separation\ of\ the\ lobes)/c \gtrsim$ few $\times\ 10^7$ yrs. Furthermore, this estimate agrees with that based on the energy argument

$\tau_{life} \gtrsim (Energy\ in\ the\ lobes)/Luminosity \gtrsim$ few $\times\ 10^7$ yrs. Note that a few $\times\ 10^7$ yrs is the typical lifetime of O stars - which is not long by the usual criteria!

The recent conference "Extragalactic Radio Sources: from Beams to Jets" (Roland, Sol and Pelletier 1992) contains an excellent series of papers reviewing and discussing the observations and theory, which I recommend.

2.3 The X-ray Spectra

Active galaxies emit strongly in X-rays. The general form of their spectra is taken to be that of a power law; under that assumption, the data can be analysed taking absorption into account. It is found that the best fit to the data has a spectral index $\alpha \sim 0.7$ (Mushotzky 1984, Turner and Pounds 1989). It is likely that the primary source of such high-energy power-law continua in compact sources is a population of non-thermal particles, radiating through Compton or synchrotron processes. Until recently, the spectral resolution of X-ray observations was so low as to prevent more detailed analysis of the spectra: many models for the details of the continuum production were suggested, with little to choose between them on the basis of the small amount of data available. These models are reviewed by Zdziarski (1992), and in the references he gives.

As the spectral resolution and sensitivity of X-ray detectors has improved in recent years, various extra features have been detected in observed X-ray spectra. A good review of this is given by Pounds (1989). At low (< 1 keV) energies, an excess of emission is often seen above the smooth power-law continuum defined by the higher energies. This soft excess can be interpreted as the blue tail of the UV (big-blue) bump (Arnaud et al. 1985, Turner and Pounds 1989, Masnou et al. 1992). Measurements of the soft X-ray absorption shows a surprisingly low column density, $N_H \sim$ several $\times\ 10^{19}\,cm^{-2}$, although some objects may have a larger column density. However, reprocessing of the hard X-rays required by some of the optical/UV observations of emission lines requires an absorbing column density, $N_H \gtrsim 10^{22}\,cm^{-2}$. In many low luminosity AGN, the radiation at these wavelengths is obscured, absorbed by intervening material of column density $N_H \gtrsim 10^{22}\,cm^{-2}$. The column density of this absorbing gas varies with time, and is likely to be a caused by changes in the ionization state of the intervening gas rather than changes in the physical column density on the line of sight (Turner et al. 1992).

At rather higher energies, particularly around 7 keV, various sharp spectral features are found. The first such observation was of Cen A, by Mushotzky et al. (1978). They found evidence for the Fe K_α line, emitted by fluorescence in cool material. Recently, the Ginga mission provided sufficiently detailed data to see this feature in several other AGN. For the iron line to be emitted by fluorescence, a significant fraction of the hard X-ray emission must reprocessed by a cool or warm medium with a large column density of singly-ionized Fe (thus implying $N_H \gtrsim 10^{23}\,cm^{-3}$) (Krolik 1990). This gas will also reprocess the hard X-ray continuum through Compton scattering, producing a broad maximum at energies between 10 and 50 keV (Lightman and White 1988; Guilbert and Rees 1988; George and Fabian 1991). The downturn at the high-energy edge of this broad warm feature has now been observed. Because of this bump, the X-ray spectral slopes observed in energy bands around 10 keV may be rather shallower than those emitted by the central energy source.

Also, in the intermediate energy band, absorption edges are observed (Nandra et al. 1990). The depth of these features is too great for them to be produced by reflection;

the ionization potential of the species producing the edges suggests that the absorbing material is fairly highly ionized. The cool column of gas is also too small to produce the depth of absorption edge. The variability of the cool column suggests that this material is in equilibrium with a warmer, photoionized component, which may be a 'warm absorber' (Pounds 1989; Turner et al. 1992). A warm absorber, in contrast to the cool material which emits the soft X-ray component, is indicated by the presence of the deep Fe K edge. It is assumed to be photoionized by the hard X-ray flux.

As it does in other wavebands, variability provides an important extra insight into X-ray emission. The X-ray emission is rapidly variable, on timescales down to ~ 100 seconds in NGC 6814 (Tennant et al. 1981). NGC 6814 is indeed an important test case: although a very low luminosity source, it is one of the closest extra-galactic X-ray sources. This has allowed some of the most detailed of the X-ray observations to be done on this object. Important amongst the results of this study are the 12 000 second periodicity found in its X-ray light curve (Mittaz and Branduardi-Raymont 1989), and the rapid response of the reprocessed iron features to continuum variations, on a timescale of 400 seconds (Turner et al. 1992). These observations give a very definite limit on the size of the emission region in this object: the well-maintained 12 000 second periodicity is particularly striking, as it suggests that some coherent structure must be modulating the X-ray emission over many dynamical timescales.

In many objects, X-ray spectra are observed to harden as the flux decreases. This may be due to the relatively long time the Compton reflection component takes to respond to the central, power-law continuum source, in a similar fashion to that in which the longer wavelength, optical to UV continuum and line spectrum responds on a yet longer timescale, and smooths out to a rather greater extent the image of the central light curve.

Interpretation of the X-ray observations is highly model dependent and will be discussed further on. However, it is worth pointing out here that most of the observations have so far been interpreted assuming that the X-rays are emitted at or very near a central black hole. Recent observations appear to require that the hard X-ray source is at least 10 light-days removed from any central source, and that it sits in front of a 'cold' reflector ($\sim 10^4 - 10^5$) (George, Nandra and Fabian 1990). This geometry is similar to that proposed by Collin-Souffrin et al. (1987), and may also be compatible with radiation from compact supernova remnants (Terlevich et al. 1992). I return to this point briefly in the last sections of this article.

2.4 Absorption Lines

Absorption lines are very common in the spectra of QSOs, particularly those at high redshift. The higher frequency of absorption features at high redshift is partly due to an increase in absorbing matter and is partly an observational feature: the most common strong absorption lines are UV resonance lines which enter the optical window only at high z. Recent observations are discussed in detail in the proceedings of the conference *QSO Absorption Lines* (Blades, Turnshek and Norman 1988) so my discussion here will be brief.

Most of the absorption lines are very narrow and their wavelengths therefore well defined. In $\lesssim 10\%$ of high-z QSOs, very broad absorption is also present. The narrow lines can usually be identified with resonant transitions of commonly occurring ions and grouped into systems of lines with a common z_{abs}. Such identifications are subject to consistency criteria based on plausible excitation conditions and abundance ratios. In very rich spectra the proposed identifications must be checked for statistical significance. This is because there are so many lines, only a small fraction of which generally fit any individually proposed z_{abs}, and because identifications must be accepted within some reasonable error limits of $\Delta\lambda/\lambda$.

Absorption line systems of all types are mostly blue-shifted with respect to the emission line redshift of the QSO. There are a small number of red-shifted systems, but all these have $z_{abs} - z_{em} \lesssim 0.01$. The dominant blue-shift is consistent with either foreground absorption, with $z_{abs} = z_{cosmological}$, or with outflowing gas within the QSO itself, in which case

$$\frac{v_{outflow}}{c} = \frac{(1 + z_{em})^2 - (1 + z_{abs})^2}{(1 + z_{em})^2 + (1 + z_{abs})^2} \tag{2}$$

There are three main categories of absorption systems: the narrow metal line systems; the Lyα forest; and the broad absorption line (BAL) systems. Almost all the narrow lines to the red of Lyα in any given spectrum can be identified and belong to one or more systems containing heavy elements: the narrow metal line systems. The lines most commonly identified (in the optical and UV) are Lyα and the doublets CIV $\lambda\lambda$ 1548,1550 Å and MgII $\lambda\lambda$ 2796,2803 Å. The reason CIV and MgII doublets are most often identified is that they are both strong, and have definite doublet ratios and separations. The 21-cm fine structure line of neutral hydrogen is often observed in absorption in the radio, and extends to low redshift the possibility of detection of hydrogen.

The CIV and MgII doublets are most easily found in different redshift ranges, the MgII dominating below $z \sim 1.5$ and CIV from $z \gtrsim 1.2$ to ~ 3.5. Lyα enters the optical window for $z_{em} \gtrsim 1.9$ and is usually identified in systems with $z_{abs} \gtrsim 1.9$. The MgII and CIV doublets are usually, but not always, accompanied by lines from (amongst others) SiII, CII, FeII, AlII, or SiIV, NV, OIV respectively. Examination of tables of ions of all the abundant elements reveals that every ion which has resonance lines in the observable window over the redshift range $0 < z < 3.5$ has been identified in at least some spectra (Perry 1979). The two doublets have given their names to these two respective systems, commonly known as MgII or CIV systems.

By and large, the lines bluewards of Lyα cannot be identified with resonant lines of heavy elements. Lynds (1971) suggested that most of these lines are Lyα in absorption in gas of small enough optical depth that no other lines, such as the CIV doublet, are strong enough to be seen. This is now the accepted explanation of the 'picket-fence' of lines found to the blue of Lyα which is now known as the Lyα forest. Often Lyβ and rarely lines of heavy elements can be identified at the same z_{abs} as a moderately strong Lyα line (Chaffee et al. 1986; Meyer and York 1987; Lu 1991).

The Lyα forest lines are extremely numerous, outnumbering all other types of absorption system by two orders of magnitude. They have low HI column densities, and are usually unaccompanied by absorption due to metals (Pettini et al. 1990; Carswell et al. 1991). Clustering, if present, seems to be weaker and over a smaller characteristic scale than the clustering of metal line systems. The number of Lyα systems is compatible with a Poissonian distribution, indicating that the systems arise in intervening clouds having a widespread cosmological distribution (Tytler 1987; Bajtlik, Duncan and Ostriker 1988; Lu, Wolfe and Turnshek 1991; Bahcall et al. 1991).

Very strong, damped, Lyα lines are rare, but very important. These high column density systems form a subset of the narrow metal line systems; at low redshift the high N_H (and low T) result in strong 21-cm absorption – which is how they were first discovered. Wolfe and his collaborators have carried out extensive and detailed studies of these systems for more than a decade; in an important recent survey Wolfe et al. (1986) determined their cosmological distribution. Pettini, Boksenberg and Hunstead (1990) and Hunstead, Pettini and Fletcher (1990) have recently found that these systems are ∼ 2 orders of magnitude under-abundant. The interpretation of these systems is discussed in Sect. 7.1.

Broad absorption lines occur in the spectra of about 3 – 10% of QSOs. Their velocity extent Δv_A is $\gtrsim 10^4$ km s^{-1}. The absorption troughs have a wide spread in appearance ranging from smooth to highly structured, attached or detached from the adjacent emission lines, which are almost always the high ionization lines such as CIV, NV or OVI. The BAL are identified with the adjacent emission lines and the resulting (high) ionization states are therefore presumably comparable. The absorbing gas must have column densities $N_H \approx 10^{20}$ and electron densities $n_e \gtrsim 10^6$ cm^{-3}. Line strengths (but not velocity widths) are observed to vary on timescales as short as 2 years. For a full review of the observations of the BAL see Turnshek (1988).

Absorption can occur, in principle, anywhere along the line-of-sight between the emitter and the observer. In practice it appears that the majority of the narrow metal-line systems and the Lyα forest arise in "intervening" material: "clouds" which accidently lie along the line-of-sight but are not directly associated with the QSO. The BAL are universally believed to be intrinsic to the QSO, since no possible intervenor which could give rise to such broad lines is known. There is a subset of narrow metal line systems which also appear to be "associated" with the QSO, and have $v_{outflow}$ of $\lesssim 3000$ km s^{-1} (Anderson et al. 1987 and references therein). The implications of the observations are discussed further in Sect. 7.

3 AGN Host Galaxies

Initially the focus of research into the physics of *activity* was almost entirely on energetic processes in the nucleus, essentially detached from the host galaxy, and reduced the rôle of the galaxy to that of a reservoir of fuel for whatever monster was driving the AGN. Over the past several years emphasis has shifted, and recently the rôle of global phenomenon in perhaps initiating and driving activity has been receiving increasing attention.

Broadly speaking, Seyfert nuclei are found in early type spirals; powerful radio galaxies are giant ellipticals; WRGs in less luminous ellipticals; and the morphological types of the hosts of QSOs and quasars are unclear. However (see below) there is increasing evidence for distorted morphology, taken to indicate interactions, in most of the host galaxies. Because Seyferts and radio galaxies are found at lower redshift than QSOs, and because their nuclei are less luminous, it is possible to resolve, image, and study the host galaxies.

Something like ∼ 80% of Seyferts are found in Sa-Sbc galaxies (Terlevich et al. 1987, Woltjer 1990). Radio galaxies, in contrast to Seyferts, are ellipticals. In a study of their morphology, Heckman et al. (1986) found that many are peculiar and show evidence of recent star formation, with extra-nuclear colours bluer than those of normal ellipticals, and clear evidence for kinematic peculiarities. They find that these peculiar structures have a higher surface brightness in radio-loud galaxies than they do in normal galaxies. In their sample, 13 of 15 optically peculiar galaxies have strong emission lines in their nuclei, as compared to 10 of 20 normal galaxies. Also 8 of the 15 are strong IRAS sources, as against only 2 of the 12 morphologically normal galaxies. They found that the peculiar galaxies are optically less luminous than normal giant E galaxies. The interpretation of the latter is unsure however, because of the effects of dust extinction and recent star formation in the peculiar galaxies. Since Seyferts are weak radio sources, and powerful radio galaxies are found in large bright ellipticals with $M_V = -23.3 \pm 0.7$ (Smith and Heckman 1989) it is usually *assumed* that quasars form in ellipticals and QSOs in spirals.

QSOs are so luminous that even a first ranked elliptical galaxy host is difficult to image without masking out the QSO. Furthermore, QSOs are found mostly at high redshift, and so require deep imaging as well as subtraction of the nucleus in order to resolve the host galaxy, rendering the results inconclusive. Nevertheless, virtually all QSOs with $z \lesssim 0.5$ studied by deep imaging are resolved and surrounded by extended nebulosity. In a study of 17 QSOs, Gehren et al. (1984) found that, assuming the extended nebulosity is a galaxy, host galaxies are very luminous and quasars are in galaxies which are 2 magnitudes brighter than are the hosts of QSOs. Malkan et al. (1984) studied 24 X-ray selected AGN, and were able to resolve 15 (and marginally 2 more). They found the images were consistent with the nebulosities being spiral galaxies. Careful imaging of 3C 273 puts the quasar in a first ranked giant elliptical galaxy – but not at the centre! (Tyson et al. 1982). It is unclear what causes the offset: is it real, or a result of projection effects, or of a distorted light distribution of the galaxy? Romanishan and Hintzen (1989) found, in a study of quasars with $0.2 < z < 0.7$, that their galactic hosts are more compact and bluer than are normal ellipticals, indicative of star formation concentrated in their nuclear regions. Boroson et al. (1985) divided the nebulosity around quasars into two groups; they found that extended steep spectrum sources were surrounded by nebulosities with flat blue continuum with strong emission lines, whereas those surrounding compact flat sources had a red continuum and weak or no emission lines. Furthermore, they find that although there is no clear-cut evidence that the quasars are in giant ellipticals, all the associated galaxies are peculiar. The study of the IR CaII triplet by E. Terlevich et al. (1990) indicates that the light in the cores of Seyfert 2 galaxies is dominated by a young stellar component.

In groups or clusters, the hosts of AGNs are the brightest galaxies, and quasars sit in giant ellipticals in which a significant amount of hot gas and stars may be present, perhaps indicative of interactions. In a study of 74 AGN, Hutchings et al. (1984) also found that radio-loud AGNs are present in nebulosities (galaxies) more luminous than those hosting radio-quiet AGN. They found that ~1/3 appear to be in interacting systems, ~1/4 lie in small groups or clusters and that ~40% show some indication of spiral structure, but they found no *positive* evidence that any of the nebulosities are elliptical galaxies.

Good-quality high-resolution spectra of the nebulosity surrounding QSOs are rare. The only one showing clear stellar spectral features – essential in establishing that such nebulosities are galaxies of stars and not just ionized gas – is the spectrum of 3C 48 (Boroson and Oke 1982, 1984). This spectrum shows the Balmer series strongly in absorption, clear indication of a relatively *young* ($\lesssim$ few $\times 10^8$ yrs) stellar population. Miller (1981) observed several quasars spectroscopically and found, because MgI was not present in absorption, that the integrated light is not consistent with the assumption that the host galaxies are giant ellipticals with *old* stellar populations. Although the direct evidence that QSOs are in galaxies is sparse, Seyferts are observed to be in spiral galaxies. The continuity in the properties of QSOs and Seyferts is used to infer the existence of host galaxies for QSOs.

New analysis of the correlation of the far-infrared to radio luminosity of late-type galaxies, Seyferts, quasars and radio galaxies (Sopp and Alexander 1991) shows that the FIR-radio plot may be a powerful discriminator between host galaxy type. This follows from the striking result that the FIR and radio luminosities correlation for radio-quiet QSOs follows the same correlation as normal star-forming galaxies and ultra-luminous IR galaxies, whereas the radio-loud quasars fall on the same correlation as radio galaxies. They conclude that both the FIR and radio emission from QSOs is from star-forming host galaxies and not from the active nucleus.

All of the correlations cited above (RG to blue colours; starbursts to interaction; AGNs to interaction; AGNs to starbursts) are such that there is a greater chance, compared to galaxies in general, of finding one if you find the other, but none of them are universal. Therefore, perhaps the relationships between AGN and starbursts and/or interactions is that there is an essential trigger required by an AGN which is often provided by a starburst and/or interaction but which can possibly be provided by other means (or the connection is indirect or delayed). Unfortunately the observations are not as clear cut as a theorist might like them to be.

Thus, although the situation is still far from clear, there are strong suggestions that the galaxies which host nuclear activity are not normal and quiescent on galactic scales. It is important to determine what causes this global abnormality: is it the presence of the active nucleus, or is the nuclear activity a coincident result of whatever causes the galactic abnormality? Exciting new research is pointing toward the possible importance of interactions between galaxies being the cause of these abnormalities.

3.1 Companions and Interactions

Yee, Green and their collaborators have conducted a series of direct imaging and multi-object spectroscopy observations of QSOs and their galactic hosts and neighbours (see e.g. Yee 1987, 1989). Imaging studies of 37 Palomar-Green QSOs spanning $0.05 < z < 0.3$ found that 40% have a companion brighter than $m_v = -19$ within a separation distance less than 100 kpc. Thus the frequency of finding a companion is roughly six times greater than for field galaxies, but is similar to that for Seyfert galaxies (Dahari 1984, 1985). In this survey they found that the companions had a similar luminosity function to local normal galaxies, although other groups have found that companions to Seyferts are themselves "active" (see Fricke and Kollatschny 1989). Furthermore, they find that environment of quasars at $z \sim 0.6$ is three times richer in galaxies than is the environment of quasars with $z < 0.5$. Their quasar-galaxy covariance amplitude at $z < 0.5$ is however still 2.7 times that of the galaxy-galaxy amplitude.

There is a paucity of quasars at low z in clusters, but for $z > 0.5$ 35% of quasars are found in clusters as rich as or richer than Abell class 1. The blue galaxy fraction of clusters with quasars may be higher than is usual, and there is some evidence that the blue galaxy fraction increases with radial separation from the quasar.

But there is clear evidence that galaxies in low-z clusters are statistically *less* likely to be active than are field galaxies. As early as 1978, Gisler found in a study of 1316 galaxies that line emission was found much more commonly in field than in cluster galaxies. Dressler et al. (1985), in a sample of 1095 galaxies in clusters and 173 field galaxies, found that whereas 31% of field galaxies had prominent emission lines only 7% of cluster galaxies did, and that active nuclei occurred in 5% of the field vs. only 1% of the cluster galaxies. Although this evidence seems - on the face of it - to contradict the interaction theory, things are not so simple. Since these studies are at low z, it could be that the clusters had more activity in the past and now contain "dead" QSOs and AGN. Alternatively, it has been suggested that activity, like morphological disturbance, is inversely proportional to the velocity of the collision. Since $v_{collision} \propto v_{dispersion}$, activity will be less common in rich clusters since their velocity dispersions are too high (now) to allow mergers – this is not so in the field. It is also possible that the triggering of activity is yet more complex and that possibly interactions may trigger activity in spirals but that cooling flows are the cause of activity in ellipticals (Romanishan and Hintzen 1989).

Other evidence pointing to the possibly important rôle of interactions is provided by observations of systems with double nuclei, several of which have one Seyfert nucleus, and all of which are extremely strong far-IR sources (Fricke and Kollatschny 1989). In a controlled imaging study of blue and red galaxies, Lavery and Henry (1988) found a statistically higher chance for blue galaxies to have "companions" than do red galaxies.

To summarize: active nuclei may be in galaxies which are firstly "bluer" than normal, indicating recent star formation; and secondly in galaxies both closer to other galaxies and morphologically more distorted than 'normal' galaxies, indicating interactions in the recent past. In turn, the degree of peculiar morphology appears to be directly proportional to the power output and incidence of close companions. However, activity appears to be anti-correlated to the local galaxy density. Although only circumstantial, this evidence linking activity to both star formation and interactions between galaxies has inspired new avenues of research attempting to understand the physical basis of this link.

Ever since the pioneering work on tidal interaction between galaxies by Holmberg (1941) (using an ingenious analogue computer built from light bulbs, photocells and galvanometers, with the gravitational force modeled by light) and Toomre and Toomre (1972) (who solved a restricted 3-body problem, see below, with digital computers) tidal tails have been recognised as symptoms of the gravitational effect of galactic encounters. Stockton and MacKenty (1987) present evidence from the morphology of extended emission around QSOs that suggests that the extended gas *may* trace such tidal tails.

It is not only disturbed morphology in spirals, however, which is now believed to result from interactions and merging between galaxies. Combes (1987) has reviewed the arguments that elliptical galaxies are the results of interactions. Observational evidence cited to support this view is that perhaps as many as 40% of ellipticals are surrounded by thin stellar 'rings' or shells, and that such shells are also found around many merging systems. An attractive theory for their formation is that they result from head-on, or nearly head-on, collisions between spirals. Combes speculates that the large population of globular clusters in ellipticals, which has been considered a major problem for the interaction hypothesis, could result from dissipative interactions between gas-rich interacting spirals. The fact that powerful radio galaxies are associated with morphologically disturbed galaxies (Heckman et al. 1986) then fits neatly into this picture.

How then could interactions between galaxies trigger nuclear activity? Tidal interactions perturb and distort the gravitational potential of the interacting galaxies (which are assumed to have been in virialized equilibrium prior to the interaction). These perturbations then accelerate both stars and the interstellar medium (ISM), causing rapid internal motion within the galaxies. In the case of the stars the interaction can produce both the long tidal tails observed in the outer regions and, by violent relaxation, the spheroidal distributions characteristic of ellipticals. The gaseous component is dissipative; interstellar clouds may collide; shocks and the resulting compression within the ISM may then trigger large scale star formation. The non-axisymmetric potential may, aided by dissipative processes, also enable the ISM to fall into the nucleus of the galaxy (or galaxies) (Noguchi 1988a, 1988b; Noguchi and Ishibashi 1986; Hernquist and Barnes 1991) either then to feed a central monster or to initiate a nuclear starburst, or both (see e.g. Lin, Pringle and Rees 1988).

Since there are both regions of current star formation (on small scales) within our own galaxy and nearby galaxies showing extensive star formation, it is helpful to turn now to the evidence on star formation.

4 Starbursts

It is worth noting the different character of the definitions given in Sect. 1 for Starbursts as against all other classes of active galaxies. Whereas the others are defined in spectral terms, thus evading precise physical definitions, starbursts are defined primarily in terms of a clear physical idea, evading a precise spectral definition. We *know* stars form, though not how; a century of observation reinforced by decades of modeling stellar structure and evolution has given us confidence that we can recognise a massive star younger than $\sim 5 \times 10^7$ years old. Thus we can conclude that when many such young, massive stars are found in a spatially confined region, a star-forming event occurred within the last 5×10^7 years. The presence of young stars is not in itself sufficient to define a starburst: the time scale of star formation must be short, about the dynamical time scale, and the rate of star formation must be such that it could not be sustained over the life of the galaxy. I recommend the proceedings of two recent conferences, *Starbursts and Galaxy Evolution*, Thuan et al. (1987), and *Massive Stars in Starbursts*, Leitherer et al. (1991).

What are the observational signatures of a starburst? Direct evidence of recent star formation includes very blue resolved UBV colours, or bright far-IR and resolved near-IR emission, a continuum rising to the blue, characteristic of OB stars, and emission lines, including e.g. the Balmer and Brackett lines of hydrogen, Lyα, HeI and II lines, the TiO bands and the CaII triplet. Starbursts only rarely show strong narrow stellar absorption features. The strong stellar absorption typical of a young population, e.g. the Balmer series and HeI lines are all filled in with the emission from the ionized gas. The exception is the CaII triplet which is visible after $\sim 10^7$ yrs of evolution (E. Terlevich et al. 1990, E. Terlevich, Diaz and R. Terlevich 1990) Indirect evidence comes from effects the young stars have on the ISM: primarily this is the strong IR re-emission by dust heated by the young stars; the nebular emission lines which are indicators of HII regions formed by massive young stars, optical, IR and radio recombination lines of hydrogen; galactic superwinds; and evidence of supernova remnants (SNR) – many SNR have been resolved and mapped in the radio, revealing both unusually high SN rates, and SNR hundreds to thousands of times more luminous than 'normal'. The X-ray continuum of SBs is modest compared to the IR. Starbursts, in contrast to AGN, are not known to be variable.

It is the dramatic scale of suspected galactic starbursts that have 'electrified the atmosphere': IRAS galaxies radiate the equivalent of up to $10^{14}\,\mathrm{L}_\odot$. The most luminous IRAS galaxies ($L_{\mathrm{FIR}} \gtrsim 10^{11}\,\mathrm{L}_\odot$) are interacting galaxies or mergers (Sanders et al. 1986); such ultraluminous galaxies form between 3 - 9% of all IRAS galaxies (Wolstencroft et al. 1987). Because the IR is reprocessed radiation, one seeks the source of the excitation: is it in fact due to a hidden starburst, or to a hidden AGN? Joseph (1987) argues that starbursts are present in all the ultraluminous interacting galaxies (although he does not exclude the presence also of an AGN nucleus in some) because (1) the IR emission is resolved in most and comes from a extended region physically too large to have been heated by only a central point source of radiation; (2) the optical line ratios are characteristic of HII regions in 2/3 of the sources; and (3) the radio emission is qualitatively consistent with the supernova activity following a starburst, rather than with AGN radio spectra. In a study of 10 ultraluminous ($L_{\mathrm{FIR}} \gtrsim 10^{12}\,\mathrm{L}_\odot$ IR galaxies) Sanders et al. (1988) found that all are strongly interacting and that their optical spectra could be modeled as a superpostion of a young starburst stellar population and a typical AGN. Rowan-Robinson and Crawford (1989) studied 227 IRAS galaxies whose IR spectra they modeled as a "cool" disc component with a superposed "warm" starburst and a Seyfert nucleus. Recently, Lonsdale, Lonsdale and Smith (1991) found evidence for hidden AGN in a sample of IRAS galaxies. Again, the observations are not unambiguous; however, whether or not a 'monster' is present, ultraluminous nuclear starbursts are almost certainly associated with interacting galaxies.

One of the more dramatic recent results of SB observations has been the discovery of a class of very luminous radio SNR. In high resolution observations of M82, Kronberg et al. (1985) mapped 40 discrete radio sources within an area 100×600 pc in size. Each source is more luminous than any in our Galaxy. M82 is not ultraluminous: its total luminosity, dominated by the far-IR, is $4 \times 10^{10}\,\mathrm{L}_\odot$. The SN appear to lie along in a flattened organised structure, which does not share the disrupted appearance of the outer regions of M82, and may indicate that massive star formation and the resulting SN "occur in a system which has been organized by the dynamics of the inner 300 pc region". Detailed study of the brightest SNR lead them to conclude that it did not undergo normal free expansion in its early phase, and that it must have expanded into a dense, high pressure ambient shell, which they conclude must have had a local density $n_{\mathrm{shell}} \sim 10^7\,\mathrm{cm}^{-3}$ and total mass $M_{\mathrm{shell}} \lesssim 20\,\mathrm{M}_\odot$.

Since starbursts occur both in nearby as well as in cosmologically distant objects, it is possible to study them in more detail than is common for QSOs. The mass is determined from rotation curves and the luminosity and spectral characteristics determined within the uncertainties introduced by extinction. These detailed observations are used to attempt to determine the following (interdependent) characteristics of the galaxies and their stellar populations:

a. The slope, α, of the initial mass function (IMF) of the burst. [The initial mass function is the number of stars of mass M born per unit $\log M$ bin, given by $dN(M)/d\ln M \propto M^{-\alpha}$; in the standard Salpeter IMF $\alpha = 1.35$.] In a number of studies of blue and starburst galaxies evidence has been presented for IMFs flatter (relatively more high-mass stars) than normal (see e.g. Olofsson 1989, Rieke 1991); however this conclusion is far from universally accepted (see, e.g. the discussion following Rieke 1991). The importance of the population of massive stars is twofold. Firstly, low mass stars are a mass sink. A given mass of gas will generate significantly more luminosity if it is turned into high mass stars only, than if a large fraction goes into low mass stars. Secondly, it is a reflection of the significant change in properties of stars at $\sim 10\,M_\odot$. Stars more massive than $\sim 10\,M_\odot$ have strong continua shortward of the Lyman edge; they are therefore important in ionizing the ISM, and their radiation drives strong stellar winds which both affect their own evolution profoundly and inject both mass and kinetic energy into the ISM. It is hardly surprising that determining the IMF in distant SB galaxies is difficult; the determination of the IMF in the solar neighbourhood is itself fraught with uncertainty (Garmany 1991; Scalo 1990). Garmany concludes that within the local group there is no evidence for any IMF differences between the Galaxy, LMC and SMC; however, Scalo's interpretation of the evidence contradicts this conclusion.

b. The upper and lower mass limits on the IMF. There is some evidence that star formation may be *bimodal* in normal galaxies, (see e.g. Mezger 1987) with *induced star formation* (which is triggered by compression in spiral arms) forming only massive stars, $M_* \gtrsim 3\,M_\odot$, in contrast to *spontaneous star formation* in which stars are formed down to $0.1\,M_\odot$ (in e.g. the inter-arm regions of spiral galaxies). Similar processes may operate in starbursts resulting in a bias towards massive stars. Rotation curves combined with detailed population synthesis are required to determine the mass limits; these are available in only a few cases, and the interpretation of the spectral results is contentious because of high, variable, ill determined extinction. Rieke (1991) concludes, in a detailed study of M82, that an IMF which favours high mass stars is required to satisfy all the observational constraints. However, he is unable to determine if this is due to a raised low mass cutoff or to a flatter IMF. Puxley (1991) modeled the extinction and dust distribution in M82 and concluded that the IMF may be deficient in low mass stars. Similarly, Joseph and his co-workers, combining high resolution rotation curves and detailed spectroscopy, find evidence for a low mass cutoff in many starbursts of $\sim 5\,M_\odot$ (Joseph 1991), and evidence that the IMF must extend to at least $\sim 30\,M_\odot$, and possibly to $60\,M_\odot$. The situation remains unclear; for a review of the difficulties involved in interpretation of the observations see Scalo (1990).

c. The star formation (SF) history of galaxies. How often do episodes of star formation occur, and how long do they last? The observed IR - UV spectral energy distribution can be compared to the results of population synthesis models to constrain possible SF histories. For example, in M82, which at ~ 3.2 Mpc is close enough to map in detail, individual giants and supergiants in the outer regions could have been resolved had they been present; of course, the presence of significant amounts of dust make this null-detection far from clear cut. The fact that they are *not* detected indicates that star formation in the outer regions ceased more than 3×10^8 years ago. Yet both the spectrum of the nuclear region, and the presence of young SNR there, indicates that star formation in the nucleus has been as recent as $\lesssim 3 \times 10^7$ years ago (Kronberg et al. 1985). Fanelli, O'Connell and Thuan (1988) find that the spectra of blue compact dwarf galaxies cannot be accounted for by continuous SF; rather they require discrete star formation episodes, typically lasting $\sim 10^7$ yrs, superimposed on an underlying old ($\sim 10^{10}$ yrs) population.

d. The efficiency of star formation. Star formation efficiency, ϵ_{SF}, is generally defined as the ratio of the mass of stars formed, to the mass going into the star-forming clouds. In giant molecular clouds in normal galaxies the observed ϵ_{SF} ranges between 1% and

10%. Shu, Adams and Lizano (1987) find that in clusters star formation must have $\epsilon_{SF} \gtrsim 30\%$. Yet in interacting galaxies it appears that bursts of star formation in the nucleus of one member of the interacting pair may occur with approaching 100 % efficiency (Wright et al. 1988) although the massive gaseous outflow may confuse this determination.

e. Are the SFRs and the IMF determined primarily by local or global processes? Because of the strong [OI] lines (usually indicative of shocked gas) in interacting SB spectra (Huchra 1987) the star formation may be shock induced, caused by the global dynamical interaction or the forbidden [OI] may occur in the superwind.

f. The spatial variation of the SFR and IMF. It is important to distinguish between a *global* IMF incorporating the entire region of the burst, and possible important spatial gradients. For example, in the star-forming region 30 Doradus (in the LMC), which is near enough that individual stars can be resolved spatially, the very massive stars are found in the core. There is a 'radial' gradient in the average mass of newly formed stars, so that the low mass stars are found in the outer regions of the burst (E. Terlevich 1987; R. Terlevich et al. 1991). If this behaviour is typical of all starbursts then it would have important consequences for models of nuclear activity in galaxies.

The theoretical understanding of star formation has made some progress in the last decade. Although far from complete, predictive models *are* beginning to emerge (Larson 1987, 1991). An eventual theory of star formation will need to encompass the following:

a. What triggers the onset of star formation and how are they formed?

b. Why does star formation cease? The process of SF is probably self-regulating: once significant numbers of O stars have 'turned on', their UV continuum heats and ionizes the surrounding collapsing cloud and may prevent further dissipative collapse; this early phase is followed by the onset of strong stellar winds and SN both of which inject kinetic energy into the cloud, which also makes further collapse unlikely (Larson 1987) unless the shocks they produce induce further star formation.

c. What determines the shape of the IMF, in particular the upper and lower mass cutoffs? Low and high mass stars are thought to form differently; low mass star formation may proceed by slow accretion, with angular momentum and magnetic field diffusing outward by ambipolar diffusion (Shu, Adams and Lizano 1987), whereas high-mass stars form in turbulent hot regions where the Jeans mass is high. Norman (1991) speculates that in starbursts, a burst of high-mass stars may be able to inhibit the slow ambipolar-diffusion-driven low-mass star formation process, leading to a 'top-heavy' IMF.

5 Cosmological Distribution and Evolution of AGN

The study of the evolution of active galaxies requires firstly that the AGN be located and counted and then that the z-distribution of their properties be determined. In parallel it is necessary, in principle, to have the same sort of information about the parent galaxies. However, as we have already discussed, AGN (in particular QSOs) significantly outshine their parent galaxies and for this reason, and because unlike galaxies they appear unresolved and are thus less affected by the sky background, they are visible out to much larger distances than are the galaxies. There is no reason why, if they exist, they could not be detected at $z \gtrsim 6$. It is not yet possible to observe the bulk of the galaxy population at redshifts $z > 0.5$. In the last few years powerful radio galaxies at redshifts $z > 3$ have been detected and studied in ever greater numbers (see e.g. Lilly 1989) as have high-luminosity galaxies in rich clusters at $z \sim 0.8$. However, the only high redshift ($z \gtrsim 2.5$) objects for which large numbers of observations exist are QSOs: therefore *most* of the observational constraints currently available on models of galaxy formation and evolution at these early times ($t \lesssim 0.2\,t_{\text{Hubble}}$) comes from a study of the QSO population. In order for these constraints to be physically meaningful, an understanding of the relationship between the evolution of 'normal' galaxies and the development of activity in the nucleus is necessary.

If there is an essential connection between nuclear activity in galaxies, star formation and galactic interaction, then that connection is expected to be reflected both in their respective space densities and in a related cosmological evolution of active galaxies and the galactic population as a whole. Increasingly the evidence supports the view that even 'ordinary' galaxies have evolved in fits and starts and have had recent episodes of significant star formation. This new evidence is challenging the conventional theories of galaxy formation and evolution which are based on the assumption that the collapse of galaxies is accompanied by an initial phase with an extremely rapid star formation rate which subsequently declines sharply. Furthermore, as discussed below, it is *probable* that nuclear activity lasts for only a small proportion of the life of the parent galaxy, which would imply that a significant percentage of all galaxies go through an active period at some time in their history. In this case AGN may be good tracers of the galactic population and their study will help elucidate galaxy formation and evolution itself.

Many aspects of the relationship between nuclear activity and normal galaxy evolution, crucial for the interpretation of the AGN and QSO data are not yet understood. In particular, the time it takes to trigger activity is an important unknown effecting our understanding of both AGN and their parent galaxies. If, as is usually assumed, supermassive black holes are needed, then what are the conditions and timescales needed to form such a massive central object? Is it related to the initial collapse of the protogalaxy, to interactions, or to some other phenomenon altogether? If interactions between galaxies were to be found to be a necessary, and not just a sufficient, condition for activity, then the time delays between galaxy formation and the observation of activity might be very long, and the observation of QSOs at $z \sim 5$ would imply that at least some galaxies formed at much higher redshift. Alternatively, if galaxies themselves are formed by the merger and collapse of systems of stellar clusters and gas clouds, as is envisioned in the cold dark matter (CDM) picture (see e.g. White et al. , 1987), then activity could be associated with this merger-driven galaxy formation. The quantification of galactic and AGN evolution, which can test to what degree the "fits and starts" scenario is valid, is one of the tasks of galaxy and AGN surveys.

We start by asking how many active galaxies there are as a function of redshift, and how that compares with the known population of all galaxies.

5.1 Finding, Counting and Measuring AGN

To count active galaxies, it is first necessary to find them in a systematic way free of obvious bias, or at least with a good understanding of any bias present. To this end, large, systematic surveys in the optical, radio, X-ray and IR have been and are being carried out. Because of the relatively small range in frequency in any particular survey, and of the combination of K-corrections (that is changes in a 'standard' spectrum as it is redshifted, and as different spectral ranges move into and out of the observational window) with the probable evolutionary changes in the intrinsic spectrum with z, it is difficult to define unambiguous criteria of "completeness". Therefore, when carrying out surveys, completeness is defined observationally: as long as the intrinsic properties of the objects are not known, it is not possible to define completeness fractions for the true population. Furthermore, most surveys are flux limited or apparent magnitude limited, so intrinsically faint objects drop out of the survey as they move out in z, leaving only the intrinsically brightest objects for study at high redshift. The faintest systematic surveys for quasars have now reached $m_B \sim 23$, thereby somewhat reducing this problem.

Surveys are carried out in a limited range of wavelengths and each spectral region has its own selection effects. Thus, it is difficult to compare populations discovered by different techniques, as the 'mixture' of objects will also be different. It is for this reason that number densities and luminosity functions are usually defined for particular populations (such as, for example, X-ray or objective-prism selected) rather than for AGN as a class. For extensive recent discussions of the achievements and inherent difficulties of observational surveys I recommend Warren and Hewitt (1990), Smith (1989) and the conference proceedings *Optical Surveys for Quasars* (Osmer et al. 1988).

Quasars were originally discovered by identifying the redshift of the emission lines in the optical counterparts of two 3C radio sources, 3C 273 (Schmidt 1963) and 3C 48 (Greenstein and Matthews 1963). The first catalogues of quasars consisted therefore of the optically identified counterparts of radio sources. Radio-quiet QSOs were discovered by Sandage (1965) in the course of a systematic two-colour photographic search for quasars, a technique developed because of the discovery that the first four quasars found all had an ultra-violet excess. This ultra-violet excess, characteristic of all the QSOs originally found, has remained an important basis of subsequent searches.

The source of the extreme difficulty in discovering QSOs and AGN can be readily understood by examining Table 1: only $\lesssim 1$ in 40 stellar images brighter than 20^m is a QSO. Therefore searches must be based on predicted spectral properties of the population in order to isolate the population in question from the stellar background.

Table 1. Approximate integral surface densities, $m < m_B$, per square degree, at high galactic latitude (Warren and Hewitt 1990)

m_B	Stars	Galaxies	QSOs All	QSOs $z_e > 3$
16	200	2	0.02	< 0.001
18	500	40	1	0.02
20	1000	400	25	0.5
22	2000	3000	100	2

The traditional method of separating QSOs from galactic stars relies on their optical broadband colours: for $z \lesssim 2.2$ the $U - B$, $B - V$ colours of QSOs are sufficiently different from those of galactic stars (the well known UV excess) for the colours to be used as a discriminant. Based on comparison with samples selected by optical variability, radio or X-ray emission, UVX samples are found to be very nearly complete to $z = 2.2$ (Veron 1983). After the initial colour selection, candidate QSO stellar images must be observed spectroscopically to determine their redshifts and confirm the original selection.

The change in colours of QSOs at $z \sim 2.2$ results from the fact that the Lyα emission line moves into the B band and the Lyα forest depresses the continuum flux in the U band. For $z \gtrsim 2.2$ more sophisticated, and complex, multi-colour selection criteria must be applied (see e.g. Warren and Hewett 1990). Because of the extensive range in the spectral properties of AGN discussed in Sect. 2, which argues against targeting "typical" quasars, these surveys identify all unusual objects in the field. Since this will throw up peculiar stars and galaxies as well as quasars, it is necessary to obtain higher-resolution data to confirm the nature of the candidates.

There has been a dramatic increase in the number of known QSOs in the past 6 years, due to new instrumentation, particularly the introduction of fibre-optic systems and machine measurement of photographic plates, and to more sophisticated multi-colour search techniques. Figure 4 (from Boyle 1992) illustrates the difference between the numbers of spectroscopically identified QSOs at faint magnitudes and $z < 3$ in 1986 and in 1991. At *really* high redshifts the difference is even more dramatic. Prior to 1986 no QSOs were known with $z > 4$; now there are more than 20.

5.1.1 AGN Counts

Complete surveys for QSO candidates result in catalogues of candidate images and their location and observed fluxes, or optical magnitude. Subsequent spectroscopic observation and confirmation of the identifications is lengthy and telescope-intensive (but significantly easier than it would be for ordinary galaxies, due to the presence of strong emission lines). Important information about the evolution and spatial distribution of the population can be obtained directly from the survey itself, prior to the spectroscopic observations, from the observed number-magnitude ($N(m)$) relation. (Usually $N(m)$ is defined as the total number of objects brighter than m observed per square degree; differential number counts – the number of objects observed between m and $m + dm$ – are also frequently plotted, as are z-limited counts in spectroscopically complete samples.) For a spatially uniform, non-evolving population in a Euclidean space $\log N(m) = k + 0.6m$. Sandage (1965) observed a slope, $d\log N/dm \sim 0.4$ for his first sample of ultra-violet excess objects brighter than 16^{m}. Normal galaxies at the same magnitude level have a slope of 0.6. Sandage therefore deduced he was observing an abnormally bright, extremely distant population for which the curvature of space causes the $N(m)$ relationship to deviate from the simple Euclidean power-law, as he had shown previously (Sandage 1961). This initial sample also showed that QSOs were far more numerous than quasars, and that their space density was only a small fraction of that of galaxies. These initial, qualitatively correct, results have been extended and quantified over the past quarter century.

Ever since the discovery of the first quasars, it has appeared from the number counts that the space density of QSOs versus z (or look-back time) had an apparent peak between $z \sim 2$ and $z \sim 3$, as shown in Fig. 5. The *comoving space density* $N(z)$ of a

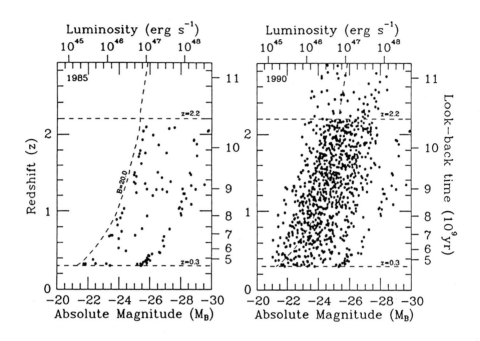

Fig. 4. (a) Absolute magnitude vs. z for all optically selected QSOs identified in complete spectroscopic surveys up to 1986. (QSOs with $z < 0.3$ are not included because of incompleteness problems). (b) As for (a) but now including all such surveys up to December 1990. (Boyle 1992)

particular class of object is defined as the total number of objects per comoving volume element at a given z, where a comoving volume element is one which expands at the same rate as the universe as a whole. Thus the comoving space density of a non-evolving population of objects is constant with time – i.e. independent of redshift. As is clear from Fig. 5, there is significant evolution in $N(z)$.

There has been little agreement between the conclusions drawn by workers from different research groups regarding the evolution of the space density of quasars beyond $z = 2$, some claiming that the space density of high-redshift quasars increases, while others cited evidence for a sudden and dramatic "cutoff" in the numbers of faint quasars beyond $z = 3$. Obscuration by dust can almost certainly be ruled out as an explanation for a observed decline in numbers of QSOs at high-z (Warren and Hewitt 1990).

Recent studies (Boyle 1992) have found that the decline in comoving space density above $z \approx 3$ may not be as dramatic as previously thought, at least at bright magnitudes ($M_B \lesssim -27$), although there are still large uncertainties in the selection effects.

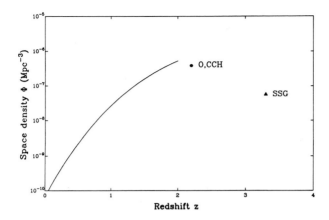

Fig. 5. Space density of QSOs brighter than $M_B = -25.9$ (Warren and Hewitt 1990), calculated from the model fit of Boyle, Shanks and Peterson (1988) - solid line. The filled circle is the space density of quasars with Lyα emission line luminosities, $L_{Ly\alpha} > 10^{38}$ J s^{-1} from the Osmer (1980) and Crampton, Cowley and Hartwick (1989) samples for $z \sim 2.2$ and the filled triangle is from the Schmidt, Schneider and Gunn (1988) third survey

5.1.2 Galaxy Counts

What about the numbers of galaxies? Figure 6 shows the space densities as a function of absolute magnitude, M_B, of normal and Seyfert galaxies (at $z = 0$) and of QSOs at $z = 0.5$ and 2. Figure 6 illustrates a number of critical points: (i) luminosity functions extend over a significant range in luminosity, (ii) the space density of galaxies today is several orders of magnitude greater than the space density of quasars at any redshift $z < 2$, (iii) quasars possess far greater luminosities than the galaxy population as a whole, and (iv) bright quasars were more common in the past than they are today.

To relate the incidence of activity in galaxies to their structure and evolution at $z > 0$ it is necessary to determine when and how galaxies formed. There are several different aspects possibly important in this context: the accumulation of the mass of the galaxy, possibly over an extended period by accretion, or in a number of discrete merging events; the formation of the several components – halo, bulge and disc in spirals; and the formation of stars – both the metal deficient population II stars and the metal rich population I stars which must have formed after the build up of the heavy elements. Since all AGN emission spectra analysed to date appear to come from gas well enriched in heavy elements, it is clear that activity must post-date several generations of star formation in the nucleus. Galaxy formation is clearly beyond the scope of these lectures; however, several aspects of the observations of the distribution and properties of normal galaxies are directly relevant, and I shall now touch on a number of them.

Observations of individual galaxies at different z provide essential clues. At high z ($2 < z < 4$) both young-looking Lyα 'bursters' and old-looking galaxies with broad-band spectra explicable by stellar populations ~ 1 Gyr old are seen (see Peebles 1989). Simi-

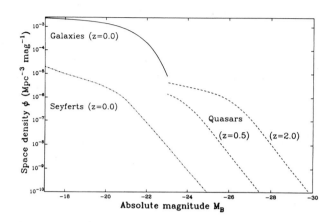

Fig. 6. The space density, ϕ, as a function of absolute magnitude, M_B, for normal and Seyfert galaxies at $z = 0$, and of QSOs at $z = 0.5$ and 2 (Warren and Hewitt 1990b)

larly, at low z (as discussed earlier) both old-looking galaxies, which seem to have formed their stars in a burst early in the universe (and which were long thought to encompass the entire galactic population), and starbursting galaxies are observed. The extent to which the latter population is 'young', and the extent to which they are 'rejuvenated' is unclear – M82 is a good example of an old galaxy undergoing a starburst. Other low redshift galaxies, e.g. intergalactic HII regions and blue galaxies, may be truly young.

In an attempt to understand the evolution of "normal" galaxies, counts of "local" faint galaxies were used to model the expected high redshift population assuming "no evolution": i.e. no episodes of unusual star formation and no change in the total number of galaxies. By using normal stellar evolution models it is possible to evolve the population backwards passively and predict expected number counts of faint blue galaxies (see Guiderdoni and Rocca-Volmerange 1991, and references therein). Compared with the results of these models, one observes an excess of faint blue galaxies by at least a factor of 2 (Broadhurst, Ellis and Shanks 1988). (Interestingly, the non-evolving – or passive evolution – models fit the number counts of red galaxies better than they do the blue counts.) Guiderdoni and Rocca-Volmerange find that density, as well as luminosity, evolution is required to fit the galaxy number counts in an $\Omega = 1$ universe.

In the period of the early 1970s to early 80s the "excess" blue images were assumed to be either protogalaxies at high z or a (burst) population of star forming galaxies at $z \sim 1$. Since then the faint galaxy redshift survey has shown that the "excess" galaxies are rather mainly at $z \sim 0.2 - 0.3$. Thus, there is a population of faint, blue, low redshift ($z \sim 0.2$) galaxies. Since this indicates that we are in fact detecting recent bursts of SF the problem remains: where are the high-z young and/or protogalaxies?

Wolfe and his collaborators (see e.g. Wolfe et al. 1986) have suggested that the damped Lyα absorption systems arise in the *disks* of young galaxies. From their statistical analysis of the frequency of such systems at $z \approx 2.4$ they conclude that the Lyα disks have a cross-section 5–6 times higher than expected on the basis of present-day spiral galaxy with the same column density limit. They conclude that they are detecting the progenitors of disk galaxies, at a time when they were still undergoing gravitational collapse and when most of their baryonic matter was still in the ISM. However, these systems have recently been found to have extensive spectral similarities to present-day HII galaxies (Pettini, Boksenberg and Hunstead 1990; Hunstead, Pettini and Fletcher 1990), and to be underabundant in the heavy elements (with respect to solar) by ~ 2 orders of magnitude, leading Pettini and his collaborators to conclude that they are more likely to be high redshift examples of HII galaxies rather than proto-disks.

Extended Lyα emission objects at high redshift are commonly assumed to be young galaxies, and the search for such objects has focused on Lyα surveys (see e.g. Smith et al. 1989; Spinrad 1989). Yet in a recent detailed study, E. Terlevich et al. (1992) found that young or unevolved galaxies exhibit only weak or absent Lyα emission consistent with emission by photoionized gas rather than a stellar continuum.

IRAS Galaxies. Far-infrared counts are dominated by starburst galaxies and AGN and the IRAS survey is almost unique in that it has full sky coverage. It is thus ideally designed to test the idea that all or many galaxies go through episodes of rapid star formation. Lonsdale (1990) and her coworkers (Lonsdale, Lonsdale and Smith 1992; Lonsdale et al. 1989) find that the IRAS number counts are incompatible with non-evolving (i.e. passive stellar evolution only) galaxy models and that either strong evolution of the population in luminosity or in density (see below) fit the counts. They find that the statistics are consistent with *all* gas rich galaxies experiencing at least one infrared-luminous phase lasting $\sim 10^8$ yrs during the Hubble time. This clearly would have an important effect on the dynamics and the star formation history of all galaxies. Lonsdale further concludes that the fact that a large proportion of FIR luminous (starburst) galaxies harbour AGN "indicates that whatever the triggering and fueling mechanisms of AGN, they tend to go hand-in-hand with far infrared activity. Since galaxy-galaxy interactions seem to be implicated in many far infrared-luminous events, one attractive possibility would be density evolution in which there is an increasing interaction rate with look back time, resulting in a higher frequency of interaction-induced starbursts, and/or AGN, and/or shock heating, without any increase in luminosity of a given event."

5.2 Luminosity Functions

The luminosity function (LF) of a population is defined as the co-moving number distribution (per Mpc^{-3}) as a function of absolute magnitude of a volume-limited sample (i.e. of a sample within a small Δz). The integral of the LF over luminosity gives the number density of the population. The LF is a parametric fit to the observed data for the entire population: only if the objects are long lived do the changes in LF with z represent evolution of individual members of the class; if the objects are short lived, then changes in the LF reflect changes in the characteristics of members of the class at different z convolved with possible changes in the numbers of such objects at the different z.

5.2.1 AGN Luminosity Functions

Schmidt first demonstrated, in 1968, that the properties of the QSO population change with redshift. He showed that the observations could be explained by a rapid change in their comoving space density. Subsequently Mathez (1976) showed that the changes could equally well be represented by an increase in the average luminosity of the population instead. These conflicting interpretations arise because the early LFs were limited to the bright end of the luminosity range, where it was found that the LF was a simple, and featureless, power-law. In that case the change with redshift can be due to a translation of the LF either in space density or in absolute magnitude, as is shown in the cartoons in Fig. 7 from Boyle (1992). It is only since faint QSOs have been able to be observed in sufficient number to determine the faint end of the LF, which fortunately shows a crucial break in the high-L power-law, that it has become possible to begin to disentangle density and luminosity evolution. Such disentangling is essential if we are to relate QSO and galaxy evolution as discussed below and in Sect. 7.

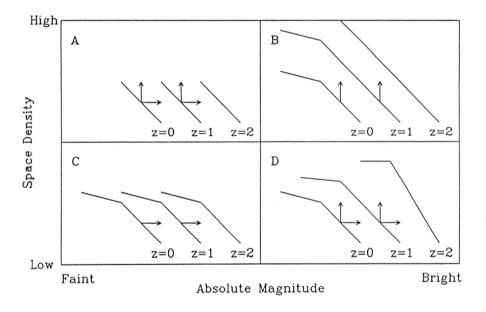

Fig. 7. A schematic representation of the QSO LF to illustrate the differences between density and luminosity evolution (Boyle 1992). (a) Pure power-law LF - as found in the initial surveys. No discrimination is possible, because as the arrows show, movement may have been along either axis. Broken power-laws, as found in the deeper surveys, may be due to (b) pure density evolution, (c) pure luminosity evolution, or (d) both density and luminosity evolution

A full description of the evolution of the LF should ideally include a description of the range and possible changes in the QSO spectral properties, and should yield the evolution of their bolometric luminosities, as well as the changes in their space density. However, because of observational limitations the blue rest-wavelength absolute magnitude $z > 3$ are shown (Boyle 1992).

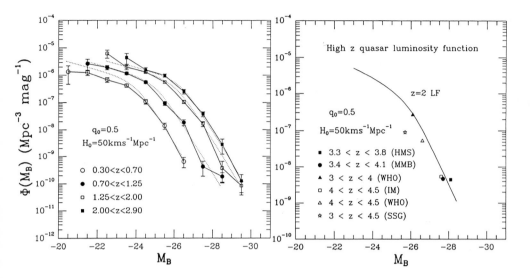

Fig. 8. The QSO luminosity functions for (a) $z < 3$ and (b) $z > 3$. In (a) the redshift bins represent equal intervals in $\log(1+z)$ so that the $(1+z)$ power law luminosity evolution witnessed at low redshifts is represented by a constant shift in absolute magnitude between successive redshift bins. The dotted lines are the Boyle, Shanks and Peterson (1988) model fits to the data. In (b) references are: HMS - Hazard et al. 1986, MMB - Miller et al. 1990, WHO - Warren et al. 1988, IM - Irwin and McMahon 1991, SSG - Schmidt et al. 1988

Boyle, Shanks and Peterson (1988) parameterised the LF by a two power-law form as shown in Fig. 8. The steep power law at high L turns over to a shallower slope at low L. This break in the LF is the crucial feature which allows the analysis of luminosity versus density evolution. Boyle, Shanks and Peterson showed that the data are consistent with a parameterization where the shape and normalization of the LF is invariant with redshift but shifts progressively to brighter magnitudes at higher z – i.e. what is known as pure *luminosity evolution*. They found that models for luminosity evolution as a power-law function of z, $L = L_o(1 + z)^{k_L}$ with $k_L = 3.2 \pm 0.1$ ($q_0 = 0.5$) provided a good fit to the low red-shift data. The strength of the evolution with z is dramatic – the characteristic luminosity in their model is a factor ~ 35 times brighter at $z = 2.2$ than at $z = 0$. They also showed that the extrapolation of the QSO LF to $z = 0$ gives a good fit to the Seyfert data – a possibly important link between the two populations. However the fact that the shift between $2 < z < 2.9$ is smaller than the shift between $1.25 < z < 2$ means that the strong luminosity evolution does not continue beyond $z = 2$. Furthermore, there is no evidence (Boyle 1992) for any decrease in the space density of QSOs between $z = 2$ and $z = 3$. The new results require the addition of a further parameter describing the LF evolution – z_{max} – the redshift at which the luminosity 'switches off': $z_{max} \sim 2$ fits the current data. The uncertainties in comparing observations with models is discussed by Boyle (1992). He concludes that the current situation can be summarized as follows: "A decline in the comoving space density of QSOs at $z > 3$ is only clearly seen at the low luminosities ($M_B > -26$) probed by the Schmidt et al. (1988) survey. Even in this case, correction for incompleteness and the adopted spectral index could significantly reduce the amount of a decline required. Models with no decrease in the comoving number density of quasars at high redshift are certainly consistent with the number density of bright ($M_B < -26$) QSOs at high redshift. ... The lack of any significant decline in the comoving number density of QSOs at high redshifts would pose significant problems for models of galaxy formation which predict a 'late' epoch of galaxy formation, in particular the Cold Dark Matter (CDM) model." Efstathiou and Rees (1988) discuss the implications of this constraint for QSOs observed at $z > 5$.

5.2.2 Galaxy Luminosity Functions

The luminosity function of galaxies, selected without specific regard to type, is well described by the Schechter luminosity function (Schechter 1976). This functional form combines an exponential decrease in number density above a characteristic luminosity L^*, and a slowly increasing power-law tail to low luminosity. Careful detailed study of the luminosity functions of different Hubble types reveals that the LF differs between them; a comprehensive recent review of galactic LFs is given by Binggeli, Sandage and Tammann (1988).

The evolution of galaxies in clusters, where the effects of interactions, in particular, can be studied in detail, are reviewed by Dressler (1984). As he points out, when detailed luminosity functions of galaxies first became available, a striking result emerged: most clusters, the field galaxies, and small groups, showed LF very similar in form, with variations of less than factors of $\sim 2 - 3$ in L^*. Several possible interpretations of this result are possible: either the processes that determined galactic luminosity were insensitive to local conditions such as ρ_{GAL}, T, or turbulence (angular momentum); or these processes varied little from protocluster to protofield regions of space; or evolution of the distribution is not substantial. However, merging, tidal stripping, and accretion in clusters are expected to cause changes in the structure and evolution of the cluster galaxies and therefore to produce some evolution of the luminosity function. If luminosity evolution is partially induced locally over a wide range in z then the fact that at high z only large bright clusters are visible may introduce a false apparent z dependence into the LF.

Whether or not environmental effects are expected to dominate depends on the model of the early universe. Depending on whether galaxies formed before or after clusters, one would, or would not, expect the LF to be dominated by local environmental effects. In order to test the importance of local environmental factors, Oemler (1974) studied 15 rich clusters. He found *weak* evidence for differences in the LF of cluster galaxies depending on whether the clusters are "spiral rich", "spiral poor", or "cD" – those with a giant elliptical at the centre. The small differences he found were confirmed and the results strengthened by the detailed statistical study of Dressler (1978).

There is a tight relationship between Hubble type and the local density of galaxies, ρ_{GAL} (Dressler 1980; Butcher and Oemler 1985, and references therein). Spiral-poor clusters are denser than spiral-rich clusters. In low density fields $80 - 90\%$ of all galaxies are found to be spirals, whereas in high density fields 80 – 90% of all galaxies are elliptical. "cD clusters" contain few other bright galaxies (Dressler 1978).

There are severe problems of selection in the study of galaxy LFs, discussed in detail in Binggeli et al. (1988). Although they state that the "jury is still out" on many of the details of the LFs of clusters, they do "open this Pandora's box" by concluding that "it is now possible to investigate the LF with sufficient morphological type resolution to show that there *cannot* be a universal LF because every type has a specific LF, varying in form from nearly Gaussian to exponential. Hence, the summed LF *must* depend on the type mixture and consequently on the environment."

The relationship of the evolution of AGN and the majority population of "normal" galaxies depends on how the birth, evolution in luminosity, and death of individual AGN are connected with the life history of the host galaxy. The number counts and luminosity functions are compatible with the picture of sporadic bursts of star formation and nuclear activity in the bulk of the galactic population which appears to evolve in fits and starts. It is, however, far from a proven case, and much more data needs to be gathered and analysed, and far more theoretical understanding of the important processes developed.

6 The Symbiosis of AGNs and Their Host Galaxies: Unpeeling the Onion

Up until this point, the evidence that interactions between galaxies may trigger nuclear starbursts has been emphasized. However, once a burst of stars – or an active nucleus – has formed, it almost certainly has a profound effect on its parent galaxy. There is a variety of ways in which the interaction can occur: for example, via the intense radiation field, via gas flows induced by the SB or AGN and via jets from the central region (Dyson and Perry 1986). The jets can interact with the host galaxy either mechanically or via the radiation they emit.

Starbursts, and most particularly their attendant massive stars and supernovae, inject both mass and mechanical energy into the interstellar medium (ISM). The ISM is both heated and compressed by the SN shocks, and it is unclear if this terminates star formation locally, or enhances it. These shocks, and those produced by radio emitting plasmons ejected from the nucleus, will be radiative in the extreme conditions of nuclear SBs and AGN. Because of this, such shocks ultimately produce cool gas which can radiate both broad and narrow emission lines. Radiative shocks in the high gas and radiation density conditions of galactic nuclei were first analysed in Perry and Dyson (1985); they are discussed in more detail in Sect. 10.1. Direct observational evidence for radiative shocks is provided by observations of the NLR – see below.

The effects of the central radiation field on the galaxy can take many forms: the most direct, the simple "illumination" of interstellar gas and clouds is the only one to date which it has been possible to explore observationally, and it is the main subject of this section. The radiation scattered by the ISM and the emission spectra of extended extra-nuclear regions, although weak, have revealed important aspects of the central radiation not otherwise accessible to observation, and have shown also that there is a strong mechanical interaction of the radio emitting plasmons with the ISM of the galaxy.

In principle, various combinations of all the interactions listed above are possible. To a large extent, theory preceded observations in the area of the effects of AGN on their host galaxies. The effects of AGN winds on their parent galaxies were first treated by Dyson, Falle and Perry (1981) and Falle, Perry and Dyson (1981), who showed that if a high-luminosity active nucleus produced a supersonic wind with a kinetic-energy luminosity of $\sim 10^{44}\,\mathrm{erg\,s^{-1}}$ that it could sweep out the galactic ISM and create a thin dense shell at tens of kpc. They suggested that such a wind could explain the associated narrow-line absorption systems (but not the BAL) seen in the spectra of many high-redshift luminous quasars (see also Perry and Dyson 1990). Such winds, representing only 0.1 – 1% of the total luminosity of the system would be expected to affect the evolution of the high redshift systems in which they occur. Winds of that strength would not be expected in low-luminosity low-redshift AGN; however, it is interesting that some starbursts appear to have very strong "superwinds" (see e.g. Heckman, Armus and Miley 1987, 1990) driven by the kinetic energy input from the supernovae and stellar winds. Begelman (1985) studied the effects of X-ray heating on the interstellar gas of the parent galaxy, and in particular the effects on interstellar clouds. He showed that long-lived QSOs could cause runaway heating and thermal expansion of the ISM. Chang, Schiano and Wolfe (1987) considered the effects of radiation pressure on the ISM gas through its coupling to interstellar dust in addition to the effects of the X-ray heating. They assumed the AGN turns on very rapidly, on timescales shorter than the dynamical timescales of the ISM. Because of the efficiency of the coupling of the gas and dust, they found that a radiatively driven shock could propagate, "damaging" the host galaxy on timescales of only a few $\times 10^7$ yrs. A through study of nuclear winds in active galaxies and their effect on the host galaxy has recently been completed by Smith (1991). Amongst the many effects on the galaxy which he analyses he finds that nuclear winds can eject the hot component of the ISM from the galaxy and alter the X-ray profile of the host galaxy. Smith's analysis is applied to elliptical galaxies, and he discusses the interaction of the winds and narrow emission line regions.

The evidence in nearby resolved active spiral galaxies – Seyferts – is that large quantities of cool gas ($\sim 10^4$ K) and dust exist extensively within the inner 1 kpc (e.g. Wilson, Baldwin and Ulvestad 1985) contrary to expectations based on models such as those outlined above. Shanbhag and Kembhavi (1988) showed that proper consideration of the mass exchange between the galactic stellar population and the ISM gas would substantially diminish the stripping of the ISM by an AGN. In addition, asymmetry and obscuration of the central continuum – as inferred by recent observations of radiation cones in extended emission line regions – reduces the effects of the central radiation on the ISM. The evidence for asymmetry and obscuration comes from spectropolarimetry and long-slit observations of the narrow-line emission, to which I now turn briefly. First, however, a note of caution. The interpretation of much of this evidence is controversial; the observations are very recent and there is not yet a statistically meaningful sample from which to draw general conclusions. Almost all of the objects discussed below are nearby, low-luminosity Seyferts or radio galaxies. These results may – or may not – be applicable also to high-z, high-luminosity objects. Only after significantly more observations will we know.

6.1 Spectropolarimetry and Resolved Narrow Emission Line Studies

Important diagnostics of some of the possible effects of the active nucleus on the host galaxy have been found since 1985 through spectropolarimetry, and since 1987 in the extra-nuclear line emission discovered in many low redshift active galaxies which are near enough to resolve spatially. These emission lines also serve as probes of the nuclear emission itself, in interesting and startling ways, in some cases revealing the presence of an otherwise undetected active nucleus. These lines come from two regions (discussed below) which have been called the EENLR (or E^2NLR) and ENLR, to distinguish them from the (often unresolved) 'classical' NLR which gives rise to the prominent narrow lines seen superimposed on the broad emission lines in most AGN spectra.

Spectropolarimetry.

The discovery by Antonucci and Miller, in 1985, that the prototype Seyfert 2 galaxy, NGC 1068, has a 'hidden' Seyfert 1 type spectrum dramatically changed the discussion of activity, and led to the development of 'unified models' for Seyfert galaxies. NGC 1068 normally shows no broad permitted emission lines; yet, when they observed it spectropolarimetrically, Antonucci and Miller found that the polarized component was dominated by a Seyfert 1 spectrum with broad permitted lines and a strong 'non-stellar' continuum. The polarization of the narrow lines is quite different; this accords with the view that they arise in a quite distinct extra-nuclear region. They interpreted these findings as showing that a Seyfert 1 nucleus was hidden from our direct view by obscuring material, but that its radiation escaped and was reflected by the extra-nuclear ISM, both polarizing it and allowing us to observe it. Miller and Goodrich (1990) have since found at least four other examples of 'hidden' AGN in Seyfert 2 galaxies. These observations led directly to the model that a thick torus of obscuring material lies within the narrow-line region, but outside the BLR, and that whether any individual object is seen to be a type 1 or 2 depends on the viewing angle. However, data exists which appears to conflict fundamentally with this model, most particularly the apparent change in the spectra of e.g. NGC 4151 from that of a Seyfert 1 to a Seyfert 2 within 20 years (Penston and Pérez 1984). Since 4151 cannot have precessed within this time, it is likely that some more fundamental difference is involved here other than only orientation. These results are not understood at present. Spectropolarimetry of several radio galaxies has revealed a polarized blue continuum which appears to be a scattered continuum from the nucleus (Jackson and Tadhunter, in preparation) and which does not appear to be integrated starlight.

Line Emission.

The EENLR - The Extremely Extended Narrow Line Region. Recently very extended narrow-line regions at up to 100 kpc from their nuclei have been discovered in some broad- and narrow-line radio galaxies and quasars (McCarthy et al. 1987, Tadhunter et al. 1987, Pérez et al. 1989). They have a rich spectrum which includes some very high ionization species. They are clearly not HII regions photoionized by the continuum of normal stars. Their line spectra are explicable, however, if they are photoionized by either a high temperature ($\sim 10^5$ K) blackbody or an anisotropic power-law central continuum

(Robinson et al. 1987). Their structures and kinematics cover the full range from regular disc/ring systems to chaotic systems for which no pattern can be discerned. The gas motions appear to be predominantly gravitational, although the three systems with the largest velocities show signs of interaction with radio jets. The EENLR are elongated along the radio axis. It appears likely that this is due to anisotropies in the radiation field, although it cannot yet be ruled out that cloud concentrations, possibly because of asymmetric perturbations in the gravitational field, are responsible for this morphology. Tadhunter et al. (1989) conclude that "if there is a single origin for the warm gas in powerful radio galaxies then accretion from an external source is it." Direct evidence for mergers or tidal interaction has been found for several of the galaxies in their sample.

The ENLR – The Extended Narrow Line Region. Within the extremely extended narrow line region and beyond the narrow-line region lies an emission region whose velocity measurements are often consistent with normal galactic rotation (Unger et al. 1987). The emitting gas is supposed to be undisturbed gas in the galaxy, but detailed observations of the kinematics do not yet exist. Detailed long-slit spectroscopy of the nearby Seyfert galaxy NGC 3516, on the other hand, provides evidence of bipolar mass outflow from the nucleus (Ulrich and Péquignot 1980, Goad and Gallagher 1988, Pogge 1989, Mulchary et al. 1992). Its line spectrum is compatible with the supposition that the gas is photoionised by the nuclear source. Detailed photoionization studies of their spectra show that they must be ionized by a continuum which is stronger than that which reaches the earth directly. These 'radiation cones' seen in narrow-band line images – direct evidence for anisotropy – are reviewed by Fosbury et al. (1992).

The NLR - The Narrow Line Region. The 'classical' narrow line region is emitted by gas with densities $n_e \approx 10^4 \, \mathrm{cm}^{-3}$ and velocities widths of the order of 300 to 500 $\mathrm{km\,s}^{-1}$. Photoionization calculations place the NLR at distances from the continuum source of between 10 pc and 1 kpc. Wilson and Heckman (1985) reviewed the NLR properties in some detail. The line widths in Seyfert galaxies are similar to, but systematically higher than, the stellar Keplerian velocities in spiral galaxies with luminosities comparable to the Seyfert galaxies. The line wings, however, extend to much greater velocities than could reasonably be expected on a gravitational origin. High resolution mapping of the NLR of several Seyferts has shown that there is a spatial correlation between NLR and the radio structure, in that the NLR is elongated along the direction of the axis of the radio emission. This suggests that both the ionizing radiation and the radio-emitting plasma may emerge from the central source in the same direction. This led Pedlar, Dyson and Unger (1985) and Taylor, Dyson and Axon (1992) to propose a model whereby the NLR arises in ISM gas compressed within radiative shocks driven by the radio-emitting plasmons; it is photoionized primarily by the nuclear source, and by precursor radiation from the shock fronts. As we shall see in Sect. 10, radiative shocks driven by supernovae may also be responsible for the formation of the clouds emitting the broad lines. We now turn to the construction of models to account for the observations outlined previously.

7 AGN Models: What Are the Observations Telling Us?

What are the primary (non-radio) observations of AGN which must be explained by any physical model of the nuclear regions? The energy source for the bolometric luminosity (ranging from $10^{43}\,\mathrm{erg\,s^{-1}}$ in LINERs up to $10^{48-49}\,\mathrm{erg\,s^{-1}}$ in the brightest QSOs); the existence of the broad emission line gas, its abundances, excitation and kinematics, including the systematic differences between the low and high ionization emission (the LIL and HIL respectively); the broad absorption lines seen in roughly 3–10% of QSOs and the (short) time variability of the line spectra. Furthermore, all of these nuclear properties appear to exist predominantly in disturbed galaxies which often show signs of recent interactions. This latter feature suggests that models which incorporate, in some fashion, effects of interactions – such as enhanced gas deposition in the nuclear regions and/or enhanced nuclear star-formation – may be the most appropriate starting point. It is necessary first to develop a self-consistent pseudo "steady-state" model explaining *all* these features, and then to develop an evolutionary model capable of explaining the number counts and luminosity evolution, including AGN formation. It is again good to bear in mind the stellar analogy: main-sequence models were successfully developed prior to evolutionary models, and star formation and stellar death are still not well understood. We shall, however, consider the implications of the observations in the inverse order and begin by discussing the constraints which can be placed on steady-state models by the evolution of the population of active galaxies (discussed in Sect. 5).

7.1 The Luminosity Functions

Although the pure luminosity evolution parameterization of the data presented in Sect. 5 is a strikingly simple description of the variation of the LF with cosmological epoch, the physical significance of this behaviour is not clear. The standard interpretation of the apparent luminosity evolution of QSOs/AGNs is that either:

(a) *Only a small fraction (of the order of 10%) of all galaxies contain active nuclei.* This is the simplest and most direct interpretation of the data. Such active nuclei must then be long lived (of the order of 10^{10} yrs). The observed LF evolution from high luminosity at high z to low-luminosity at $z = 0$ then represents the slow dimming, with a timescale of 3 Gyr, of long-lived active nuclei born before $z = 2$. The fading, under the conventional BH hypothesis, would be the result of a dwindling fuel supply. The fuel could be provided through mass loss from young massive stars, or by stellar collisions in a dense star cluster and through tidal disruption of stars by the central BH. For example, theoretical studies by Duncan and Shapiro (1983) and David, Durisen and Cohn (1987) of such processes found fueling rates of the right order of magnitude which also decay with time. However, these models decay as $(1 + z)$ whereas the observations require $(1 + z)^2$.

This long-lived hypothesis implies that supermassive black holes must exist in $z \approx 0$ descendants of high-z quasars, i.e. in the nuclei of Seyferts. Most other galaxies today would, by implication, have considerably smaller, if any, central massive-objects. The minimum mass in Seyfert nuclei can be estimated directly from the present-day luminosity, as the sum of the accreted mass, over the redshift interval $0 < z < 2.2$, plus the minimum mass associated with the initial luminosity at $z = 2.2$ (obtained by assuming that QSOs initially radiate at their Eddington limit). It is found (see below) that the typical central masses would then have to be 10^9 to $10^{10}\,M_\odot$.

Observational constraints on the masses of central objects in nearby galaxies are difficult to obtain. However, there is some evidence that masses of Seyfert nuclei do not greatly exceed $\sim 10^8\,M_\odot$, too small for the hypothesis that quasars are long lived (Rokaki, Boisson and Collin-Souffrin 1990; Clavel et al. 1987; Gaskell 1988). We must then consider an alternative picture in which a large fraction of all galaxies go through an active but short-lived phase. The evolution of the LF then reflects behaviour of the population rather than of individual AGN.

(b) *Most galaxies go through a phase in which their nuclei are active.* This may occur only once, or be recurrent. These episodes probably have short active lifetimes (perhaps $\tau_{active} \sim$ few $\times\, 10^7$ to 10^8 yrs). Under the BH hypothesis most galaxies would have to now have nuclear BHs of $\sim 10^8\,M_\odot$; under the pure starburst hypothesis these are not required, and not expected to be present. If this short lifetime interpretation is correct, then the luminosity evolution is of the population, not of individual sources, and presumably represents a decrease in the 'strength' of the episodes – perhaps related to the amount of fuel available or to the number or efficiency of star formation.

This picture fits both the timescale, $\sim 10^8$ yr, associated with the galaxy-galaxy inter-actions which have been proposed as "triggers" for activity (see e.g. Hernquist 1989) and the lifetimes of massive stars in a starburst stellar cluster. In this scenario activity would be recurrent. Evidence that many currently inactive galaxies may have been through an earlier active phase comes from the observation of very low-level QSO-like activity in the nuclei of many spiral galaxies (Filippenko and Sargent 1985), and from evidence for central objects with masses $\sim 10^7\,M_\odot$ in currently inactive galaxies (e.g. Dressler and Richstone 1988, Kormendy 1988).

The fact (discussed in Sect. 3) that at high redshifts quasars lie preferentially in galaxy clusters, while at low z they are found in less rich environments supports this interpretation. Since the analogues of the higher-z AGN are not found at low z, they could have died out in the intervening time.

It is usually assumed that some form of recurrent activity occurs. The possibility, however, cannot be excluded that activity is a single short-lived event. In a model in which activity is short-lived, one expects the LF evolution to result from the combined effects of density evolution due to the change in the frequency of galaxy-galaxy interactions with cosmological epoch, and luminosity evolution due to changes with time of the luminosity of individual members of the population. In the black-hole accretion picture, this would probably be due to a change in the fuel supply rate. This might be a result of changes with cosmological epoch and of the size and mass of the nuclear starburst triggered by interactions, and of the evolution of the nuclear SB stellar cluster (Williams and Perry 1992); or of changes in the rate of the tidal disruption and collisions of stars near the black-hole; or of the depletion of the extra-nuclear sources of gas (Shlosman, Begelman and Frank 1990). In the pure SB model, luminosity evolution is a direct result of the expected decrease in the core star formation rate and of normal stellar evolution of a coeval core cluster (Terlevich and Boyle 1992). At first, the short-lived hypothesis appears more contrived than the long-lived picture, implying a conspiracy of the variations with time of the luminosity of individual objects and their space density in order to match the simple pure luminosity evolution of the LF. However, if interactions control both the space density and the luminosity, these factors are linked and probably reflect directly the spatial and velocity distribution of the galaxy population in an expanding universe. In this case, the dimming of active nuclei may be a result of the decreasing mean density of the universe. These very speculative ideas need exploration before it can be concluded that the short-lived hypothesis is contrived. One model of recurrent interactions was suggested by Cavaliere et al. (1988), which qualitatively reproduces the behaviour of the LF at low z. However, in their model the space density of low-luminosity quasars increases with time, while the observations suggest that the space density remains constant or even declines (Boyle 1992).

Heckman (1991) recently summarized the evidence for the relative contributions of AGN and starbursts to the energy density of the universe. He concludes that the luminosity functions of AGN and starburst galaxies are remarkably similar over the narrow range of $L_{\mathrm{Bol}} \approx 2 \times 10^{10}$ to $5 \times 10^{11}\, L_{\odot}$ where both are well-determined. The total amount of radiant energy produced by AGN over the history of the universe is roughly 10% of that produced by massive stars in the nucleosynthetic production of the heavy elements. He then asks if this similarity in their energetic significance implies a fundamental link between 'true' AGNs and starbursts.

Clearly, more data about the contribution of interactions to activity, and the measurement of the masses of the nuclei of a substantial number of local galaxies, both active and normal, are needed to unravel the connection between galactic evolution and activity.

7.2 Absorption

Intervening Absorption. It is now generally accepted that the vast majority of the narrow-line absorption systems are intervening. Therefore, the study of their distribution and properties has become a study of the distribution of matter in the universe. Their z-distribution appears broadly consistent with the known distribution of galaxies in the universe, if their halos are ~ 30 to 50 kpc in size. There was a considerable controversy when they were first discovered (reviewed in Perry, Burbidge and Burbidge 1978) between advocates of the intervening hypothesis and those who believed they arose in gas ejected from the QSO itself. At the time little independent evidence supported the contention that galactic halos were large enough to account for their frequency. That has now changed, and it is generally accepted that the MgII systems originate in normal galactic halos (Bergeron and Boissé 1991) or star forming regions (Yanny, York and Williams 1990). The origin of the CIV systems is still unclear, but in a recent study Peng and Weisheit (1991) suggest that they arise in clouds strongly ionized by OB stars, presumably in young galaxies.

The damped Lyα systems are usually interpreted as young galactic disks (Wolfe et al. 1986). From the frequency with which they are observed in the spectra of $z \sim 2.4$ QSOs, Wolfe et al. deduce that they must have cross-sections larger than those of present-day disks by factors of more than three. They thus deduce that these systems are disks in the process of collapsing. This is at a time when the first generation of QSOs are already "old" and would therefore place the QSO phenomenon firmly in the epoch of continuing star and galaxy formation. Under this scenario, one is tempted to speculate that their further collapse may lead to the formation of Seyfert galaxies. However, as a result of their very detailed study of the spectral characteristics of these systems, Pettini, Boksenberg and Hunstead (1990) and Hunstead, Pettini and Fletcher (1990) conclude that it is more likely that they are dwarf or HII galaxies.

The abundances of the heavy elements in the different categories of absorption system appear to be significantly different one from another, although abundance determinations are very difficult and the results to date only tentative. The absence of heavy element lines in the Lyα systems led Boksenberg (1978) to suggest that they are significantly under-abundant with respect to solar as is expected for primordial material in intergalactic space. Their low abundances are in accordance with the theory of nucleogenesis in galaxies. Considerable controversy still surrounds the interpretation of their distribution and of their physical properties.

The narrow heavy-element systems which are thought to arise in galactic haloes or in intracluster gas, the MgII and CIV systems, also appear to be under-abundant with respect to solar (Pettini, private communication) particularly at high z. Again, this is consistent if these systems arise in the haloes of young or forming galaxies. As mentioned previously, the damped Lyα systems are under-abundant by ~ 2 orders of magnitude. In their far-reaching spectral analysis of several individual damped Lyα systems Pettini et al. (Pettini et al. 1990; Hunstead et al. 1990) find that the dust-to-gas ratio is also lower than usual in these high-z galaxies and that the star formation rates seem to be below 10 $M_\odot$ yr^{-1}.

The BAL, in contrast to the intervening systems, are 'normal' or even over-abundant with respect to solar (Turnshek 1988); their abundances are equal to or higher than in the emission line regions. If the abundance estimates for these absorption line systems are confirmed, then they will be important constraints on physical theories of active galaxies. These results, taken together with the apparently normal abundances in the emission line gas, would seem to indicate that although most of the detected *gas* at high redshift has not yet been enriched significantly, gas in the nuclei of active galaxies – at all z – has. It is worth noting that the abundance of interstellar material is more rapidly enriched than the population of stars. This *requires* that a significant number of massive stars with their strong winds and resulting supernovae – the only sources of enrichment – precede or be coeval with nuclear activity, at all z up to ~ 5.

Broad and Associated Absorption. The broad absorption lines, which arise within the nucleus or circumnuclear region of the QSO itself, map the velocity field and distribution of gas in the immediate environment of the QSO. The study of the BAL therefore is part of the study of the structure of the ISM or circumquasar flows within the active galaxy. In the BH model, the necessity for accretion, combined with the observation of outflow, implies a non-spherical geometry, although if the accretion is in a thin disc the opening angle for outflow may be substantial. BALs are observed only in moderate to high-z QSOs, and not in Seyferts or radio galaxies. They are thus a reflection, perhaps, of either luminosity or epoch dependent changes in the structure of active galaxies, or of some essential difference between QSOs and Seyferts. The BAL provide direct evidence for gas flows towards the observer of up to $\gtrsim 0.1c$. It is unclear if these velocities are part of a global nuclear or galactic outflow (Weymann et al. 1982; Begelman, deKool and Sikora 1991) or are locally concentrated high-velocity regions, such as supernova remnants (Perry and Dyson 1992). Models based on global outflow have difficulty explaining both the appearance of BAL in only 5 to 10% of all QSOs and the very great (highly supersonic) velocity *width* required within a thin layer of cool gas. If an expanding supernova remnant lies along the line-of-sight both of these difficulties are overcome; I discuss the SNR model for the BALs in Sect. 9.2.1. The "associated" narrow-line absorption systems require 'outflow' velocities of only $\sim 3000 \mathrm{~km\,s^{-1}}$ when the systems are observed. If, however, they represent a thin shell formed by a galactic wind acting as a 'snow plough', then the initial velocity of the wind must have been considerably larger than is observed directly (Falle et al. 1981, Dyson et al. 1981). Although intrinsic absorption systems are important diagnostics of the ISM kinematics they are not yet well understood, and deserve far more attention. It is again worth considering the stellar analogy: atmospheric structures, and thus stellar appearance, depends on stellar type and stage of evolution; the incidence – or not – of strong stellar winds depends also on evolutionary stage and can occur from radiatively or convectively supported stars. There is no reason to expect less diversity in the structure of active galactic nuclear regions.

7.3 The Continuum

The fact that AGN emit broad non-thermal continua indicates that they are not even in approximate thermal equilibrium if the emission is due to a single coherent object. Current models for the radiation processes giving rise to the continuum assume that a wide variety of mechanisms conspires to produce the broad observed spectra: pair creation and annihilation, synchrotron radiation and Compton scattering, dust re-radiation, thermal emission from an accretion disc, radiative shocks and perhaps young stars are amongst the ingredients invoked. The details of the radiation processes are outside the scope of these lectures, and I refer the reader to the excellent review by Blandford (1990) which treats emission mechanisms particularly in the context of the black hole model. In the pure starburst model proposed by Terlevich and his collaborators, the broad continua are assumed to be the integrated spectra of the SB stellar cluster and its compact supernova remnants (CSNR) (R. Terlevich 1990, Filippenko 1992b).

The bolometric luminosity,

$$L_{\mathrm{Bol}} = \int_0^\infty L(\nu)\mathrm{d}\nu \tag{3}$$

corresponds to an equivalent mass consumption rate

$$\dot{M} = L_{\mathrm{Bol}}/(\eta c^2) = 17.5\, L_{47}/\,\eta_{0.1}\ \mathrm{M_\odot\,yr^{-1}}, \tag{4}$$

where $L_{47} \equiv L_{\mathrm{Bol}}/(10^{47}\,\mathrm{erg\,s^{-1}})$ and $\eta_{0.1}$ is the efficiency of conversion of mass to luminosity measured in units of 0.10. In BH accretion models, η represents the efficiency of radiation of accreted mass; it is taken to be 10% implying $\eta_{0.1} = 1$. In pure SB models, η reflects the conversion of mass to energy in nuclear burning averaged over the entire cluster, and depends strongly on the stellar IMF and the star formation rate. It is normally taken to be of the order of 1%.

The total bolometric luminosities, as inferred from the observed redshifts combined with extrapolations of the observed, limited wavelength coverage, fluxes lie between $\sim 10^{43}$ and $\sim 10^{49}$ erg s^{-1}. This implies that mass consumption rates vary between

$$\text{few} \times 10^{-4} < \dot{M} < \text{few} \times 10^3\,\mathrm{M_\odot\,yr^{-1}} . \tag{5}$$

The conversion of thousands of $\mathrm{M_\odot\,yr^{-1}}$ places severe constraints on models of high luminosity QSOs, in contrast to Seyferts, whose luminosities require less than 1 $\mathrm{M_\odot\,yr^{-1}}$.

Lifetimes. No reliable estimates of the lifetimes of individual objects exist. Minimum lifetimes of extended radio sources seen concurrently with active nuclei in some sources can be estimated from the light travel time from the nucleus to the lobes, and are of the order of a few $\times 10^5$ to $\lesssim 10^7$ yr. The evolution of the LF, as discussed above, is compatible with either short (few $\times 10^7$ yr) or long ($\sim 10^9$ yr) lifetimes.

Mass Estimates. The minimum mass of a radiating object can be calculated by assuming that it radiates at it Eddington limit. The Eddington Luminosity, L_{Edd}, is the theoretical maximum luminosity which can be sustained over a long time by any given central mass, M. It is calculated by equating the gravitational force with the force on matter due to electron scattering radiation pressure:

$$L_{Edd} \equiv \frac{4\pi G M m_p c}{\sigma_T} \approx 1.3 \times 10^{46} M_8 \ \mathrm{erg\,s}^{-1} \tag{6}$$

and the equivalent mass consumption rate is

$$\dot{M}_{Edd} = 2.3 \frac{M_8}{\eta_{0.1}} \ \mathrm{M_\odot \, yr}^{-1} \tag{7}$$

where $M_8 \equiv M/(10^8 \, \mathrm{M_\odot})$ and σ_T is the Thomson electron scattering cross-section. Thus, the luminosity to mass ratio (in solar units) of a body radiating at the Eddington limit is $\sim 3 \times 10^4$. If this radiation results from accretion onto a BH, then the BH grows continuously in mass; unless the mass supply also grows the effective luminosity to mass ratio will fall in time. Therefore, to model the luminosity evolution requires modeling *both* the evolution of the BH and of the mass supply. The challenge for BH models is how to channel the required mass supply down to the BH – particularly for high luminosity objects. An attractive possibility – particularly considering the circumstantial evidence linking starbursts as well as AGN to galaxy interactions – is a young, nuclear, starburst stellar cluster which is a reservoir of fuel which releases mass locally (Perry and Dyson 1985; Dyson, Perry and Williams 1992).

Even if very luminous ($L_{47} = 100$) QSOs were as short lived as a few 10^7 yr, masses of the order of $10^{10} \, \mathrm{M_\odot}$ must have been converted into radiation; to this must be added their initial minimum Eddington masses of $8 \times 10^{10} \, \mathrm{M_\odot}$. Thus, minimum 'relict' nuclear masses left behind by the most luminous QSOs should be of the order of $10^{11} \, \mathrm{M_\odot}$. These minimum estimates of the mass apply to both BHs and to starburst stellar clusters. Massive stars radiate at close to their Eddington limit, with effective luminosity to mass ratios observed to be as high as 2.5×10^4; however, the integrated luminosity to mass ratio of a stellar cluster depends strongly on the IMF and on its age (see Sect. 8) and is expected to be significantly below this maximum.

Under the long-lived hypothesis, the luminosity evolution of individual objects is modeled using the Boyle, Shanks and Peterson (1988) parameterization, the minimum mass accreted is found to be $M \approx 10^{13} \, \mathrm{M_\odot}$ for objects with $M_B = -28$, to which must be added their minimum mass at $z = 2.2$ of $M_{Edd} = 3 \times 10^{11} \, \mathrm{M_\odot}$. The long-lived hypothesis applies only to BH models, since the lifetimes exceed by a couple of orders of magnitude the lifetimes of starburst stellar clusters. In this scenario, only very few relict BHs this massive are expected.

Central Supermassive Black Holes? There is no *unambiguous* observational evidence for the existence of supermassive black holes in AGN, although there is evidence from stellar dynamics in the centres ($R \lesssim 10$ pc usually) of several nearby galaxies that large amounts of dark matter reside in very small nuclear volumes (Dressler 1989). The observed rotation curves and central luminosity require luminosity to mass ratios less than 0.05, much less than that of an old stellar population . For example, Dressler and Richstone (1988) find that $\sim 8 \times 10^6$ $M_\odot$ of dark mass is required in the central 100 pc^3 of M32. They argue that this high *dark mass* density requires either a black hole or a cluster of collapsed stars. Recently, Kormendy and Richstone (1992) found evidence for a central dark object, possibly a black hole, in NGC 3115 of 10^8 to 2×10^9 $M_\odot$.

Masses of black holes in AGN are inferred from models of the big-blue-bump and the kinematics of the emission lines (Burbidge and Perry 1976; Malkan and Sargent 1982; Gaskell 1988; Rokaki, Boisson and Collin-Souffrin 1992) and will be discussed in detail in Collin-Souffrin's lectures (this volume). In general these kinematic arguments give BH masses of the order of $10^{7.5}$ to $\sim 10^{8.5}$ $M_\odot$.

7.3.1 X-ray Constraints

Observations of X-rays give important constraints on the structure of the innermost regions of AGN. The X-ray producing regions must have a higher energy density than those producing lower energy radiation. Since AGN spectra are flat in νF_ν, the minimum size for the emitting region, for thermal emission, decreases at shorter wavelengths. Non-thermal emission is more likely to be a source of X-rays than infrared or optical emission, so the X-ray emitting region can be even smaller than this. Also, as discussed above, the rapid variability of the observed X-rays suggests that the X-ray emitting region is very small. Many individual sources, e.g. supernovae, can produce rapid variability, each flare corresponding to a localized event. (This idea would, however, suggest that the X-ray light curve should have certain statistical properties, related to those of a Poisson distribution.)

The picture that emerges from the X-ray observations is of a soft, rapidly varying, central X-ray source which emits out to high energies. This source probably irradiates a cool, optically thick component, covering a large solid angle about the source, which generates fluorescent emission lines and a Compton reflection component which varies more slowly due to a spread of light travel-times from the central source to the reflector and then to the observer. This structure may well be the inner regions of an accretion disc. Outside this, but at least in part inside the BLR, photoionized material produces absorption features in the X-ray spectrum: warm material produces Fe absorption edges, and cooler material obscures low energy X-rays, at least in fainter sources (Pounds 1989, George and Fabian 1991, Turner et al. 1992).

7.4 The Broad Emission Line Gas

The observationally defining characteristic of AGN is the appearance in the spectra of broad emission lines - resonance lines from ions of the heavy elements with line widths characteristically $\gtrsim 3000 \, \text{km s}^{-1}$, but ranging up to $\sim 10\,000 \, \text{km s}^{-1}$ in some objects. The line spectra contain the most detailed quantitative information available about the central regions of AGN (in addition to the continuum). However, the mass of gas *directly* involved in line emission is but a tiny proportion of the gas present. From the flux in the lines, masses of the emitting gas can be estimated directly, and range from $\lesssim 1\,\mathrm{M}_\odot$ in low luminosity Seyferts to only $\lesssim 100\,\mathrm{M}_\odot$ in even the most luminous QSOs. The ISM in the region contains at least $10^4\,\mathrm{M}_\odot$ (Perry and Dyson 1985, hereafter PD). Thus the gas we observe directly is a tracer, not a determinant, of the local dynamics. The breadth of the lines means that very high velocity (almost certainly supersonic - see below) gas is present. This gas must have been accelerated, probably by either gravity, radiation pressure or explosions, or some combination of these.

Any serious model of the BLR must account for the physical conditions in the gas, and its kinematics. Further, the actual origin of the emitting gas must be both plausible and consistent with the physical and kinematical properties. It would seem logical that both the properties and presence of the BLR are closely linked both with the central engine *and* the expected nuclear components (stars, gas, etc.) of the parent galaxy. Finally, since gas flows in the nuclear regions seem inevitable as gas is liberated from stars and is subjected to energy and momentum transfer from the intense central radiation field and the nuclear stellar cluster, it again would seem logical to expect some connection between the flows and the BLR.

The details of the excitation state of the gas, and of the emission processes and of photoionization equilibrium are covered in detail in Collin-Souffrin's lectures (this volume) and I will not cover that material here. I summarize and comment upon only the main results which are relevant to the construction of global dynamical models.

The Standard Model. From the very first investigations of the BLR spectrum, photoionization analysis has had a remarkable success in accounting for overall properties of the spectrum. These models required that the broad line clouds be photoionized by the broad observed (and extrapolated) continuum. Its state is then determined by the ionization parameter – the ratio of the (nominal) radiation pressure to electron gas pressure,

$$\Xi \equiv \frac{L_{\mathrm{ion}}}{4\pi r^2 cnkT} \cdot \tag{8}$$

Here L_{ion} is the ionizing luminosity, r the distance from the source, and n and T the electron density and temperature respectively. If the continuum is radiated anisotropically, then the flux in the relevant direction must be substituted for L_{ion}/r^2. Ξ (introduced by Krolik, McKee and Tarter 1981, hereinafter KMT) is the ionization parameter most suited to dynamical discussions. Because co-spatial gas sees the same ionizing flux, pressure equilibrium implies equality of Ξ. Analysis of the line spectra shows that the line emitting gas must have a high density ($n \sim 10^9$ to $10^{11}\,\mathrm{cm}^{-3}$, be located at $r \sim L_{47}^{1/2}$ pc, and exist as very small "clouds". It was realized very early on that since the clouds must be physically small, they would expand at their sound speed and dissipate in less than a dynamical crossing time of the BLR if they were not "confined". Because early calculations relied on spectra as hard as that of 3C 273, it was found that a two phase equilibrium was possible (see Fig. 9), so that small cold clouds could, in principle, be pressure confined by a hot ISM existing at the Compton temperature of the continuum $kT_{\mathrm{c}} \equiv \langle h\nu \rangle$, where $\langle h\nu \rangle$ is the mean energy of the radiation field.

Problems with the Standard Model. The 'standard' model suffers from a number of difficulties, and these will be discussed in full by Collin-Souffrin. Briefly, (a) the required

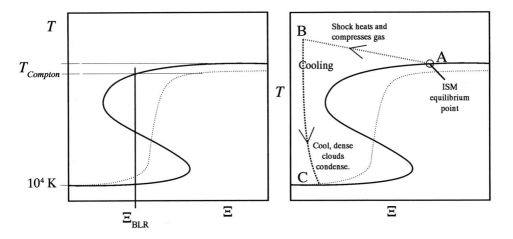

Fig. 9. The equilibrium state of gas photoionized by the broad continuum of AGN: the "S" curve (KMT). To the left of the curve, cooling exceeds heating, to the right heating exceeds cooling. If the spectrum resembles that of 3C 273 then the curve has an S form as indicated by the solid line, and a two phase equilibrium is possible. For more common AGN spectra, however, the equilibrium curve is likely to resemble the dotted curve. In the case of an "S" curve, a possible two-phase equilibrium is indicated on the figure by Ξ_{BLR} in (a). In (b) the effects of strong shocks are illustrated. These are discussed in Sect. 9

confining ISM would be optically thick (AD) and would therefore smooth out any rapid variability in the continuum, in contrast to the observations; (b) most continua are not as hard as that of 3C 273 and the equilibrium curves calculated for 'average' AGN spectra do not have the S form which allows a two-phase equilibrium (Collin-Souffrin 1987); and (c) unless the clouds move subsonically with respect to the ISM they will be disrupted (PD). The latter requirement means that the ISM must be both supersonic and have very large velocity gradients within the narrow confines of the BLR. These constraints are difficult to meet hydrodynamically.

The standard model is appealing in its simplicity and the success of photoionization models is usually invoked in its defence. Yet it is important to note that the *recombination time* is orders of magnitude shorter than any other time scale of relevance to the BLR (flow times, heating and cooling times, etc) so that the observation of a photoionization equilibrium emission line spectrum tells us only that clumps of cold moving gas are exposed to the continuum radiation field – it tells us nothing about their origin, lifetime, or dynamical history. The emission line spectrum is thus merely a *snapshot* of the gas. For this reason, it is not surprising that photoionization models give good agreement with observed line intensities and ratios independent of the dynamical model in which they are embedded; they cannot, by themselves, yield a global, self-consistent model for the BLR. That is the task of dynamics, which must explain the supply of the gas, and the creation and life history of the gas, under the constraints imposed by the line spectra.

Alternatives to the Standard Model. If the standard model of pressure-confined clouds fails, what are the alternatives? Clearly either a new confinement mechanism is required, or the broad line clouds must be continuously produced to replenish those that dissipate. Gravity is perhaps the most obvious alternative confinement mechanism, and the atmospheres of both the accretion disc (see Collin-Souffrin's lectures, this volume) and of stars (Scoville and Norman 1988) in the BLR have been proposed as alternative sights for broad line emission. However, no confinement is actually necessary: for example, SN shocks in galactic nuclei seem perfectly capable of producing and replenishing line emitting cool gas in exactly the right quantities. I discuss such models in Sects. 8 and 9.

Line Profiles. Different geometrical and velocity distributions should be distinguished by different line profiles, and evolution of the spatial and velocity distribution of line emitting material should be distinguishable in the temporal behaviour of the line profiles. For example, in a spherically or conically symmetric decelerating outflow illuminated by a changing ionizing continuum, the blue wing should appear to respond with a shorter lag than does the red wing. In a spherically symmetric chaotic velocity field or in the case of rotation in a disc both wings evolve simultaneously but the core-to-wing ratios may change. The observations clearly reveal that the line intensities change in response to changes in the continuum, and that the profiles also vary. However, to date, no clear systematic trends have emerged in the behaviour of the line profiles. Sometimes there appears to be no change in the profile shape, sometimes the wings vary faster than the cores, sometimes the blue wings vary faster, sometimes slower, than the red wings.

Despite this complex behaviour, one systematic trend has emerged (see Sect. 2). This is the clear separation of the kinematics of the LIL and the HIL. The systematic blue shift of the HIL with respect to the LIL, combined with the differences in the conditions required for the photoionization equilibrium in the two regions, has led to the conclusion that a *two-component model* is necessary to successfully describe the broad line emitting region. Collin-Souffrin's lectures cover these points in detail; in the next section I will address some possible dynamic explanations for the appearance of these two regions.

After many years of observation, however, it appears that no *obviously* universal velocity field has made its presence clear. Given the complexity of the nuclear region, and the expected diversity in, for example, the angular momentum of the system or the possibility of both winds and accretion flows, it would seem logical to conclude that there can be a wide diversity in the detailed motions and distribution of the line emitting gas.

Line Variability. Variations in the broad emission line strengths, in AGN with continuum variability, are a characteristic feature of AGN spectra. This behaviour is most naturally explained as the response of the emitting gas to the changes in the exciting continuum radiation which it is reprocessing. Because of finite light travel times, the observed response of the emitting gas should be delayed with respect to the continuum, and should reflect the spatial distribution of the gas. Therefore, in principle, the study of variability should allow observers to "map" the BLR (Blandford and McKee 1982, Gaskell and Sparke 1986). Reverberation mapping (see e.g. Peterson 1988), as the study of the delay (lag) has become known, arose out of the standard model. If the emitting gas exists in a shell thin in comparison to its distance from the central source, then the measured lag is a direct measure of the distance of the gas from the ionizing continuum. On the basis of the first analyses of AGN spectra which showed remarkably short lags, it was concluded that the emitting gas is significantly closer to the continuum source than the distances determined from photoionization calculations. However, as first pointed out by Penston (1986) reverberation mapping is far more complicated than these simple assumptions would imply. In a series of papers modeling the response for more realistic distributions of emitting gas, Pérez and Robinson and their collaborators (Robinson and Pérez 1990; Pérez, Robinson and de la Fuente 1992, and references therein) have shown that the lag tends to be heavily biased towards the **inner boundary** of a distributed emission line region. They show also that the centroid of the cross-correlation function corresponds, in principle, to the *luminosity-weighted radius*. Their detailed study illustrates some of the dangers in interpreting variability data – they show that even the ideal of measuring the inner radius and luminosity-weighted radius may not be achieved in practice since the results are very dependent on sampling techniques. Furthermore, reverberation mapping *assumes* a stable cloud population. If the emitting gas condensations are ephemeral this would introduce additional complications into the analysis; yet there are good hydrodynamic reasons for supposing that exactly this condition may be common. Thus, although line variability studies place constraints on models, it is not yet clear exactly what the physical interpretation of the observations are.

7.5 Extranuclear Emission

The observations discussed in Sect. 6 have revealed directly that (at least) low-redshift Seyfert and radio galaxies have anisotropic continuum radiation fields. It is unclear to what extent this is due to 'beaming' of the radiation, and to what extent it is due to geometrical shadowing in the circumnuclear regions. Direct evidence for anisotropy is provided by the 'radiation cones' seen in narrow-band line images of Seyfert galaxies (see the review by Fosbury et al. 1992, and references therein).

7.6 Unified Empirical Models

We appear still to be far from our goal of finding an AGN equivalent of the stellar HR diagram; it is not yet clear just how diverse a phenomenon activity really is. Are all the various observations of activity really different manifestations of a single common type of object which is simply observed from different perspectives? Do the apparently different objects represent, rather, different evolutionary stages of the same phenomenon, so that e.g. QSOs evolve into Seyferts? Or is there, in fact, a plethora of active nuclei - perhaps all sharing some similar underlying physics but with a wide variety of structures? It is important to distinguish the primary from the secondary (or superficial) parameters controlling the appearance of active galaxies. By analogy with stars we would expect that the total mass must be a primary factor. Yet clearly the rate at which mass is supplied and consumed may be just as important, and could dependent more on e.g. the galactic structure than on the mass of a central BH.

In an attempt to create order out of chaos many unified schemes have been suggested, founded in the evidence reviewed earlier for anisotropy. A sketch of a synthesis of such schemes (reviewed by Woltjer 1991) is shown in Fig. 10. (See also Blandford 1991)

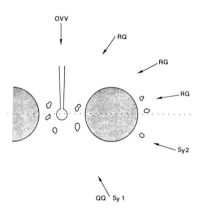

Fig. 10. Sketch of the Unified Model (from Woltjer 1991). The scheme is described in the text

The underlying assumption is that the central region containing an accreting BH and a circumnuclear BLR is surrounded by an opaque, dusty torus. Interestingly, the possibility of such a torus was predicted by Lynden-Bell (1969). In ~ 10% of sources, the BH system powers radio jets ejected along the axis. Most sources, however, are radio quiet. This division is indicated schematically by considering the top half of the figure to apply to radio-loud sources, the bottom half to radio quiet ones. Under these scenarios, OVVs and BL Lacs are sources seen when the observer views them nearly face-on. In that case, relativistic boosting means the continuum source is unusually strong, and the broad lines appear to be weak or absent. When viewed from a larger angle, but still within the opening angle of the torus, a radio-loud quasar will be observed. From viewing angles closer to the plane of the torus, the core is obscured but some of the broad line region is still visible; in this case a broad-line radio galaxy is observed. From closer to the plane yet, a radio galaxy with only narrow lines is observed. Since in all cases the narrow line spectrum is similar, the torus must lie within the narrow line region, allowing the narrow lines to be observed independent of viewing angle. A similar scheme is used to order the radio quiet objects, with QSOs and Seyfert 1s observed if the source is viewed nearly face on; whereas Seyfert 2s are observed when the source is viewed edge-on. The real situation is likely to be far more complex than this simple picture, but, for example, the unification of radio loud QSOs and radio galaxies is supported by the similarity of the form of their LFs (Peacock and Miller 1988).

Evidence which contradicts such a simple universal picture comes, as pointed out before, from the observation of changes in several Seyfert galaxies which are observed at some time to look like Seyfert 1s and within years to look like Seyfert 2s (Penston and Pérez 1984, Véron-Cetty and Woltjer 1990). There are a plethora of evolutionary changes which could effect the appearance of active sources, most of which are far from understood. One of the great advantages of unified phenomenological schemes is that they provide a framework for ordering the data; even if they turn out to be quite wrong they serve a useful purpose while the underlying physics is under exploration.

8 Relating AGNs and Nuclear Starbursts

It is now clear that as some galaxies evolve they form stars in bursts of short duration throughout the history of the universe. Evidence is mounting that there is an intimate connection between starbursts and AGN (indeed, some astronomers maintain that they are in fact one and the same phenomenon); nevertheless, it remains important to keep in mind the observed distinction between them. All the energy output of a starburst can be related *directly* to star formation, whereas that is not **clearly** the case with AGN. Two fundamentally different physical processes may be operating. It is possible that AGN are a subset of all starbursts and that the additional component, which may be a nuclear supermassive black hole, plays a more significant rôle in, say, the more luminous objects.

There are two categories of starburst model for activity in galactic nuclei: those which equate the two phenomena, modeling AGN through a nuclear starburst, with no central black hole or other "monster"; and those which, while accepting the importance of a central supermassive black hole, account for a substantial body of observations in terms of a nuclear starburst stellar cluster surrounding, and feeding, the black hole. I will call these two theories the SB theory and the SB-BH theory. In this section I review briefly these suggested connections between starbursts and AGN and, since the supernovae associated with starbursts play a central rôle in both of the most comprehensive theories for starburst AGN - both with and without central engines - I deal with them in more detail in the next section.

8.1 AGN as Nuclear Starbursts Without Black Holes

Early SB Theories. The earliest suggestions that nuclear activity and starbursts were related were made by Shklovskii (1960) and Field (1964). Shklovskii, in a far ranging analysis of extended radio sources, suggested that the most probable source of the relativistic particles ejected from the nuclei of radio galaxies is supernovae outbursts, and collisions between stars. He estimated that in the first billion years of the evolution of galaxies, when most of the nucleosynthesis was occurring and the gas density was high, the supernova rate would have been two orders of magnitude higher than in normal present epoch galaxies. He concluded that a radio galaxy is a more or less short-term phase early in the evolution of a "normal" galaxy. Field, shortly after the discovery of quasars, extended these suggestions to quasars. He proposed that a high angular momentum gas cloud, collapsing to form a galaxy, would form a small, high density core which would then rapidly form a coeval stellar cluster with masses ranging up to 100 $M_\odot$. He suggested that the resulting stellar and supernova luminosities would be adequate to power quasars of 10^{46} erg s^{-1}. Field pointed out, however, that "a major problem for the interpretation proposed here is the observed light variation." Colgate (1967) also invoked a stellar cluster to explain the luminosity of QSOs, but he proposed that they were the result of the *end-point* of the evolution of a self-gravitating cluster of stars, when a rapid coalescence of stars led to the formation of massive stars which then rapidly became supernovae. Larson (1974, 1977) modeled the star formation in collapsing protogalaxies, and found that as the system evolved, residual gas became more and more condensed at the centre, so that star formation in the nucleus continued long after that in the halo had ceased. He speculated that this continuing star formation could explain QSOs on the basis of Field's models. The common thread through all these early proposals was that the bolometric luminosity could be explained by a high rate of supernovae, and that the high-energy electrons could account for the radio emission observed.

Enter Black Holes. Attention turned away from stellar cluster theories as the arguments for a gravitational origin for the energy generation of QSOs (Lynden-Bell 1969) – through accretion onto BHs – gained favour, and because these early SB models failed to account for the rapid variability of the continuum, for the broad and narrow emission lines, or for their radio emission. All of these features seemed to find a natural explanation in the BH models or, in fact, to form a basis for believing that BHs must be present. The rapid variability is most directly interpreted in light travel-time terms as the maximum radius of the emitting volume. This implies that the emitting volumes are at least as small as the solar system. For the minimum masses estimated earlier it then follows that the only stable configuration is a black hole (Lynden-Bell 1969). Lynden-Bell's suggestion that such a BH could accrete matter through formation of a parsec size accretion disc and thereby generate the observed radiant energy – and (broadly) its correct spectrum – has led to over 20 years of creative and imaginative research into the physics of BH and their immediate environment (see e.g. Rees 1984). For much of this time the excitement and fascination of the "monster" led to an attempt to explain everything solely in terms of the BH, reducing the involvement of the parent galaxy to that of an onlooker.

The parent has been reclaiming much of her lost prominence as attention has shifted back to the rôle of galaxies in the activity at their centres. To a large extent recent discussion of starburst-AGN models has focused on explaining the line emission *as well as* the continuum, since the line emission serves as a direct probe of the region where the stellar cluster, if present, would reside. The first suggestion that a young stellar cluster *surrounding* a massive BH could be important in explaining the origin of the broad emission line clouds was made by Dyson and Perry (1982). Developed in detail by Perry and Dyson (1985, hereafter PD), their model invokes the radiative SN shocks, which result when SN explode in a high gas and photon density environment, to account for the production of cool BEL clouds. These clouds are photoionized by the central continuum and respond to its temporal variations. PD showed that the BLR was produced *self-consistently* with the fueling of the continuum by the mass loss from the stellar cluster. We return to these models below, and in the next section.

Recent SB Theories. At about the same time Terlevich and Melnick (1985) revived the idea that AGN could be explained by starbursts without the necessity for central massive BHs. In their model "nuclear activity is the *direct* consequence of the evolution of a massive young cluster of coeval stars in the *high metal abundance* environment of the nuclear region of early type galaxies." They postulate that this occurs in the final stages of the formation of the central part of the spheroid of a young normal galaxy. They first modeled the early stages of evolution of such a cluster as it passes from being a normal HII region to its appearance as Seyfert 2 galaxy due to the development of "warmers" – bare-core post-main-sequence stars with $T_{\rm eff} \gtrsim 10^5$ K. More recently, Terlevich et al. (1992) have extended their model to include supernovae evolving in a high density region where the SN shocks are radiative. The observed non-stellar UVX continuum draws its energy from the kinetic energy of the SN explosion which is thermalized and radiated as the SNR shocks against dense circumstellar material ejected by their metal rich progenitor stars. The cool broad-line emitting "clouds" are formed in the radiative SNR shells by a mechanism similar to that proposed by PD. In order to explain the inferred bolometric luminosities they require a SN rate of $\sim 1 \text{ yr}^{-1}$ for $L_{\text{Bol}} \sim 3 \times 10^{44} \text{ erg s}^{-1}$ in a SB stellar cluster of $5 \times 10^9 \, M_\odot$ or ~ 300 SN yr^{-1} for $L_{\text{Bol}} \sim 10^{48} \text{ erg s}^{-1}$ in a cluster of $\sim 10^{12} \, M_\odot$. In their models the bolometric luminosity comes from spatially extended regions. The apparent inverse correlation of variability to bolometric luminosity implies that the emitting volume is larger for more luminous sources, and, as they point out, spatial resolution of high luminosity QSOs would be a necessary test of the validity of the model. They show that their model can explain many features of radio-quiet AGN, including – most importantly – the rapid variability in the UVX continuum which would be due to cooling instabilities in the expanding SNR shells (Terlevich et al. ,in preparation). The correlated line-continuum variability is a natural result of the proximity of the line emitting shell to the continuum source. The structure of these supernova shells is described in Sect. 9. Excellent recent critical reviews of the connections between pure starbursts and AGN have been given by Heckman (1991) and Filippenko (1991).

Before turning to proposed SB-BH models we consider several basic properties of starburst stellar clusters, particularly if they surround a nuclear black hole.

8.2 Evolution of Young Stellar Clusters

Modeling the detailed evolution of a young stellar cluster requires knowing the IMF and the SFR. To illustrate the relative luminosities of the various cluster components (main-sequence and post-main-sequence stars and SN) and the rôle of an accreting BH, an example showing the evolution of a coeval stellar cluster formed instantaneously, and a coeval cluster with a decaying star formation rate, is shown in Fig. 11. Both clusters shown obey a Salpeter IMF and have upper and lower mass limits of 1 to 50 $M_\odot$ (Williams and Perry 1992). Results for a wide variety of IMFs and mass limits show similar features to those illustrated here (Williams and Perry, in preparation); the main differences are in the relative length of the SN epoch and the effective L/M of the cluster. It is clear from Fig. 11 that if a nuclear BH exists – and can accrete the mass lost by the cluster and radiate 10% of the accreted rest mass – then the bolometric luminosity so generated will dominate the underlying cluster by 2 orders of magnitude – an increase in the effective L/M of the cluster of a factor of ~ 100. In the case of the "instantaneous burst" cluster, $10^9 \, M_\odot$ generates $L_{\text{Bol}} \sim 10^{47} \text{ erg s}^{-1}$ at $t \sim 3 \times 10^6$ yr, which decays to $L_{\text{Bol}} \sim 10^{46} \text{ erg s}^{-1}$ by $t \sim 3 \times 10^7$ yr. The phase during which SN play an important rôle in the evolution of the cluster and its ISM lasts for 3×10^7 to $\sim 10^8$ yr, depending on the IMF. This time span fits well the proposed lifetime for recurrent activity (Sect. 7.1).

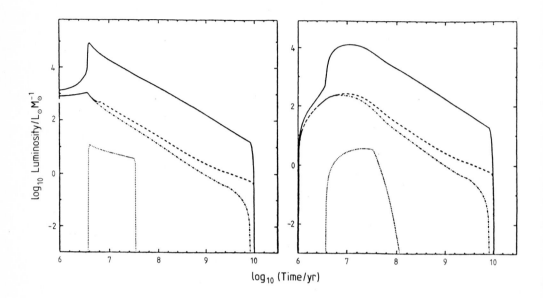

$\log_{10}$ (Time/yr)

Fig. 11. Luminosity per solar mass of a 1-50 $M_\odot$ cluster with a Salpeter IMF (Williams and Perry 1992). The curves shown are: *solid:* accretion of all stellar mass loss with 10% of the rest-mass radiated; *dashed:* total stellar luminosity of the cluster; *dash-dotted:* luminosity of the main-sequence stars only; *dotted:* luminosity of supernovae, if each produces a mechanical energy 10^{51} ergs which is efficiently radiated when the ejecta shock against the ISM. (The luminosity from normal galactic supernovae would be 100 times smaller.) (a) Instantaneous star formation, (b) star formation spread over $\sim 10^7$ yr

8.3 AGN as Black Holes Surrounded by Nuclear Starburst Stellar Clusters

Stellar Motions. If a BH exists at the centre of a galaxy then the Keplerian orbital velocities are

$$v_{\rm orbit} \approx 660(\,M_8/r_{\rm pc})^{1/2}\,{\rm km\,s}^{-1}\;. \tag{9}$$

The sound speed in the ISM depends on temperature; assuming that it is at the Compton temperature of the observed non-stellar continuum, $T_{\rm ISM} = 10^7\,T_7$ K, then

$$v_{\rm sound} = 380\;T_7^{1/2}\;{\rm km\,s}^{-1}\;. \tag{10}$$

Shocks in the ISM. From the above, we see that *all* stars within $r < 3(\,M_8/\,T_7)$ pc move through the ISM supersonically (Perry 1992). Therefore they drive strong bow-shocks into the ISM, and their atmospheres are likewise subjected to strong shocks. Therefore *all stellar mass injection into the ISM – whether through stellar winds, supernovae, tidal stripping or stellar collisions – involves shocks.* This would be true even in a static ISM; a static ISM is, however, dynamically improbable (PD) and we expect that flow velocities in the region will range from $v_{\rm ISM} \sim 3v_{\rm sound} \approx 1100\,T_7\,{\rm km\,s}^{-1}$ in a thermal wind, to perhaps 30 000 km s^{-1} (the width of the extreme wings of some emission lines). Shocks convert kinetic energy into thermal energy; the post-shock gas is heated and compressed, and in the limit of strong shocks the post-shock temperature, $T_S \approx 10^8 \omega^2$ K, (where the shock velocity, $v_{shock} \equiv 3000\omega$ km s^{-1}, is the *relative* velocity between the flowing ISM and the star or SNR and is the vector sum of v_{Kepler} and v_{ISM}). The post-shock density is 4 times the pre-shock density. The effects of such shocks on the ISM can be understood by reference to Fig. 9b. The post-shock ISM is heated to temperatures in excess of T_c and is no longer in equilibrium with the radiation field: its T and Ξ are increased and the point corresponding to the shocked ISM moves up and to the left in the Ξ–T diagram. The gas will then cool along a path determined by the local pressure gradients until it returns to thermal equilibrium. The cooling path depends on the size of the shock; if the gas is trapped within a large, effectively isobaric, post-shock region it will cool along the path **B-C**; if the shock is small, such as those around normal stars, the gas will return to T_c far downstream retracing a path such as **B-A**. We discuss these points again in Sect. 9.

The SB-BH model as originally proposed by PD envisions that a young dense stellar cluster (such as those described above) surrounds a central black hole. The non-stellar luminosity is generated by accretion onto the central black hole. Broad emission line clouds are formed by compression in that subset of the radiative shocks in the ISM which are large enough, so that gas is trapped within them long enough, for cooling to $T \approx 10^4$ K to occur. Such shocks are formed by SN Type IIs and are described further in Sect. 9.

The LIL cannot be explained by these dynamic pressurization models. Collin-Souffrin (1986, 1987) showed that the observed soft X-ray continuum up to a few keV is insufficient to account for the LIL and that hard X-rays must also be absorbed in the emitting region. One condition is that the emitting region must have a very high column density ($\gtrsim 10^{25}$ cm^{-2}). She therefore proposed that the LIL is emitted from the outer regions of an accretion disc. In this case the LIL velocity dispersion reflects Keplerian motions.

At about the time that Collin-Souffrin proposed the formation of the LIL in accretion discs, the evidence for a systematic blue shift of the HIL with respect to the LIL strengthened the case for the necessity of a *two-component* model of the BLR. Collin-Souffrin, Dyson, McDowell and Perry (1987, hereafter CDMP) then combined the model of PD for the formation of broad-line clouds, which they identified as the HIL, with disc models for the LIL. They showed that the systematic differences *and* the correlations between the HIL and the LIL are the natural consequences of the interactions between the nuclear stellar cluster and the central BH. As Collin-Souffrin covers most of the details of the line emission in her lectures (this volume) I will only discuss the dynamics of the ISM, and do so in Sect. 9.

Norman and Scoville (1987) considered a similar, but alternative scenario for the relationship between a nuclear starburst stellar cluster and AGN. They postulate a nuclear starburst stellar cluster of $\sim 4 \times 10^9$ M$_\odot$ formed as the result of a collision of gas-rich galaxies, surrounding a *seed* BH of $\sim 10^6$ M$_\odot$. The stellar mass loss feeds the BH, building it up to the required mass for the accretion to create the observed luminosity. They propose that the broad emission lines are formed in the radiatively heated and ionized stellar atmospheres of red giant stars. However, the number of such stars which is required to account for the equivalent widths of the lines is so large that the accompanying

SN would, by the shock compression scenario outlined above, produce an order of magnitude more broad-line emission than the red giant envelopes. Thus, although I believe that stellar envelopes may well radiate some emission lines, I do not believe that they can represent the major contribution to line emission, particularly not within the first $\sim 10^8$ yrs of the lifetime of a starburst cluster.

Sanders et al. (1988) studied 10 ultraluminous infrared galaxies ($L(8 - 1000 \ \mu\mathrm{m}) \gtrsim 10^{12} \, \mathrm{L}_\odot$) and found that nearly all were interacting, contained typically 0.5–2 $\times 10^{10} \, \mathrm{M}_\odot$ of H_2 and had optical spectra which indicated a mix of SB and AGN energy sources. They proposed that these ultraluminous IR galaxies are the initial, dust-enshrouded stages of quasars. They propose that once the nuclei shed their obscuring dust, through the effects of radiation pressure and SN explosions, that the AGN cores will shine through. They conclude that the origin of quasars is the merger of molecular gas-rich spirals and that this accounts for both the increased number of high luminosity quasars at large redshifts when the universe was smaller and gas more plentiful, and the "redshift-cutoff" of quasars which represents the epoch when the first galaxy collisions were occurring. This model thus provides an observationally based scenario for the initial evolutionary stages of evolution leading to the stage where a "pseudo steady-state" model such as those of PD or Terlevich et al. would be operative.

8.4 Triggering of AGN and Galactic Structure

The essential requirement for a SB-AGN model is a massive, coeval, nuclear starburst and in the SB-BH models a seed black hole. Rees (1984) has discussed the formation of such black holes. The major development in recent years in the study of possible formation scenarios for nuclear starbursts has been the study of galaxy-galaxy interactions and their ability to dump the gaseous content of the galaxy or galaxies into the nuclear regions. Because the gas is dissipative, it can reach the nucleus in a dynamical timescale; while the stellar content of the galaxy is disturbed and forms large scale structures it cannot concentrate in the nucleus. Hernquist (1989) and Hernquist and Barnes (1991) discuss such models as do Noguchi and his collaborators (Noguchi 1988b, and references therein). These models are very exciting and as they are explored we will learn far more about the importance of local galactic density, impact parameters and velocity of interaction.

There can be only small differences between the nuclear ($r \lesssim 10$ pc) ISM (and probably also the nuclear stellar clusters) of radio loud (RL) and radio quiet (RQ) AGN, as there is no *significant* difference between the emission line spectra of the two. Yet a strong correlation of radio luminosity and morphology appears to exist with galaxy type, with RSs in ellipticals and RQs in spirals. It is unlikely that the central engine can be strongly determined directly by the outer galaxy. Therefore, it may be that the features of the central engine controlling its radio properties (probably the angular momentum of the core) are determined by, or simultaneously with, the *formation* and/or *evolution* of the outer structure. Since collisions between galaxies appear to be linked to both starbursts and AGN it may be that the impact parameter and strength of the collision (or the galaxy types of the interacting pair, i.e. spiral-spiral or spiral-elliptical) determine simultaneously both the outer structure and the eventual character and strength of the nuclear activity. In particular the evidence that elliptical galaxies may be formed by a series of interactions between spirals, coupled with the association of strong radio sources with elliptical galaxies only, is suggestive but unexplored.

8.5 Modelling Cosmological Evolution

The time lags involved in the formation of AGN are a significant constraint on theories both of AGN, and of the development of structure in the universe. Strong metal lines are observed in the highest redshift objects: at least some gas must have been processed in stars and subsequently released through SN explosions even at these early times in the universe. Equally, to form the observed objects at least $10^8\,M_\odot$, and perhaps as much as $10^{12}\,M_\odot$, of material must have collapsed into a tiny volume in the available time, from initial conditions assumed to be almost smooth.

The strength of these constraints can be judged from Fig. 12 – the observation/history diagram (Perry and Williams 1991). In this figure, time lags are plotted against redshift for a standard ($H_0 = 50$, $\Omega = 1$) cosmological model. The redshift of an observation, z, is plotted horizontally. The various curves translate this into the redshift, z_i, at which an object formed, for various time lags between formation and observation: the formation redshift is plotted vertically.

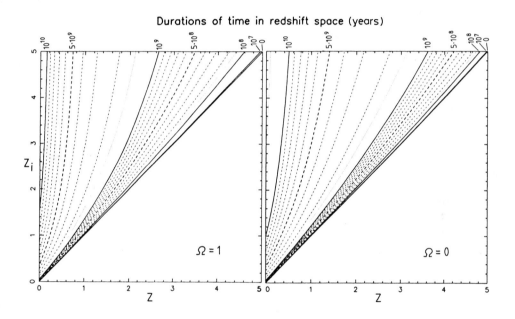

Fig. 12. The observation/history diagram. The diagonal line is for zero delay; above this, the solid lines show lags of 10^7 yr (almost hidden in the diagonal line), $10^8, 10^9\,and\,10^{10}$ yr, the dotted lines intermediate times

For instance, consider an object which is observed at redshift 2.5. Lags of 10^7 or even 10^8 years do not take us back by a significant redshift interval. A lag of 10^9 years, however, requires the object to have formed at a redshift of almost 4.5. The constraints are stronger still for higher observed redshifts, as can be seen from the figure, and for universes with low density, $\Omega < 1$. As discussed earlier, such early formation times may be in conflict with e.g. the CDM theory of formation of structure in the universe. The non-linearities inherent in construction of number counts and LFs can be demonstrated graphically. In the Appendix, I describe how different scenarios for AGN evolution can be visualized easily, and how the observation/history diagram can be used to generate $N(z)$ and LFs for the different cases.

9 Supernovae in Nuclear Stellar Clusters

Having discussed the circumstantial evidence for the association of starbursts with AGN it is now time to consider the detailed dynamical consequence of this association. It is well known that the primary source of energy in the ISM of normal galaxies is the population of SN. SNR are the primary sources of the hot phase of the ISM ($T \gtrsim 10^6$ K) and they largely determine the ISM pressure. They generate ionizing photons and cosmic rays, and are thought to play a rôle in triggering star formation. In a system with orders of magnitude more SN per year in a smaller volume, it is would not be surprising if the SN played a very significant role in the dynamics and evolution of the ISM. This is independent of the possible presence of a central black hole. A black hole, however, increases the L/M of the cluster (as discussed in Sect. 8) and the models I personally favour (and have been instrumental in developing) include a central black hole as an essential ingredient. The observational evidence for a top-heavy IMF, although inconclusive, is sufficiently suggestive (as discussed earlier) to warrant serious consideration. None of the models depend on a top heavy IMF; although a top heavy IMF, because it increases the L/M of the cluster, is an energy and mass efficient way of creating an AGN.

9.1 "Normal" Supernovae

In the ISM of normal galaxies the hydrodynamic interaction of a single SN with its environment is, in principle, a well posed problem. Yet the phenomenon is far from fully understood. One of the difficulties is that the environment depends on the the state of the ISM and on the details of the pre-supernova evolution of the precursor star. An interesting, brief review of the hydrodynamics of SNR has been given by Dyson (1986). The evolution of Type II SN is of particular interest in the AGN context, as these SN are associated with massive ($M \gtrsim 8\,M_\odot$) short-lived stars typical of young starburst populations. Normally born in dense molecular clouds, their OB progenitors produce both large numbers of ionizing photons and strong stellar winds. These processes have competitive effects on the immediate environment into which the SN explodes. The ionizing photons alone would cause an expanding HII region to evacuate the region; however the winds create various configurations of high density shells depending on whether they are fast or slow winds. This in turn depends on their initial mass. Thus the details of the evolution of SNR even in normal galaxies are varied and complex. Nevertheless it is possible to describe a basic three phase evolution in an idealized smooth ISM and to then consider the modifications required by more realistic circum-SN environments.

In the first phase of the expansion the ejecta are not yet decelerated by the ISM and they move out in essentially free expansion. They drive a forward shock into the ambient gas, and a reverse shock moves into the ejecta and heats it. An (unstable) contact discontinuity separates the two shocked regions: the ISM is heated to high temperatures

($T \gtrsim$ few $\times 10^8$ K) and cannot cool; however, the shocked ejecta is heated to only moderate temperatures in the early stages and can cool well, producing soft X-rays. The flow decelerates, so the contact discontinuity between the cool ejecta and the hot shocked ISM is Rayleigh-Taylor unstable, producing a turbulent region which may generate radio emission.

Once a remnant has swept up its own mass of material, and therefore has decelerated significantly, it forgets its early history and enters Phase II – the Sedov-Taylor phase. In the (relatively) low density environment of normal galaxies radiative losses are low and the expansion is adiabatic. The inner shock disappears and the pressure in the SNR bubble is almost uniform. This phase lasts until the cooling time drops below the dynamic time scale, at which point catastrophic cooling occurs and a thin shell is formed in the swept-up material. The presence of density and pressure gradients in the ISM can have profound effects on the evolution once shell formation occurs.

Phase III is the final phase, during which radiative losses dominate. In the classical snow plough model momentum conservation is assumed; however the hot SN bubble can continue to exert pressure and drive the outer shell, with a pressure which drops adiabatically – this is the modified snow plough model. The shell velocity eventually drops below the general dispersion velocity of the interstellar clouds and the shell merges into the general ISM.

This basic interaction is well understood and has been quite successful in explaining the overall pattern of the observations of SN. However, in fitting it to individual supernova, modifications have to be made in the details to take into account, in particular, variations in the circum-SN environment which can lead to e.g. formation of dense filaments and condensations, and multiple reflected shocks within the region. Amongst many of the deviations from the "classical" picture are the observations of e.g. clumpy optical emission accompanied by extended, hard X-ray, diffuse emission.

In the high gas and photon density environment of galactic nuclei the evolution of a supernova is considerably modified from the above picture. Most importantly, the Sedov-Taylor phase does not occur, as radiative losses set in by the time the SN has swept up its own mass. It is these *radiative SN* which are the subject of the rest of this section. I first discuss the rôle they play in models of AGN which are centred on a nuclear black hole, and then turn to the more radical proposals for AGN without BHs.

9.2 Black-Hole Starburst Models

If a starburst stellar cluster surrounds a BH, then when a supernova explodes within the inner several parsecs it does so while moving at a velocity which is supersonic relative to the local ISM (see Sect. 8.3). Its *initial* expansion will be quasi-spherical – so long as the ejecta velocity, (initially $v_{\mathrm{ejecta}} \approx 10\,000$ to $15\,000$ km s^{-1}), has not been decelerated to below the systematic velocity relative to the ISM. However, the shock in the ISM will rapidly open out and develop into a "classic" bow-shock. The high gas and photon density have as their result that the ISM shock rapidly becomes radiative, as I now discuss. The development of a SN in a galactic nucleus is sketched in Fig. 13.

Radiative SN and Production of the HIL.   Photoionization studies suggest that the HIL form in small gas clumps at a radius $r \lesssim L_{47}{}^{1/2}$ pc from the central continuum

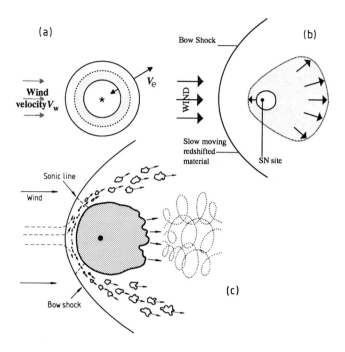

Fig. 13. Sketch of the evolution of a supernova in a high density wind. In the initial phase of the expansion (a) a fast forward shock moves outwards into the ISM at $v_{\text{shock}} \approx v_{\text{ejecta}}$; shocked ISM gas reaches high temperatures ($T \approx 10^8$ to 10^9 K). A reverse shock moves into the ejecta, initially at a very low speed, but accelerates with time. This sketch applies to both the models of PD for SN near BHs, and to those of Terlevich et al. for SB only AGN models. SN which occur in a situation where the pre-supernova moves supersonically with respect to the ISM (as in the models of PD) will soon be surrounded by open bow shocks and their further evolution follows (b) and (c). In (b) the structure of a supernova remnant in a wind at the stage when BAL absorption will be observed is shown. This phase is discussed below. In (c) the structure of the remnant and shock during the HIL phase is shown. This phase is described in detail in the text

source. These clumps have a temperature of 10^4 K and a density around $10^9\,\text{cm}^{-3}$. How can a passage through a shock *cool* the ISM gas temperature from $\sim 10^7$ K to 10^4 K?

The most important physical parameter for the gas is neither the density nor the temperature but their product: the pressure – which determines locally the value of Ξ (Sect. 8). The ISM of AGN must be optically thin – X-ray observations suggest it has extremely low column density – so it will be at the Compton equilibrium temperature of around 10^7 K. HIL clouds cannot be pressure confined by the ISM; however, bowshocks increase the local pressure to the *stagnation pressure* of the flow. PD showed that optically thin flows in galactic centres with mass fluxes about equal to the accretion rates required to feed the central BHs, $\dot{M}_{\text{ISM}} \approx \dot{M}_{\text{acc}}$, will have stagnation pressures about equal to the pressures observed in the broad emission line gas. Thus, if this gas is shocked, and if the post-shocked gas has time to cool isobarically, it will fragment to form HIL clouds. Fig. 9b shows how the process works. ISM gas exists at the Compton temperature (point **A**) until it encounters a shock. The shock heats (T increases) and compresses (Ξ decreases) the gas, to point **B**. The gas flows through the shock before re-expanding into the ISM downstream. A schematic structure for an HIL producing SN shock is shown in Fig. 13. As the gas accelerates downstream the pressure drops, and Ξ consequently increases. If the shock is spatially extended, the region within which the gas is subsonic (with respect to the system in which the shock is at rest), and effectively isobaric, is correspondingly large, and the gas will cool along the locus **B-C**, first by Compton scattering the incident radiation, then by bremsstrahlung and heavy-element line cooling. The latter is thermally unstable: cool dense clouds condense out of the flow. If the gas reaches this stage an evanescent population of HIL clouds is continually produced behind the bow shocks, and then continually destroyed as they expand back into the global, low-pressure flow. Thus the crucial question is whether or not the gas remains at roughly the stagnation pressure of the flow long enough to complete its cooling. To answer that question we must determine the size of the shocks.

How large are the shocks? The size of the shock in the ISM depends on the size of the obstacle causing the shock to form: the effective size of the star, or of the supernovae bubble. These obstacles are all deformable bodies whose effective size is determined self-consistently by the requirement for approximate hydrostatic equilibrium at the contact surface between the shocked ISM and shocked stellar material.

A star will be compressed by the shock down to that point in its atmosphere where the pressure is p_{stag}. The bowshock will be very small, and the cooling time for gas from $T \gtrsim 10^7$ K is so long that the gas cannot cool before it re-enters the ambient ISM.

A supernova, (or disrupted star), on the other hand, injects a prodigious amount of energy into the region and it expands until it reaches approximate pressure equilibrium with the shocked ISM. The pressure of the fully expanded bubble is

$$p_{\text{SNR}} \approx E_{\text{SN}}/V , \tag{11}$$

where E_{SN} is the energy injected ($E_{\text{SN}} \equiv 10^{51} E_{51}$ erg), and V is the volume of the bubble. Hence $V \approx E_{\text{SN}}/p_{\text{stag}}$ giving

$$r_{\text{SN}} \approx 0.09(E_{51}/p_{0.01})^{1/3} \text{pc} , \tag{12}$$

where $p_{\text{stag}} \approx (n_{\text{ISM}}/10^5)\omega^2 \equiv 10^{-2} p_{0.01}$ dynes (remember that $\omega \equiv v/(10^{-2}c)$).

Behind the shock $nT \approx p_{\text{stag}}/k \sim 8 \times 10^{13} p_{0.01}$ (in agreement with photoionization modeling of observed HIL spectra which find that $nT \approx 10^{14}$). Detailed calculations by Innes and Perry (1992, in preparation) show that (as indeed estimated by PD) bow shocks of the size given by (12), near an AGN continuum source, cool well. Catastrophic cooling then results in the continuous production of "cold" line emitting gas. The gas fragments into "clouds" with a spectrum of column densities whose maximum is $N_{\text{H}} \sim 10^{22} \text{cm}^{-2}$ (PD).

There is a natural maximum radial distance from the continuum source within which cooling is effective: the radiation field, and probably the ISM density, decrease $\propto r^{-2}$, and with them the cooling. PD found that $r_{\text{max}} \sim L_{47}^{1/2}$ pc, in agreement with distances deduced from photoionization estimates. It is important to note that clouds are formed by this process *throughout the region* $r < r_{max}$ and that the small reverberation distances observed are therefore compatible with this model (see Sect. 7.4).

Cloud Production, SN Rates and L_{Bol}.   A conservative estimate for the yield of cold gas by the shocks is found by assuming that only gas entering the shock within about $r_{SN}/2$ of the stagnation line cools isobarically. Then $\dot{M}_{cold} \approx (\pi/4)r_{SN}^2\rho_{ISM}v_{ISM}$ per SN. The lifetime of the cold gas (between cooling and downstream re-expansion and re-heating), is considerably greater than the post-shock cooling time, $t_{life} > t_{cool}$. The mass of cold gas produced per supernova is therefore $M_{cold} \gtrsim \dot{M}_{cold}t_{cool}$. The observed equivalent widths, which are remarkably constant from object to object, yield directly the ratio M_{cold}/L_{Bol}. This in turn determines directly the supernova rate required in the cluster, $\dot{N}/L_{47} \lesssim 0.7\,\mathrm{yr}^{-1}$ (PD). The stellar cluster simulations (Sect. 8.2) show – assuming a central BH accretes and radiates the mass-loss – that $\dot{N}/L_{47} \lesssim 0.6\,\mathrm{yr}^{-1}$ after about 5×10^7 yr (in remarkable agreement with the rate deduced above). For a bright QSO with $L_{47} \sim 1$, a cluster mass of $M_{cluster} \sim 2 \times 10^9\,\mathrm{M}_\odot$ is required. If accretion onto the BH is approximately Eddington, a BH mass of $M_{BH}/L_{47} \sim 8 \times 10^8\,\mathrm{M}_\odot$ is required. This scenario produces the *minimum mass starburst stellar cluster* required to produce – self-consistently – both the bolometric luminosity and the HIL broad line region.

HIL Ionization Parameter. An important feature of the observed HIL spectrum is the near constancy of the observed ionization parameter, $\Xi \sim 0.5$. The HIL is observed when cooled shocked gas is in the process of rejoining the flow, so $v_{shock} \approx v_{ISM}$ (PD). Then (Perry 1992)

$$\dot{M}_{ISM} \approx 4\pi r^2\,\rho_{ISM}\,v_{ISM}\ . \tag{13}$$

Behind shocks,

$$nkT \approx \rho_{ISM}\,v_{ISM}^2\ , \tag{14}$$

and thus

$$\Xi_{shock} \approx L_{ion}/(\dot{M}_{ISM}\,v_{ISM}c)\ . \tag{15}$$

Since the luminosity is produced by accretion from the ISM we can write $L_{ion} = 0.1f_{Bol}f_{acc}\dot{M}_{ISM}c^2$, where $f_{Bol} \equiv L_{ion}/L_{Bol}$, $f_{acc} \equiv \dot{M}_{acc}/\dot{M}_{ISM}$ and $\dot{M}_{acc} \equiv L_{Bol}/(0.1c^2)$. Therefore

$$\Xi \approx 0.1f_{Bol}f_{acc}(c/v_{ISM})\ . \tag{16}$$

From observations $f_{Bol} \sim 0.1$; the hydrodynamics and mass conservation through the region implies $f_{acc} \sim 1$ and the width of the broad lines implies $v_{ISM} \sim 0.02c$. Thus $\Xi \sim 0.5$, in agreement with observations.

   As discussed in Sect. 7, these radiative shocks cannot produce the observed LIL. However, in the situation outlined above, with the cluster mass-loss feeding a nuclear BH, the LIL arise naturally in the outer regions of the accretion disc and a two-component emission line region is created, as shown in Fig. 14.

The two-component model of AGN proposed by CDMP, is illustrated in Fig. 14 and also shown in the accompanying "flow chart", Fig. 15, which outlines the process schematically. An accretion disc feeds a central black hole. In the innermost regions, non-thermal processes dominate the accretion: the rapidly time varying power-law continuum is generated in the very central regions. Outside this region, gas is driven away over a wide solid angle; the continuum escapes freely outside a central region possibly optically thick to Compton scattering.

Slightly further out in the disc, emission becomes thermal or photoionized. Energy input into the disc from viscous processes and back-scattered continuum radiation is radiated away. This radiation takes several forms. In the hot central, fully ionized regions thermal processes produce the UV bump. Radiation pressure can also drive a strong wind from the disc surface. At larger radii yet, the X-ray continuum incident on the disc surface produces an emission-line spectrum typical of high density, low-ionization gas. The LIL spectrum observed in AGN agrees well with that predicted by these models.

A particularly appealing feature of this model is its natural explanation for the systematic blue-shift of the HIL relative to the LIL. The LIL has little or no radial motion, whereas the HIL motion contains a systematic ISM flow component. If this is outflow, as is likely, then obscuration by the disc of the receding component of the flow will result in the observed systematic blue-shift.

Are the BALs SNR?   Recently, Perry and Dyson (1992) analysed the evolution of SN which lie somewhat outside the BLR where the ISM density has dropped to perhaps $n_{ISM} \sim 10^4 \, cm^{-3}$. They find that SNRs naturally produce regions of cool very fast moving gas which would – if it happened to lie on the line-of-sight – explain the appearance of the BALs seen in $\sim 10\%$ of QSOs. In Fig. 14, a schematic representation of the BAL has been added to the original diagram of CDMP.

In the early stages of the SNR expansion, a reverse shock moves into the high speed ejecta, accelerating back towards the explosion centre. The ISM is shocked to temperatures exceeding $10^8 \, K$. However, the temperature of the shocked ejecta, T_{se}, is initially low, since it goes through the reverse shock at a low relative velocity; T_{se} increases with time as the relative velocity between the ejecta and the shock increases. About 10% of the post-shocked ejecta has temperatures less than a few $10^4 \, K$ while still moving at velocities greater than $8\,000 \, km\,s^{-1}$ (Perry and Dyson 1992). The ejecta slows down in $t_{stop} \sim 3(M_{ej}/n_4)^{1/3}V_{ej}^{-1}$ yr ($n_4 \equiv n_{ISM}/(10^4 \, cm^{-3})$, M_{ej} is measured in units of $M_\odot$, and V_{ej} in 0.1c.) Consider a SN which ejects 5 $M_\odot$ at $V_{ej} \approx 0.5$. Then $t_{stop} \sim 10$ yr; for a QSO at $z \sim 2$ this appears as ~ 30 yrs in the observer's frame. Such very fast moving cool ejecta does not exist for any appreciable time in the broad *emission* line region where $n_4 > 10$, explaining why the BAL are broader than the emission lines. Ejecta moving toward the central source slow down far more quickly and expand to a smaller radius than ejecta moving away from the source provided only that the ISM is outflowing locally – even in a modest thermal wind. Thus only blue-shifted troughs will be observed. The BAL may provide a discriminant between starburst models for AGN with and without a "central engine", since the BAL producing region must cover the non-thermal continuum source. In pure starburst models, which require a coeval starburst of over $10^{11} \, M_\odot$ for luminous QSOs, it is difficult to imagine a mechanism which can create a thin shell with the required velocity dispersion which can cover a continuum-producing region of at least several hundred parsecs in extent.

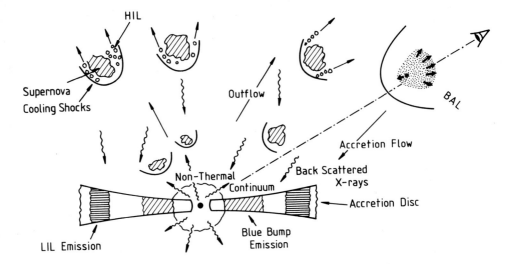

Fig. 14. A possible structure for AGN. The sketch shows schematically the two-component model of AGN proposed by CDMP. The radial coordinate is 'logarithmic' to allow physical processes responsible for observed AGN properties to be shown at various radii. The model is described in the text

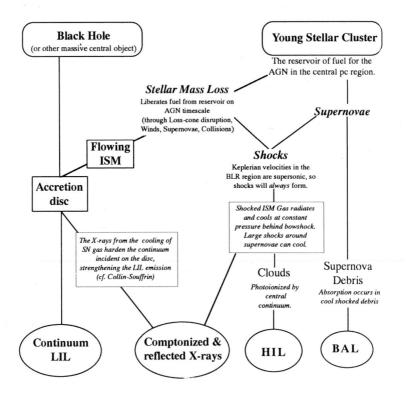

Fig. 15. Flow chart of CDMP model

9.3 Pure Starburst Models

The circumstantial evidence linking interactions, nuclear starbursts and the observed phenomena of *activity* naturally suggests the possibility that AGN are simply extreme starbursts without requiring any additional component, such as a supermassive black-hole. Terlevich and his collaborators have considered this scenario in great detail (briefly reviewed in Sect. 8.1). A critically important component of their models is the behaviour and appearance of SN which explode in a very high gas density environment. In a nuclear SB the ISM pressure will be orders of magnitude higher than in the normal galactic ISM (Chevalier 1991 estimates that $(nT)_{SB} \sim 10^7$ cm^{-3}K). Supernovae Type II are known to have progenitors with strong winds – whether these winds are fast or slow depends on the mass of the star. In the high pressure ISM of a nuclear SB such winds cannot expand freely and form dense shells into which the SN explodes. If the shell has a density $n_{shell} \gtrsim 10^7$ cm^{-3} the post-shocked gas cools efficiently by bremsstrahlung, cooling setting in within about one year of the explosion. Figure 16 shows such a compact radiative SNR.

The SN in these dense ISM develop into rapidly radiating remnants, with two massive, and geometrically thin, concentric shells moving at large speeds and bounded by radiative shock waves. A small fraction of the swept-up mass lies between the two shocks and remains at very high temperatures (about 10^8 K). The wide range of gas temperatures in the cooling region results in a 'power-law' like spectrum. The hot cavity emits by free-free in the hard X-ray region; most of the luminosity of the (lower temperature) reverse shock is in the soft X-ray and UV. The maximum luminosity is reached about 2.5 years after blast, the total X-ray and bolometric luminosities rapidly decaying thereafter. Most of the broad line luminosity comes from the two shells and the ejecta, as shown in Fig. 16. The model predicts long-term variations of a few hundred days with total energies of about 10^{51} erg, with superposed flares of 20- to 50-day duration and energies of about 10^{49} erg. The flares should have a much softer spectrum than the long-term component. This model has a large number of very attractive features; not all aspects of the evolution of SN in these environments have yet been worked out, so it is too early to decide if it can actually explain all the observed features – even of low luminosity AGN. The models make no claim to explaining radio loud AGN, which, Terlevich proposes, may be a separate class of object. Filippenko (1992b) has given an extensive and fair evaluation of the strengths and weakness of the current version of the model and I refer the reader to his article.

10 Concluding Remarks

Activity in galaxies may turn out to be "normal", in the sense that it may be a natural phase in the life of all galaxies, intimately connected with the formation of stars and the galaxy itself. It used to be thought that stars and galaxies were formed early in the history of the Universe (at $z \lesssim 20$) and thereafter aged passively. We now know that 'young' galaxies exist even at $z < 1$, that galaxies run into each other at all z down to the present and that the slow waltz they perform together causes their shapes to change and accelerates their interstellar gases which then can accumulate in their nuclei. There is mounting evidence that these motions can instigate new star formation, sometimes of

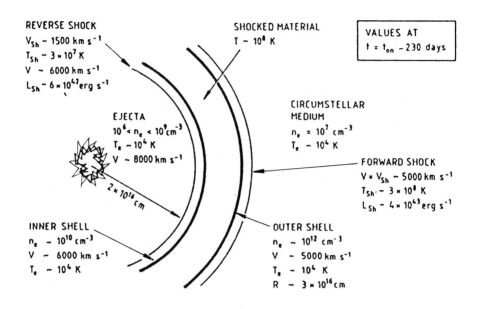

Fig. 16. A compact SNR at the onset of broad line emission. (Terlevich et al. 1992).There are three sources of ionizing radiation: the two radiative shocks and the hot central cavity. These ionize four cold regions: the ejecta, the inner and outer shells, and the circumstellar medium. The inner shell radiates the HIL and a small fraction of the LIL. The outer shell radiates the LIL and the circumstellar medium the narrow lines

very massive stars which last only several million years. Careful imaging of AGN and QSOs has revealed that they are often found associated with interacting galaxies at all epochs that can be observed. A circumstantial case is gradually developing associating activity in galactic nuclei with interactions, and the starbursts they initiate. The resulting supernovae may play a significant rôle in much of the observed AGN phenomenon - either in concert with a central massive black hole or alone. Thus we see that progress in the understanding of activity in the nuclei of galaxies depends on understanding how galaxies formed and have interacted; how stars and possibly supermassive black holes form and evolve; and upon the symbiosis of all these components.

Acknowledgements

I thank the organizers of the school, Tom Ray and Aage Sandqvist, for all the work they did to make it such a rewarding experience, for their hospitality in Dublin, their patience in waiting for these lecture notes, and for their insistent demands that they be written up. Alan Watson, Amanda Baker, Robin Williams, Roberto Terlevich, John Dyson and Brian Boyle all read individual sections – or all – of the manuscript critically and made helpful scientific, editorial and literature suggestions for which I am grateful. Robin Williams and I prepared the material for the appendix together, and he also helped assemble many of the figures. Richard Sword most kindly produced drawings and photographs at short notice with good humour. Ernst van Groningen and Uppsala Observatoriet were most hospitable during two visits when I spent long evenings finishing sections of this manuscript on their VAX. My husband, Martin Sohnius, was very supportive and helpful during the critical (and for families intolerable) final preparation of the manuscript.

Appendix: Cosmology with a Ruler

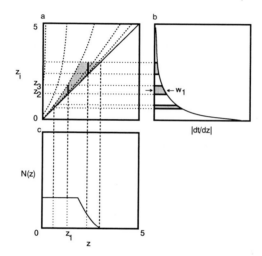

Fig. A1. Cartoon showing the use of the OHD to generate number counts (for an absolute magnitude limited sample) for a hypothetical population of AGN which "turns-on" instantaneously and remains bright for $\sim 10^9$ yr. This population has a formation rate which is constant in cosmological time (but not in historical redshift) as shown in (b) upper right. Despite this uniformity of life-time and formation rate, the non-linearities in the space-time geometry cause the number counts to be non-uniform, as can be seen in (c) lower left

In Sect. 5, I reviewed the observed number counts $N(z)$ and luminosity functions (LFs) and in Sect. 8.5 showed how space-time non-linearities could be visualized easily using the observation/history diagram (OHD). Here I review, very briefly, how the OHD can be used to generate the $N(z)$ and LF (Perry and Williams 1991). $N(z)$ can be derived by considering 'regions of visibility'. These are times during an object's evolution when it falls within the selection criteria of any particular survey. The regions shaded in Fig. A1(a) correspond to an *absolute* magnitude limited sample of objects. Consider a sample of objects drawn from a hypothetical population that evolves in brightness *independently of their formation time,* so they are brighter than the magnitude limit for the same length of proper time whatever their formation time. The equivalent redshift interval, however, is dependent on the initial redshift, as can be seen from the OHD and Fig. A1(a).

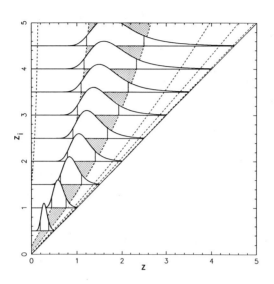

Fig. A2. Luminosity evolution of a hypothetical population of objects which brighten only slowly after formation, becoming brighter than an absolute magnitude limit only after $\sim 10^9$ yr. They then continue to brighten, as shown, for $\sim 2 \times 10^9$ yr, after which they fade, becoming dimmer than the survey limit at $\sim 3 \times 10^9$ yr. Luminosity functions are constructed by taking vertical slices at given observational redshifts. The number of objects corresponding to any given track are found from weighting functions specifying the formation rate, as in Fig. A1(b).

Fig. A1(b) shows the formation rate per unit historical redshift interval when it is constant in cosmological time. In this case the formation rate is just equal to $|dt/dz|$. In general the formation rate (in historical redshift) is the product of $|dt/dz|$ and the rate of formation per unit cosmological time. Note that in this figure, the historical redshift is on the vertical axis, the formation rate on the horizontal one. Because of the nonlinearity of $|dt/dz|$ each part of the shaded area of Fig. A1(a) corresponds to far more objects at later epochs than at earlier ones. Similarly, low historical redshifts appear more often in the population than the area they occupy in the OHD implies. Each point in the OHD plane that falls within the region of visibility should be weighted by the value shown in Fig. A1(b) at its historical redshift.

The derived number count, $N(z)$, is shown in Fig. A1(c). The dotted lines show how it is derived from the curves above. For any observational redshift (e.g. z_1) a vertical line is plotted on the OHD: it corresponds to the population at cosmological time $t(z_1)$. This cuts through the region of visibility at a range of historical redshifts $z_2 - z_3$, so objects that formed between these limits are observable at z_1. $N(z_1)$ is simply the area under the $\mathcal{F}(z_i)$ curve between these limits (shown shaded). This is, roughly, the product of the width in redshift $(z_3 - z_2)$ and a characteristic value of $\mathcal{F}(z_i)$ (e.g. $\mathcal{F}_1$) within this range. Repeating the procedure for other values of z_1 yields the entire $N(z)$ curve (Fig. A1(c)).

The dotted lines drawn for observational redshifts in the $N(z)$ tail show how the cut-off occurs. The section through the region of visibility at the observational z becomes limited not by the end of the observable period of objects, as it did previously. Rather, here even the earliest objects to form have not evolved through their observable life. The solid line in Fig. A1(a) ends at a historical z (horizontal line) corresponding to the redshift at which formation commenced. Points just above this line correspond to delays shorter than the object lifetime, but from epochs at which no objects formed. The shaded area in Fig. A1(b) is correspondingly smaller. $N(z)$ becomes zero at an observational redshift corresponding to the epoch when the first objects formed.

Similarly, the LFs can be understood by considering Fig. A2. Here, the luminosity history of a hypothetical population of AGN is shown. The objects "turn-on" very slowly after they form, reaching an absolute luminosity brighter than the survey limit after $\sim 10^9$ yr. They then remain bright enough to observe for a period of $\sim 2 \times 10^9$ yr. The LF at any z is derived by taking a vertical slice through the OHD diagram, and weighting each luminosity by the number of objects along its track, found from the formation rate per unit historical redshift, as described above. The LF at any z shows the luminosity distribution of the population existing at that observed z. For long-lived populations, such as those shown in Figs. A1 and A2, it is clear that this population is not coeval. This introduces inherent complexities in deconvolving the LFs. The parameterizations of the LF and its evolution discussed in Sect. 5 describe the empirical behaviour of the ensemble, and *not* necessarily the evolution of individual objects. With care, it is possible to use the OHD in conjunction with formation rate graphs to give reasonably accurate number count or LF model predictions when exploring models solely by graphical techniques (Perry and Williams 1991).

References

Anderson, S.F., Weymann, R.J., Foltz, C.B., Chaffee, F.H. (1987): *Astron. J.*, **94**, 278

Antonucci, R.R.J., Miller, J.S. (1985): *Astrophys. J.*, **297**, 621

Arnaud K.A., Brauduardi-Raymont, E., Culhane, J.L., Fabian, A.C., Hazard, C., McGlynn, T.A, Shafer, R.A., Tennant, A.F., Ward, M.J. (1985): *Mon. Not. R. astr. Soc.*, **217**, 105

Bahcall, J.N., Jannuzi, B.T., Hartig, G.F., Bohlin, R. (1991): *Astrophys. J. (Lett.)*, **377** L5

Bajtlik, S., Duncan, R.C., Ostriker, J.P. (1988): *Astrophys. J.*, **327**, 570

Baldwin, J.A. (1977): *Astrophys. J.*, **214**, 769

Baldwin, J.A., Wampler, E.J., Gaskell, C.M. (1989): *Astrophys. J.*, **338**, 630

Begelman, M.C. (1985): *Astrophys. J.*, **296**, 492

Begelman, M.C., de Kool, M., Sikora, M. (1991): *Astrophys. J.*, **382**, 416

Bergeron, J., Boissé, P. (1991): *Astron. Astrophys.*, **243**, 344

Binggeli, B., Sandage, A., Tammann, G.A. (1988): *Ann. Rev. Astron. Astrophys.*, **26**, 509

Blades, J.C., Turnshek, D.A., Norman, C.A. (eds) (1988): *QSO Absorption Lines* (C.U.P., Cambridge)

Blandford, R.D. (1990): in *Active Galactic Nuclei*, Saas-Fee Advanced Course 20, eds. Courvoisier T.J.-L., Mayor M. (Springer-Verlag, Berlin), p. 161

Blandford, R.D., McKee, C.F. (1982): *Astrophys. J.*, **255**, 419

Boksenberg, A. (1978): *Phys. Scripta*, **17**, 205

Boroson, T.A., Oke, J.B. (1982): *Nature*, **296**, 397

Boroson, T.A., Oke, J.B. (1984): *Astrophys. J.*, **281**, 535

Boroson T.A., Persson, S.E., Oke, J.B. (1985): *Astrophys. J. (Lett.)*, **293**, L35

Boyle, B.J. (1992): in *Proc. Texas-ESO/CERN Conference on Relativistic Astrophysics 1990*, in press

Boyle, B.J., Shanks, T., Peterson, B.A. (1988): *Mon. Not. R. astr. Soc.*, **235**, 935

Broadhurst, T.J., Ellis, R.S., Shanks, T. (1988): *Mon. Not. R. astr. Soc.*, **235**, 827

Burbidge, G. (1992): *Scientific American*, **266**, No. 2, 120

Burbidge, G., Hewitt, A., Narlikar, J.V., Das Gupta, P. (1990): *Astrophys. J. Suppl.*, **74**, 675

Burbidge, G., Perry, J.J. (1976): *Astrophys. J. (Lett.)*, **205**, L55

Butcher, H., Oemler, A. (1985): *Astrophys. J. Suppl.*, **57**, 665

Carswell, R.J., Lanzetta, K.M., Parnell, H.C., Webb, J.K. (1991): *Astrophys. J.*, **371**, 36

Cavaliere, A., Giallongo, E., Padovani, P., Vagnetti, F. (1988): in *Optical Surveys for Quasars*, eds. Osmer P.S., Porter A.C., Green R.F., Foltz C.B. (ASP Conference Series No. 2), p. 311

Chaffee, F.H., Foltz, C.B., Bechtold, J., Weymann, R.J. (1986): *Astrophys. J.*, **301**, 116

Chang, C.A., Schiano, A.V.R., Wolfe, A.M. (1987): *Astrophys. J.*, **322**, 180

Chevalier, (1991): in *Massive Stars in Starbursts*, eds. Leitherer C., Walborn N.R., Heckman T.M., Norman C.A. (C.U.P., Cambridge), p. 169

Clavel, J., Altamore, A., Boksenberg, A., Bromage, G.E., Elvius, A., Pelat, D., Penston, M.V., Perola, G.C., Snijders, M.A.J., Ulrich, M.-H. (1987): *Astrophys. J.*, **321**, 251

Colgate, S.A. (1967): *Astrophys. J.*, **150**, 163

Collin-Souffrin, S. (1987): *Astron. Astrophys.*, **179**, 60

Collin-Souffrin, S. (1992): in *Physics of Active Galactic Nuclei*, eds. Duschl W.J., Wagner S.J. (Springer Verlag, Heidelberg),

Collin-Souffrin, S., Dumont, S., Joly, M. and Tully, J. (1986): *Astron. Astrophys.*, **166**, 27

Collin-Souffrin, S., Dyson, J.E., McDowell, J.C., Perry, J.J. (1987): *Mon. Not. R. astr. Soc.*, **232**, 537

Combes, F. (1987): in *Starbursts and Galaxy Evolution*, eds. Thuan, T.X., Montmerle, T., Tran Thanh Van, J., (Editions Frontières), p. 325

Crampton D., Cowley, A.P., Hartwick, F.D.A. (1987): *Astrophys. J.*, **316**, 505

Czerny, B., Evis, M. (1987): *Astrophys. J.*, **321**, 305

Dahari, J. (1984): *Astron. J.*, **89**, 966

Dahari, J. (1985): *Astrophys. J. Suppl.*, **57**, 643

David, L.P., Durisen, R.H., Cohn, H.N. (1987): *Astrophys. J.*, **316**, 505

Dressler, A. (1978): *Astrophys. J.*, **223**, 765

Dressler, A. (1980): *Astrophys. J.*, **236**, 351

Dressler, A. (1984): *Ann. Rev. Astron. Astrophys.*, **22**, 185

Dressler, A. (1989): in *Active Galactic Nuclei*, IAU Symposium No. 134, eds. Osterbrock D.E., Miller J.S. (Kluwer Acad. Publ., Dordrecht) p. 217

Dressler, A., Richstone, D.O. (1988): *Astrophys. J.*, **324** 701

Dressler A., Thompson I.B., Shectman, S.A. (1985): *Astrophys. J.*, **288**, 481

Duncan, M.J., Shapiro, S.L. (1983): *Astrophys. J.*, **268**, 565

Dyson, J.E. (1986): in *RS Ophiuchi and the Nova Phenomenon* ed. Bode M.F., (VNU Science Press, Utrecht), p. 155

Dyson, J.E., Falle, S.A.E.G., Perry, J.J. (1981): *Mon. Not. R. astr. Soc.*, **191**, 785

Dyson, J.E., Perry, J.J. (1982): in *Third European IUE Conference*, eds. Rolfe E., Heck A., and Batrick B. (ESA, Noordwijk), p. 595

Dyson, J.E., Perry, J.J. (1986): in *Workshop on Model Nebulae*, ed. Péquignot (Observatoire de Paris, Paris), p. 23

Dyson, J.E., Perry, J.J., Williams, R.J. (1992): in *Testing the AGN Paradigm*, Proc. 2nd Annual October Conf. in Maryland, eds. Holt S. et al. (AIP Publishing), in press

Efstathiou, G., Rees, M.J. (1988): *Mon. Not. R. astr. Soc.*, **230**, 5p

Ellingson, E. (1989): in *Evolution of the Universe of Galaxies*, ed. Kron R.G., (A.S.P. Conf. Series), p. 334

Espey, B., Carswell, R.F., Bailey, J.A., Smith, M.G., Ward, M.J. (1989): *Astrophys. J.*, **342**, 666

Falle, S.A.E.G., Perry, J.J., Dyson, J.E. (1981): *Mon. Not. R. astr. Soc.*, **195**, 397

Fanelli, M.N., O'Connell, R.W., Thuan, T.X (1988): *Astrophys. J.*, **334**, 665

Ferland, G.J., Rees, M.J. (1988): *Astrophys. J.*, **332**, 141

Field, G.B (1964): *Astrophys. J.*, **140**, 1434

Filippenko, A. V. (1992a): ed., *Relationships between Active Galactic Nuclei and Starburst Galaxies* (Astr. Soc. Pac. Conf. Series)

Filippenko, A. V. (1992b): in *Physics of Active Galactic Nuclei*, eds. Duschl W.J., Wagner S.J. (Springer Verlag, Heidelberg), in press

Filippenko, A.V., Sargent, W.L.W. (1985): *Astrophys. J. Suppl.*, **57**, 503

Foltz, C.B., Chaffee, F.H., Hewett, P.C.,MacAlpine, G.M., Turnshek, D.A., Weymann, R.J., Anderson, S.F. (1987): *Astron. J.*, **94**, 1423

Foltz, C.B., Chaffee, F.H., Hewett, P.C., Weymann, R.J., Anderson, S.F., MacAlpine, G.M. (1989): *Astron. J.*, **98**, 1959

Fosbury, R.A.E., Morganti, R., Robinson, A., Tsvetanov, Z. (1992): in *Extragalactic Radio Sources: from Beams to Jets*, eds. Roland J., Sol H., Pelletier G. (C.U.P., Cambridge) in press

Francis, P.J., Hewitt P.C., Foltz, C.B, Chaffee, F.H., Weymann, F.J., Morris, R.J. (1991): *Astrophys. J.*, **373**, 465

Fricke, K.J., Kollatschny, W. (1989): in : *Active Galactic Nuclei*, IAU Symposium No. 134, eds. Osterbrock D.E., Miller J.S. (Kluwer Acad. Publ., Dordrecht) p. 425

Garmany, C.D. (1991): in *Massive Stars in Starbursts*, eds. Leitherer C., Walborn N.R., Heckman T.M., Norman C.A. (C.U.P., Cambridge), p. 115

Gaskell, C.M. (1982): *Astrophys. J.*, **263**, 79

Gaskell, C.M. (1988): *Astrophys. J.*, **325**, 114

Gaskell, C.M., Sparke, L.S. (1986): *Astrophys. J.*, **305**, 175

Gehren, T., Fried, J., Wehinger, P.A., Wykoff, S. (1984): *Astrophys. J.*, **278**, 11

George, I.M., Nandra, K., Fabian, A.C. (1990): *Mon. Not. R. astr. Soc.*, **242**, 28

George, I.M., Fabian, A.C., (1991): *Mon. Not. R. astr. Soc.*, **249**, 352

Gisler, G.R. (1978): *Mon. Not. R. astr. Soc.*, **183**, 633

Goad, J.W. and Gallagher, J.S. (1987): *Astron. J.*, **95**, 948

Greenstein, J.L., Matthews, M. (1963): *Nature*, **197**, 1041

Guiderdoni, B., Rocca-Volmerange, B. (1991): *Astron. Astrophys.*, **252**, 435

Guilbert P.W., Rees M.J., (1988): *Mon. Not. R. astr. Soc.*, **233**, 475

Hazard, C., McMahon, R.G., Sargent, W. (1986): *Nature*, **322**, 38

Heckman, T. (1991): in *Massive Stars in Starbursts*, eds. Leitherer C., Walborn N.R., Heckman T.M., Norman C.A. (C.U.P., Cambridge), p. 289

Heckman, T., Armus, L., Miley G. (1987): *Astron. J.*, **93**, 276

Heckman, T., Armus, L., Miley G. (1990): *Astrophys. J. Suppl.*, **74**, 833

Heckman, T., Smith, E.P. Baum, S.A., et al. (1986): *Astrophys. J.*, **311**, 526

Hernquist, L. (1989): *Ann. N.Y. Acad. Sci.* **571**, (Proc. 14th Texas Conference on Rel. Astrophysics, ed. E.J. Fenyves), p. 190

Hernquist, L., Barnes, J.E. (1991): *Nature*, **354**, 210

Hewett, P.C., Foltz, C.B., Chaffee, F.H., Francis, P.J., Weymann, R.J., Morris, S.L., Anderson, S.F., MacAlpine, G.M. (1991): *Astron. J.*, **101**, 1121

Holmberg, E. (1941): *Astrophys. J.*, **94**, 385

Huchra, J.P. (1987): in *Starbursts and Galaxy Evolution*, eds. Thuan, T.X., Montmerle, T., Tran Thanh Van, J., (Editions Frontières), p. 199

Hunstead, R.W., Pettini, M., Fletcher, A.B. (1990): *Astrophys. J.*, **356**, 23

Hutchings, J.B., Crampton, D., Campbell, B. (1984): *Astrophys. J.*, **280**, 41

Irwin, M., McMahon, R.G. (1991): *Gemini*, **30**, 6

Irwin, M., McMahon, R.G., Hazard, C. (1991): in *Space Distribution of Quasars* ed. Crampton, D. (ASP **21**,) p. 117

Joseph, R.C. (1991): in *Massive Stars in Starbursts*, eds. Leitherer C., Walborn N.R., Heckman T.M., Norman C.A. (C.U.P., Cambridge), p. 259

Joseph, R.C. (1987): in *Starbursts and Galaxy Evolution*, eds. Thuan, T.X., Montmerle, T., Tran Thanh Van, J., (Editions Frontières), p. 293

Kormendy, J. (1988): *Astrophys. J.*, **325**, 128

Krolik, J.H. (1990a): in *The Interstellar Medium in External Galaxies* eds. Thronson H.A., Shull J.M. (Kluwer Acad. Publ., Dordrecht), p. 239

Krolik, J.H. (1990b): in *High Resolution X-ry Spectroscopy of Cosmic Plasmas* ed. Gorenstein P., Zombeck M., (IAU Colloquium 115), p. 264

Krolik, J.H., McKee, C.F., Tarter, C.B. (1981): *Astrophys. J.*, **249**, 422 (KMT)

Kronberg, P.P., Bierman, P.L., Schwab, F.R. (1985): *Astrophys. J.*, **291**, 693

Landau, R. et al. (1986): *Astrophys. J.*, **308**, 78

Larson, R.B. (1974): *Mon. Not. R. astr. Soc.*, **166**, 585

Larson, R.B. (1977): in *The Evolution of Galaxies and Stellar Populations*, eds. Tinsley B.M., Larson R.B., (Yale Univ. Obs., New Haven), p. 77

Larson, R.B. (1987): in *Starbursts and Galaxy Evolution*, eds. Thuan, T.X., Montmerle, T., Tran Thanh Van, J., (Editions Frontières), p. 467

Larson, R.B. (1991): in *Fragmentation of Molecular Clouds and Star Formation* eds. Falgarone E., Boulanger F., Duvert G., (Kluwer Acad. Publ., Dordrecht), p. 261

Lavery, R.J., Henry J.P. (1988): *Astrophys. J.*, **330**, 596

Lightman A.P., White T.R. (1988): *Astrophys. J.*, **335**, 57

Lilly, S.J. (1989): in *The Epoch of Galaxy Formation*, eds. Frenk C.S., Ellis R.S., Shanks T., Heavens A.F., Peacock J.A. (Kluwer Acad. Publ., Dordrecht), p. 63

Lin, D.N.C., Pringle, J.E., Rees, M.J. (1988): *Astrophys. J.*, **328**, 103

Leitherer, C., Walborn, N.R., Heckman, T.M., Norman, C.A., (1991): *Massive Stars in Starbursts* (C.U.P., Cambridge)

Lonsdale, C.J. (1990): in *Windows on Galaxies*, eds. Fabbiano G., Gallagher J.S., Renzini A. (Kluwer Acad. Publ., Dordrecht), p. 121

Lonsdale, C.J., Hacking, P.B., Conrow, T., Rowan-Robinson, M., (1989): *Astrophys. J.*, **358**, 60

Lonsdale, Colin J., Lonsdale, Carol J., Smith, H.E. (1992): *Astrophys. J.*, in press

Lu, L. (1991): *Astrophys. J.*, **379**, 99

Lu, L., Wolfe, A.M., Turnshek, D.A. (1991): *Astrophys. J.*, **367**, 19

Lynden-Bell, D. (1969): *Nature*, **223**, 690

Lynds, C.R. (1971): *Astrophys. J. (Lett.)*, **164**, L73

Malkan, M.A., Margon, B., Chanan, G. (1984): *Astrophys. J.*, **280**, 66

Malkan, M.A., Sargent, W.L.W. (1982): *Astrophys. J.*, **254**, 122

Masnou, J.-L., Wilkes, B.J., Elvis, M., McDowell, J.L., Arnand, K.A. (1992): *Astron. Astrophys.*, **253**, 35

Mathez, G. (1976): *Astron. Astrophys.*, **53**, 15

McCarthy, P.J., van Breugel, W., Spinrad, H., and Djorgovski, S. (1987): *Astrophys. J. (Lett.)*, **321**, L29

McDowell, J.C., Elvis, M., Wilkes, B.J., Willner, S.P., Oey, M.S., Polomski, E., Bechtold, J., Green, R.F. (1989): *Astrophys. J. (Lett.)*, **345**, L13

Mezger, P.G. (1987): in *Starbursts and Galaxy Evolution*, eds. Thuan, T.X., Montmerle, T., Tran Thanh Van, J., (Editions Frontières), p. 3

Meyer, D.M., York, D.G. (1987): *Astrophys. J. (Lett.)*, **315**, L5

Miller, J.S. (1981): *Publ. Astr. Soc. Pac.*, **93**, 681

Miller, J.S., Goodrich, R. (1990): *Astrophys. J.*, **355**, 456

Miller, L., Mitchell, P.S., Boyle, B.J. (1990): *Mon. Not. R. astr. Soc.*, **244**, 1

Mittaz, J.P.D., Branduardi-Raymont, G. (1989): *Mon. Not. R. astr. Soc.*, **238**, 1029

Mulchary, J.S., Tsvetanov, Z., Wilson, A.S., Pérez-Fournon, I. (1992): *Astrophys. J.*, in press

Mushotzky, R.F., Serlemitsos, P.J., Becker, R.H., Boldt, E.A., and Holt, S. S. (1978): *Astrophys. J.*, **220**, 790

Mushotzky, R.F. (1984): *Adv. Space Res.*, **3**, 157

Nandra, K., Pounds, D.A., Stewart, G.C., George, I.M., Hayashida, K., Makino, F., Okashi, T. (1991): *Mon. Not. R. astr. Soc.*, **248**, 760

Noguchi, M. (1988a): *Astron. Astrophys.*, **203**, 259

Noguchi, M. (1988b): *Astron. Astrophys.*, **201**, 37

Noguchi, M., Ishibashi, S. (1986): *Mon. Not. R. astr. Soc.*, **219**, 305

Norman, C.A. (1991): in *Massive Stars in Starbursts*, eds. Leitherer C., Walborn N.R., Heckman T.M., Norman C.A. (C.U.P., Cambridge), p. 271

Norman, C.A., Scoville, N.Z. (1988): *Astrophys. J.*, **332**, 124

O'Brien P.T., Harries, T.J. (1991): *Mon. Not. R. astr. Soc.*, **250**, 133

Oemler, A. (1974): *Astrophys. J.*, **194**, 1

Olofsson, K. (1989): *Astron. Astrophys. Suppl. Ser.*, **80**, 317

Osmer, P.S. (1980): *Astrophys. J. Suppl.*, **42**, 523

Osmer, P.S., Porter, A.C., Green, R.F., Foltz, C.B. (1988): *Optical Surveys for Quasars*, (ASP Conference Series No. 2)

Peacock, J.A., Miller, L. (1988): in *Optical Surveys for Quasars*, eds. Osmer P.S., Porter A.C., Green R.F., Foltz C.B. (ASP Conference Series No. 2), p. 194

Peebles, P.J.E. (1989): in *The Epoch of Galaxy Formation*, eds. Frenk C.S., Ellis R.S., Shanks T., Heavens A.F., Peacock J.A. (Kluwer Acad. Publ., Dordrecht), p. 1

Pedlar, A., Dyson, J.E., Unger, S.W. (1985): *Mon. Not. R. astr. Soc.*, **214**, 463

Penston, M.V., Pérez, E. (1984): *Mon. Not. R. astr. Soc.*, **211**, 33p

Pérez, P., Gonzalez-Delgado, R., Tadhunter, C., Tsvetanov, Z. (1989): *Mon. Not. R. astr. Soc.*, **241**, 31p

Perry, J.J. (1979): in *Active Galactic Nuclei* eds. Hazard C., Mitton S. (C.U.P., Cambridge) 139

Perry, J.J. (1992): in *Relationships between Active Galactic Nuclei and Starburst Galaxies*, ed. Filippenko A. V. (Astro. Soc. Pac. Conference Series), in press

Perry, J.J., Burbidge, E.M., Burbidge, G.R. (1978): *Publ. Astr. Soc. Pac.*, **90**, 337

Perry, J.J., Dyson, J.E. (1985): *Mon. Not. R. astr. Soc.*, **213**, 665 (PD)

Perry, J.J., Dyson, J.E. (1990): *Astrophys. J.*, **361**, 362

Perry, J.J., Dyson, J.E. (1992): in *Testing the AGN Paradigm*, Proc. 2nd Annual October Conf. in Maryland, eds. Holt S. et al. (AIP Publishing), *in press*

Perry, J.J., Ward, M.J., Jones, M. (1987): *Mon. Not. R. astr. Soc.*, **228**, 623

Perry, J.J., Williams, R.J.R. (1991): Inst. of Astronomy, preprint

Peterson, B.M. (1988): *Publ. Astr. Soc. Pac.*, **100**, 18

Peterson, B.M., Balonek, T.J., Barker, E.S., et al. (1991): *Astrophys. J.*, **368**, 119

Pettini, M., Boksenberg, A., Hunstead, R.W. (1990): *Astrophys. J.*, **348**, 48

Pettini, M., Hunstead, R.W., Smith, L.J., Marr, D.P. (1990): *Mon. Not. R. astr. Soc.*, **246**, 545

Pogge, R.W. (1989): *Astron. J.*, **98**, 124

Pounds, K.A., (1989): in *The Proceedings of the 23rd ESLAB Symposium, Two topics in X-ray astronomy. Part 2 AGN and the X-ray Background*, eds. Hunt J., Battrick B., p. 753

Pounds, K.A., Nandra, K., Stewart, G.C., George, I.M, Fabian, A.C. (1990): *Nature*, **344**, 132

Puxley, P.J. (1991): *Mon. Not. R. astr. Soc.*, **249**, 11p

Rees, M.J. (1984): *Ann. Rev. Astron. Astrophys.*, **22**, 471

Rieke, G.H. (1991): in *Massive Stars in Starbursts*, eds. Leitherer C., Walborn N.R., Heckman T.M., Norman C.A. (C.U.P., Cambridge), p. 205

Robinson, A., Binette, L., Fosbury, R.A.E., Tadhunter, C.N. (1987): *Mon. Not. R. astr. Soc.*, **227**, 97

Rokaki, E., Boisson, C., Collin-Souffrin, S. (1992): *Astron. Astrophys.*, **253**, 57

Roland, J., Sol, H., Pelletier, G. (eds.) (1992): *Extragalactic Radio Sources: from Beams to Jets* (C.U.P., Cambridge)

Romanishan W., Hintzen P. (1989): *Astrophys. J.*, **341**, 41

Rowan-Robinson, M., Crawford, J. (1989): *Mon. Not. R. astr. Soc.*, **238**, 523

Sandage, A. (1961): *Astrophys. J.*, **133**, 355

Sandage, A. (1965): *Astrophys. J.*, **141**, 1560

Sanders, D.B., Phinney, E.S., Neugebauer, G., Soifer, B.T., Matthews, K. (1989): *Astrophys. J.*, **347**, 29

Sanders, D.B., Scoville, N.Z., Young, J.S., Soifer, B.T., Schloerb, F.P., Rice, W.L., Danielson, G.E. (1986): *Astrophys. J. (Lett.)*, **305**, L45

Sanders, D.B, Soifer, B.T., Elias, J.H., Madore, B.F., Mathews, K., Neugebauer, G., Scoville N.Z. (1988): *Astrophys. J.*, **325**, 74

Scalo, J. (1990): in *Windows on Galaxies*, eds. Fabbiano G., Gallagher J.S., Renzini A. (Kluwer Acad. Publ., Dordrecht), p. 125

Schechter, P.L (1976): *Astrophys. J.*, **203**, 297

Schneider, D.P., Schmidt, M., Gunn, J.E. (1989): *Astron. J.*, **98**, 1507

Schneider, D.P., Schmidt, M., Gunn, J.E. (1991a): *Astron. J.*, **101**, 2004

Schneider, D.P., Schmidt, M., Gunn, J.E. (1991b): *Astron. J.*, **102**, 837

Schmidt, M. (1963): *Nature*, **197**, 1040

Schmidt, M. (1968): *Astrophys. J.*, **151**, 393

Schmidt, M., Schneider, D.P, Gunn, J.E. (1988): *Optical Surveys for Quasars*, eds. Osmer P.S., Porter A.C., Green R.F., Foltz C.B. (ASP Conference Series No. 2), p. 87

Scoville, N., Norman, C.A. (1988): *Astrophys. J.*, **332**, 163

Scoville, N., Soifer, B.T. (1991): in *Massive Stars in Starbursts*, eds. Leitherer C., Walborn N.R., Heckman T.M., Norman C.A. (C.U.P., Cambridge), p. 233

Shanbhag, S., Kembhavi, A. (1988): *Astrophys. J.*, **334**, 34

Shields, G.A., (1978): *Nature*, **272**, 706

Shklovskii, I.S. (1960): *Sov. Astron. AJ*, **4**, 885

Shlosman, I., Begelman, M.C., Frank, J. (1990): *Nature*, **345**, 679

Shu, F.H., Adams, F.C., Lizano, S. (1987): *Ann. Rev. Astron. Astrophys.*, **25**, 23

Smith, E.P., Heckman, T.M. (1989): *Astrophys. J.*, **341**, 658

Smith, M.G. (1989): in *Active Galactic Nuclei*, IAU Symposium No. 134, eds. Osterbrock D.E., Miller J.S. (Kluwer Acad. Publ., Dordrecht)

Smith, S.J. (1991): Ph.D. Thesis, University of California at Los Angeles

Sopp, H.M., Alexander, P. (1991): *Mon. Not. R. astr. Soc.*, **251**, 14p

Spinrad, H. (1989): in *The Epoch of Galaxy Formation*, eds. Frenk C.S., Ellis R.S., Shanks T., Heavens A.F., Peacock J.A. (Kluwer Acad. Publ., Dordrecht), p. 39

Stockton, A., MacKenty, J.W. (1987): *Astrophys. J.*, **316**, 584

Tadhunter, C.N., Fosbury, R.A.E., Binette, L.A., Danziger, I.J., Robinson, A. (1987): *Nature*, **325**, 504

Tadhunter, C.N., Fosbury, R.A.E., Quinn, P.J., (1989) *Mon. Not. R. astr. Soc.*, **240**, 225

Taylor, D., Dyson, J.E., Axon, D.J. (1992): *Mon. Not. R. astr. Soc.*, **255**, 351

Tennant, A.F. Mushotzky, R.F., Boldt, E.A., and Swank, J.H., (1981): *Astrophys. J.*, **251**, 15

Tenorio-Tagle, G., Terlevich, R., Franco, J., Melnick, J. (1992): in *Relationships between Active Galactic Nuclei and Starburst Galaxies*, ed. Filippenko A. V. (Astro. Soc. Pac. Conference Series), in press

Terlevich, E. (1987): *Mon. Not. R. astr. Soc.*, **224**, 193

Terlevich, E., Díaz, A.I., Pastoriza, M.G., Terlevich, R., Dottori, H.(1990): *Mon. Not. R. astr. Soc.*, **242**, 48p

Terlevich, E., Díaz, A.I., Terlevich, R. (1990): *Mon. Not. R. astr. Soc.*, **242**, 271

Terlevich, E., Díaz, A.I., Terlevich, R., Vargas, M.L.G. (1992): *Mon. Not. R. astr. Soc.*, in press

Terlevich, R. (1990): in *Windows on Galaxies*, eds. Fabbiano G., Gallagher J.S., Renzini A. (Kluwer Acad. Publ., Dordrecht), p. 87

Terlevich, R., Boyle, B.J. (1992): in preparation

Terlevich, R., Melnick, J. (1985): *Mon. Not. R. astr. Soc.*, **213**, 841

Terlevich, R., Melnick, J., Moles, M. (1987): in *Observational Evidence for Activity in Galaxies*, eds. Kachikyan E.Ya., Fricke K.J., (Reidel, Dordrecht), p. 488

Terlevich, R., Portal, M.S., Días, A.I., Terlevich, E. (1991): *Mon. Not. R. astr. Soc.*, **249**, 36

Terlevich, R.T., Tenorio-Tagle, G., Franco, J., Melnick, J. (1992): *Mon. Not. R. astr. Soc.*, in press

Thuan, T.X., Montmerle, T., Tran Thanh Van, J. (1987): *Starbursts and Galaxy Evolution*, (Editions Frontières, Gif sur Yvette)

Toomre, A., Toomre, J. (1972): *Astrophys. J.*, **178**, 623

Turner T.J., Done C., Mushotzky, R., Madejski, G., Kunieda, H. (1992): *Astrophys. J.*, in press

Turner T.J., Pounds K.A. (1989): *Mon. Not. R. astr. Soc.*, **240**, 833

Turnshek, D.A. (1988): in *QSO Absorption Lines*, eds. Blades J.C., Turnshek D.A., Norman C.A., (C.U.P., Cambridge), p. 17

Tytler, D. (1987): *Astrophys. J.*, **321**, 69

Tytler, D., Fan, X.-M. (1992): *Astrophys. J. Suppl.*, **79**, 1

Tyson, J.A., Baum, W.A., Kreidl, T. (1982): *Astrophys. J. (Lett.)*, 257 L1

Ulrich, M.-H., Péquignot, D. (1980): *Astrophys. J.*, **238**, 45

Unger, S.W., Pedlar, A., Axon, D.J., Whittle, M., Meurs, E.J.A., and Ward, M.J. (1987): *Mon. Not. R. astr. Soc.*, **228**, 671

van Groningen, E., de Bruyn, A.G. (1989): *Astron. Astrophys.*, **211**, 293

Véron, P. (1983): in *Quasars and Gravitational Lenses*, ed. Swings J.P., (Inst. d'Astrophysique, Liege), p. 210

Véron-Cetty, M.P., Woltjer, L. (1990): *Astron. Astrophys.*, **236**, 69

Wampler, E.J., Gaskell, C.M., Burke, W.L., Baldwin, J.A. (1984): *Astrophys. J.*, **276**, 403

Warren, S.J., Hewett, P.S. (1990): *Rep. Prog. Phys.*, **53**, 1095

Warren, S.J., Hewett, P.S., Osmer, P.S. (1988): in *Optical Surveys for Quasars*, eds. Osmer P.S., Porter A.C., Green R.F., Foltz C.B. (ASP Conference Series No. 2), p. 98

Weymann, R.J., Scott, J.S., Schiano, A.V.R., Christiansen, W.A. (1982): *Astrophys. J.*, **262**, 497

White, S.D.M., Davis, M., Efstathiou, G., Frenk, C.S. (1987): *Nature*, **330** 451

Williams, R., Perry, J.J. (1992): in *Physics of Active Galactic Nuclei*, eds. Duschl W.J., Wagner S.J. (Springer Verlag, Heidelberg), in press

Wills, B.J., Netzer, H., Wills, D. (1985): *Astrophys. J.*, **288**, 94

Wilson, A.S., Baldwin, J.A., Ulvestad, J.S. (1985): *Astrophys. J.*, **291**, 627

Wilson, A.S., Heckman, T.M. (1985): in *Astrophysics of Active Galaxies and Quasi-Stellar Objects*, eds. Miller J.S. et al. (University Science Books, Mill Valley), p. 39

Wolfe, A.M., Turnshek, D.A., Smith, H.E., Cohen, R.D. (1986): *Astrophys. J. Suppl.*, **61**, 249

Wolstencroft, R.D., Unger, S.W., Pedlar, A. et al. (1987): in *Starbursts and Galaxy Evolution*, eds. Thuan, T.X., Montmerle, T., Tran Thanh Van, J., (Editions Frontières), p. 269

Woltjer, L. (1990): in *Active Galactic Nuclei*, Saas-Fee Advanced Course 20, eds. Courvoisier T.J.-L., Mayor M. (Springer-Verlag, Berlin), p. 1

Wright, G.S., Joseph, R.D., Robertson, N.A., James, P.A., Meikle, P.A. (1988): *Mon. Not. R. astr. Soc.*, **233**, 1

Yanny, B. (1990): *Astrophys. J.*, **351**, 396

Yanny, B., York, D.G., Williams, T.B. (1990): *Astrophys. J.*, **351**, 377

Yee, H.K.C. (1987): *Astron. J.*, **94**, 1461

Yee, H.K.C. (1989): in *Evolution of the Universe of Galaxies*, ed. Kron R.G., (A.S.P. Conf. Series), p. 322

Zdziarski, A.A. (1992): in *Extragalactic Radio Sources: from Beams to Jets*, eds. Roland J., Sol H., Pelletier G., (C.U.P., Cambridge)

Zel'dovich, Ya.B., Novikov, I.D. (1964): *Sov. Phys. Dokl.* **158**, 811

Observations and Their Implications for the Inner Parsec of AGN

Suzy Collin-Souffrin

DAEC, Observatoire de Paris–Meudon, and
Institut d'Astrophysique, Paris, France

Abstract: These lectures aim at giving a description of the inner parsec of a "standard AGN", based on an almost direct interpretation of the observations. After a brief description of the overall spectrum of a quasar or a Seyfert 1 nucleus, we examine what we can learn about it: from the Broad Line intensities (through photoionization models and line formation processes), Broad Line profiles and variability; from the UV and ultra soft X-ray continuum (the accretion disc, its physics, structure and emission); from the X-ray continuum (the hot medium and the cool reflecting medium). We then discuss the different methods used to determine central masses and accretion rates: do they give a clue for understanding the physical and cosmological evolution of AGN? In the conclusion we mention briefly the "unified 3 parameters scheme": can it explain the various classes of AGN and what is the influence of the viewing angle?

1 Overall Spectrum and Emission Mechanisms

This section gives a description of the "standard spectrum" of an AGN, considering only Seyfert 1 nuclei and quasars, and excluding blazars. The spectrum consists of emission and absorption lines, superimposed on a continuum extending from the radio to the gamma-ray range. It was impossible, without performing an arbitrary selection, to refer to the hundreds of contributions devoted to the subject. So I decided to give only a very few important recent references, and I make my apologizes to the authors of other papers. For people who want to know more about this field, they can consult Osterbrock's book (1990) and Osterbrock's review (1991), where the observational aspects are extensively reviewed.

1.1 The Continuum

The spectral distribution of the overall continuum is quite complex and displays several bumps and breaks. For people entering the subject now, it is interesting to realize that for 20 years the AGN continuum was essentially matched by a power law with an index close to unity, extending from the millimeter to the gamma-ray range. The bulk of the emission was thus attributed to the synchrotron mechanism, while this mechanism is

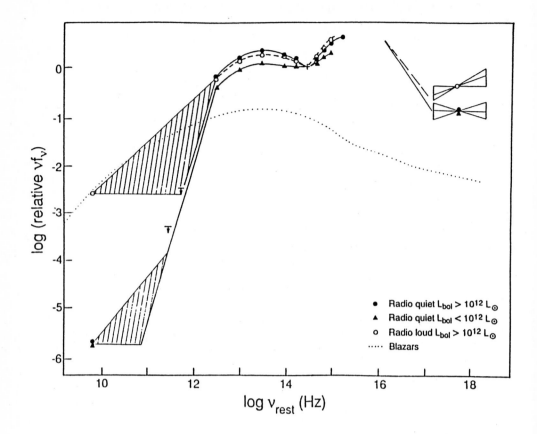

Fig. 1. Log νF_ν versus log ν, after Sanders et al. (1989)

presently considered as a minor contributor, and even by some people as being completely absent except in the radio range! An excellent review of the subject is given by Bregman (1990).

A plot of the power per logarithmic bandwidth of a typical AGN is displayed in Fig. 1. It shows that the bulk of the bolometric luminosity is probably emitted in the range $14 \leq \log\nu \leq 17$. The bolometric luminosity however may be dominated by hard X-ray photons if the spectrum extends up to a few hundreds keV with the same spectral index as in the 10 keV range.

1.1.1 The Radio Continuum

Since this is the main topic of P. Biermann's lectures, it is mentioned only very briefly here.

Let us recall first that there are two kinds of AGN: *radio loud* and *radio quiet* objects. Radio loud objects constitute about 10% of all quasars, and a small proportion of Seyfert galaxies (those called *Broad Line Radio galaxies*, always elliptical galaxies). Even in radio loud objects the overall radio luminosity represents only a very small fraction of the total luminosity.

Radio loud objects emit about 3 times more X-rays than radio quiet objects. This indicates the existence of an X-ray emission mechanism linked with the radio emission. The radio structure is complex. It consists generally of a *core*, a *jet*, and two large radio *lobes* in symmetric positions with respect to the galaxy. The jet is observed at all scales, from the parsec to the Megaparsec ones (at the parsec scale, it is called "compact radio source", and sometimes displays "superluminal" motions, which are due to bulk motions at relativistic velocities). It implies the existence of a privileged direction which is maintained over millions of years.

Radio quiet objects are not strictly "quiet": they contain small and weak radio cores and small jets with similar spectral characteristics. Their radio fluxes are about 10^{-3} to 10^{-4} that of radio loud ones.

The spectral distribution of the radio continuum displays various shapes. It can be a simple power law with an index varying from 0 (*flat spectrum*) to 1 (*steep spectrum*), or be more complex, with a negative spectral index and several inflexions. Flat spectral indices are correlated with large values of the core to lobe flux density ratio.

Radio emission is produced by the synchrotron process. Maxima and inflexions in the spectrum are due to the presence of several sources, some of them being self-absorbed.

A most important fact to keep in mind for the following is that all AGN, radio loud or not, present **a privileged direction**.

1.1.2 The Infrared Continuum

The spectrum rises very sharply in the 100–1000 μm range and peaks in the 10–100 μm range. About a third of the bolometric luminosity is emitted in this band. There is a controversy on the best fit for the infrared continuum. Some authors claim that it is a power law with a spectral index close to 1.2, on which is often superimposed a "5 μm bump". Others, such as Sanders et al (1989), prefer a fit with a "broad infrared bump". The issue is clearly to discriminate between a thermal and a non-thermal emission mechanism. The shape of the "break" between the radio and the submillimeter range can be a discriminator: it seems steeper than a power law with an index equal to -2.5, which would be the signature of a self-absorbed synchrotron mechanism. Variability could also be a powerful test for a thermal or non-thermal process: unfortunately it is still not clear if the infrared flux is variable. Variations have been detected only in the near infrared range (1 μm), where the spectrum displays a clear minimum, due to a change in the dominant emission mechanism. So this variability is probably linked with the visible part of the spectrum. The idea which generally prevails is that the infrared continuum is a mixture of non-thermal (synchrotron) and thermal (dust emission) processes. Possibly radio quiet AGN are dominated by thermal emission and radio loud AGN by non-thermal emission. The dust emission could be simply that of the whole host galaxy, the dust being heated by the central continuum, or it could be associated with the Narrow Line Region. The "5 μm bump" is probably due to hotter dust located inside or near the Broad Line Region.

1.1.3 The Optical-UV Continuum

In this range, the continuum is also a mixture of various components.

In low luminosity AGN, the visible continuum is dominated by starlight, and observations with a high spatial resolution are required to correctly subtract this contribution. The other components are (cf. Fig. 2):

- an underlying ν^{-1} power law continuum
- the "big bump": a quasi-black-body continuum corresponding to a temperature of a few 10^4 K, which is generally considered as the signature of an accretion disc
- the "small bump": a mixture of FeII blends and Balmer continuum, both produced by the Broad Line Region.

The UV continuum steepens on the blue side of Lyα, its spectral index reaching a value of 2 near the Lyman limit. An important question is to know if this steepening is intrinsic and implies a relatively low temperature of the black-body-like emission, or if it is extrinsic (due to the Lyα forest of intergalactic clouds). One should also mention the absence of any *intrinsic* Lyman discontinuity, either in emission or in absorption (Antonnucci et al. 1989).

In recent years variability of the optical–UV continuum has been intensively studied in several monitoring campaigns. Its characteristics are:

- an increase of amplitude with frequency
- an increase of characteristic variation time scale with luminosity, from a few days (Seyfert 1) to a few months (quasars)
- a flattening of the UV continuum when the flux increases.

The existence of a time lag between variations observed at different wavelengths is difficult to assess. In Fairall 9, Clavel et al. (1989) have observed a time lag increasing with the wavelength between the UV and the infrared flux variations, which they did attribute to the emission of hot dust located at increasing distances from the central source. In any case the time lag between the optical and the UV flux variations, if any, is very short ($\lesssim$ a few days).

1.1.4 The X-ray Continuum

The X-ray continuum is observed from 0.2 keV to a few tens keV. It can be divided in two ranges, the soft and the hard X-ray range, corresponding to two different emission mechanisms.

The Hard X-ray Range, from 2 keV to a Few Tens keV.

The continuum is well fitted by a power law. The average value of the spectral index is equal to 0.7 (but there is a large range of values around this mean, with a standard deviation of the order of 0.3). One does not know if this power law extends up to a few hundreds keV, as is the case for the three AGN observed so far in the gamma-ray range (NGC4151, 3C273 and MCG.11.11). For these objects the hard X-ray range dominates the bolometric luminosity.

One is used to link the X-ray flux to the UV flux through a spectral index α_{ox}, defined as the slope of the continuum between 2000 Å and 2 keV. This index is close to 1.5 in radio quiet objects and to 1.2 in radio loud ones. The difference is attributed to

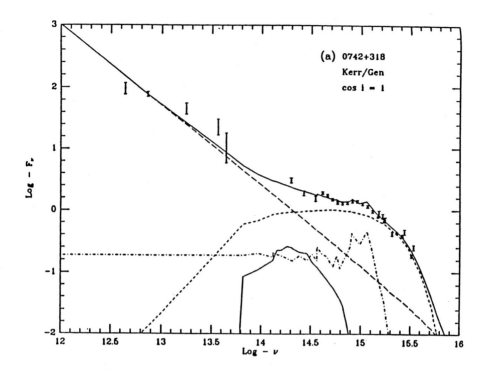

Fig. 2. $\mathrm{Log}F_{\nu}$ versus log ν: an example of decomposition of the optical-UV continuum, after Sun and Malkan (1989). Solid line: stellar emission; dot-dashed line: Broad Line emission; small dashed-line: disc emission; large dashed-line: power-law continuum

an additional X-ray mechanism in radio loud objects (probably inverse Compton in the compact radio core).

X-ray measurements with Ginga allow us to define the shape of the spectral distribution (Pounds et al. 1990). A composite spectrum, obtained by summing 12 AGN, is displayed in Fig. 3. The spectrum is correctly matched by the sum of a power law continuum with a spectral index close to 0.8 on which is superimposed a dip in the range 7–10 keV, an emission line at 6.3 keV, and a "hump" extending above 12 keV. The emission line and the hump are identified with the FeK fluorescence line and the reflection continuum due to a *cold* medium at a temperature lower than 10^{6} K, seen from the central source with a large covering angle of the order unity. The dip might be due to the absorption by a "warm" medium where all atoms are stripped (so it does not absorb soft X-ray photons, cf. below).

Variability measurements provide important clues on the X-ray emission mechanism. Variations at the 10% level were detected in the shortest accessible time (100 s) in several AGN, but a typical time scale for large amplitude variations is a few hours, implying a size of the X-ray source of the order of 10^{15} cm at most. The (dominant) power-law continuum is then consistent with a non-thermal mechanism, such as the synchrotron

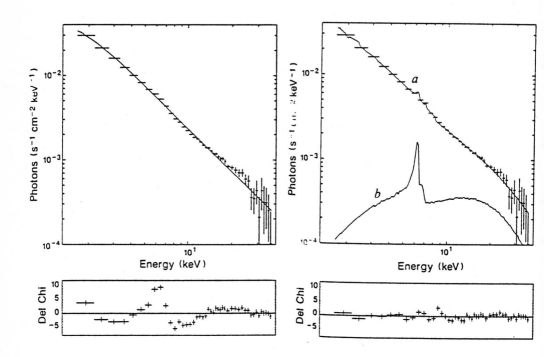

Fig. 3. Observed composite spectrum of AGN fitted by a power law alone (left panel), and by a power law plus reflection and warm absorber model (right panel), together with residuals, after Pounds et al. (1990)

self-Compton, i.e. the Compton upscattering of synchrotron infrared or UV photons by the relativistic gas itself. However several other thermal mechanisms can also well account for the shape of the spectrum and the small size of the source, such as comptonization of soft photons in a hot optically thick gas (either in a spherical stream or in an accretion disc, cf. Maraschi and Molendi 1990, Wandel and Liang 1990), or simply thermal emission of a hot gas with a distribution of temperatures.

The Soft X-ray Range, from 0.2 to 2 keV.

The shape of the continuum in this range is strongly variable from one object to another. The spectral index is steeper than in the hard X-ray range, and may reach values of 2 or 3 (it is determined by taking account of interstellar absorption, cf. Wilkes and Elvis 1987). In fact these large spectral indices correspond to a continuous steepening of the continuum towards small frequencies: it is called the *soft X-ray excess* and can account for a large part of the bolometric luminosity (cf. Fig. 1). It is commonly believed that this excess is the extrapolation of the UV-bump, although it is not clear if they are directly linked: as we have seen indeed, the UV continuum near 1000 Å is already very steep and may not fit the extrapolation of the soft X-ray continuum. Radio quiet objects have a steeper soft X-ray continuum than radio loud ones, again indicating a larger flux in the hard X-ray ray range for the latter objects.

An important result from the soft X-ray measurements is the relatively small column density of the absorbing "cold" matter located along the line of sight to the X-ray source: 10^{21} cm^{-2} to 10^{23} cm^{-2}, this last value being reached in low luminosity objects. As we shall see later, it is much smaller than the typical column density of the Broad Line Region.

Finally it is worth mentioning the luminous quasar HS 1700+6416 where the UV continuum has been observed with IUE down to 500 Å (Reimers et al. 1989). It displays a large EUV excess, which would argue in favour of a very high black-body-like temperature emission. This high temperature raises some problem for the standard accretion disc model (note, however, that the shape of the continuum is strongly dependent on the intergalactic correction).

1.1.5 The Gamma-Ray Continuum

Among AGN, only 3C273 has been observed at over 100 MeV (with CosB). The spectrum is an extension of the power law continuum up to a few MeV, with a flat spectral index of 0.5. It breaks at this energy, and displays then a power law continuum with a spectral index close to 1.5 (we recall that spectral indices given in this chapter all refer to flux and not to photon number). Thus the major part of the bolometric luminosity is emitted around a few MeVs in this object. It may, however, not be representative of all AGN. Several mechanisms are invoked for the gamma-ray emission: including extension of the X-ray mechanism, pair production and annihilation, Penrose Compton scattering.

1.2 The Line Spectrum
1.2.1 Absorption Lines

High redshift quasars display a rich absorption line spectrum, consisting of the Lyα forest and in systems of lines of heavy elements. These lines are very narrow, and are due to intervening intergalactic matter located along the line of sight to the objects and present a smaller redshift i.e. protogalactic clouds and halos of galaxies. These lines are often studied in a cosmological context, but do not relate to the topic of this chapter. We have seen, however, that they are important for the determination of the shape of the quasar continuum in the far UV range. Some of the absorption redshifts are close to the emission redshift of the quasar, and a few are larger. They are probably due to protogalactic clouds or galaxies in the same supercluster, but this interpretation is sometimes disputed, and the lines are attributed in this case to high velocity clouds in the quasar's own system.

Broad absorption lines are undoubtedly formed in the quasar's own system. They are either completely detached from the corresponding emission lines, (which in some cases can even be completely absent), or they display a P Cygni profile on the blue side of the emission wing. They correspond to very high ionization lines, such as OVI, NV, or SiIV. These systems are present in 9% of quasars, which are then called "Broad Absorption Line" or BAL quasars. BALs have been extensively studied (cf. for instance Turnshek et al. 1988). They are attributed to outflowing highly ionized dilute matter, which are most probably conical structures surrounding or embedding the Broad Line Region, which can be seen only when the viewing angle is favourable.

Similar but narrower absorption lines are observed in the blue wings of high ionization resonance emission lines in a few low luminosity Seyfert galaxies (in Lyα, CIV... of NGC 4151, 3516...). The same interpretation holds for these lines, except that the outflowing velocity is only a few 10^3 km s^{-1} and not a few 10^4 km s^{-1} as in BAL quasars. In NGC4151 absorption lines may have also been identified in the wings of subordinate lines, namely in Hβ and in the HeIλ5875 line, but it is difficult to distinguish between complex emission profiles and absorption components. If the existence of these absorption components is confirmed, it would imply that they are produced in a dense and highly excited neutral medium, i.e. in the Broad Line Region itself.

1.2.2 Emission Lines

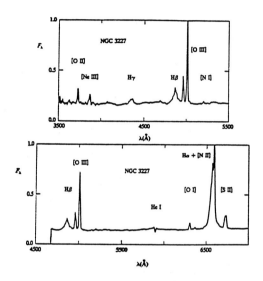

Fig. 4. Line spectrum of the Seyfert 1 galaxy NGC3227, after Osterbrock (1989)

Figure 4 shows the line spectrum of a Seyfert 1 galaxy. As can be seen, there are two kinds of emission lines: "Narrow Lines" whose Full Width at Half Maximum (FWHM) is equal to a few 10^2 km s^{-1}, and "Broad Lines" whose FWHM is equal to a few 10^3 km s^{-1}. The origin of the broadening had been amply discussed in the past, and several intrinsic mechanisms such as electron scattering or thermal broadening have been eliminated. The only possible mechanism seems to be Doppler broadening due to large relative velocities of individual emitting clouds or to large differential velocities in a continuous medium. An immediate consequence is that Narrow Lines and Broad Lines are produced in different media, whose dispersion velocities are different. These media are called the Broad Line and the Narrow Line Regions (BLR and NLR). Their respective proportion or their physical conditions should vary with luminosity, as the ratio of Broad Line to Narrow Line intensities increases with luminosity (although observational bias are not excluded, due to varying spectral resolution and range of wavelength observed). Also the ratio decreases continuously from Type 1 to Type 2 Seyfert galaxies (cf. Sect. 4).

Narrow Lines.

Narrow lines are mainly *forbidden lines*, the most intense ones being nebular transitions in the fundamental level of p^n ions (all processes described in this section are very extensively discussed in Osterbrock 1989). Some examples are: [OI]6300, [OII]3727–29, [OIII]4959–5007, [NeIII]3868, [NeV]3345–3425, [NII]6548–83, [SII]6716–30. Auroral and even transauroral lines of the same ions are also observed, but are fainter (one important example is [OIII]4363). So the Narrow Line spectrum is quite similar to a planetary nebulae spectrum, but it covers a larger range of ionization, including for instance [FeX]6374. Also the relative intensities are not completely identical to those of planetary nebulae.

These lines are formed by collisional excitation in a low density medium. So the auroral to nebular ratios are strongly dependent on the temperature and density of the emitting gas, and constitute excellent diagnostics for the physical conditions of the NLR. One finds that the temperature is equal to $1–2 \times 10^4$K, and the density to 10^4 to 10^6 cm^{-3}.

Some broad lines, but not all (for instance FeII lines), also have a narrow component located on top of the broad profile. Some of the most intense narrow permitted lines are Balmer lines and the HeI5548 line: these lines are formed mainly by recombination, as in planetary nebulae, but in contradistinction with the broad line components. So the Balmer decrement is close to case B, and consequently is a good reddening indicator: compared to the case B $H\alpha/H\beta$ ratio ~ 3, the $H\alpha/H\beta$ ratio is generally larger in the NLR, indicating an important reddening.

As recombination lines, Balmer and helium lines do not depend much on the temperature. Once the density is known, they allow one to determine accurately the mass of emitting gas. A useful formula is:

$$L(H\beta) \sim 10^{-25} \ V_{\text{eff}} \ n_e^2 \quad \text{ergs cm}^{-3} \text{ s}^{-1} \tag{1}$$

where V_{eff} is the effective emission volume (which can be spread in a much larger volume) and n_e is the electron (i.e. the total) numeric density in cm^{-3}. This equation leads to a value of the order of 10^6 M$_\odot$, which represents an underestimation of the NLR mass since it corresponds only to the ionized fraction of the NLR.

An important implication of the relatively low temperature of the NLR is that it is **not collisionally but radiatively ionized.** Indeed collisional ionization would require a temperature of the order $5 \ 10^4$ K to get large fractional abundances of O^{+2} or Ne^{+2} ions, and 10^5 K for Ne^{+4} ions. This is why the so-called *photoionization models* have been intensively developed for AGN since the beginning of the seventies. In these models the emission region is assumed to be ionized by a central source of UV and X-ray continuum (cf. Sect. 2). Since the ionization level depends strongly on the distance to the central source, photoionization models lead to a value of the "radius" of the emission region (in fact a typical dimension):

$$R(\text{NLR}) \sim 30 \ \sqrt{L_{44}} \ \text{pc} \tag{2}$$

where L_{44} is the ionizing luminosity expressed in 10^{44} ergs s^{-1}. This dimension agrees with measurements from direct imaging in nearby Seyfert galaxies. It allows one to determine the ratio of the effective to the total volume of the NLR, called the "*filling factor*":

$$f = \frac{<n_e^2>}{<n_e>^2} \sim \frac{V_{\text{eff}}}{R^3} \sim \frac{10^{-2}}{\sqrt{L_{44}}} \tag{3}$$

The small value of the filling factor implies a filamentary or cloud structure.

We have mentioned the existence of lines from highly ionized species: several [FeVII] lines, [FeX]6374, [FeXVI]5303... The origin of these lines have been the subject of much debate. Their FWHMs are larger than those of other forbidden lines, so these lines are generally attributed to an *intermediate medium*, located between the NLR and the BLR. Moreover, they show a loose inverse correlation between their FWHMs and the ionization potential of the corresponding ionic species. This correlation can be interpreted as a decrease of the ionization level with increasing distance, owing to the increasing dilution of the ionizing radiation. As we shall see below, the BLR is much closer to the central engine. All these facts argue for *a continuous decrease of the dispersion velocity with increasing distance* from the central engine, as it would be the case in a gravitationally bound medium.

A very important and still open problem concerns the kinematics of the NLR. Is it dominated by rotational or radial motions? And if the motions are radial, do they correspond to an inflow or an outflow? Several constraints are obtained through the study of line profiles. For instance the existence of a blue asymmetry in a majority of objects can be interpreted as a radial motion in the presence of absorbing dust. But the direction of the motion is not known. Either the dust is located *inside* the clouds, in the neutral region; the lines are then emitted preferentially by the illuminated face of the clouds, and the blue asymmetry implies an inflow. Or the dust is located *outside* the clouds, in the surrounding medium. Emission from the part of the NLR located further from us is then more absorbed, and the blue asymmetry implies an outflow.

Several recent observations can also give important clues for the structure and the kinematics of the NLR: the existence of a link between the structure and the kinematics (line widths) of the NLR and the nuclear radio source, the presence of "Extended Narrow Line Regions" (ENLRs) made of dilute gas far from the nucleus and ionized by the central nuclear continuum, which sometimes display "conical" structures, and finally the possible connection between the NLR and a starburst activity, etc. All these observations could be replaced in the "unified scheme" of AGN to lead to a globally consistent picture. We will try to perform this task in the last section.

Broad Lines.

Broad lines consist only in permitted lines (except a few "semi-forbidden lines"). It means that the density of the BLR is much larger than the critical density for collisional de-excitation of forbidden lines, i.e. $\gtrsim 10^8$ cm^{-3}. The most intense lines are:

- in the UV range: Lyα, NV1238, CIV1549, HeII1640, CIII]1909 (intercombination line), MgII2800, many FeII blends.

- in the optical range: Hα and Hβ, HeI5875, HeII4686, many FeII blends.

- in the infrared range: Paschen and Brackett series of HI, OI8446, CaII triplet near 8540 Å, HeI10830.

An impressive result is **the almost complete independence of the line spectrum on the luminosity, spanning a range of five decades in luminosity**: indeed line ratios and equivalent widths are remarkably similar in Seyfert 1 galaxies and in luminous quasars. Line intensities are actually proportional to the overall luminosity of the objects. However a noticeable exception, for which several explanations have been given (cf. later), is the so-called "*Baldwin effect*", namely an inverse correlation between the equivalent width of CIV1549, and to a lesser degree of Lyα, with the underlying continuum intensity. The slope of the correlation is small (about 0.2) and corresponds to an average difference of only a factor 2 in the line equivalent width between Seyfert nuclei and luminous AGN.

FeII blends represent an important fraction (about a fourth) of the total energy radiated in the broad lines. This fraction reaches even 50% in a few AGN. Since these lines are heavily blended, the extraction of the underlying continuum is a very difficult task, which explains why its exact shape and the existence of the UV bump was only recently established.

Several easy diagnostics can be used to determine, at least roughly, the physical conditions prevailing in the BLR. The presence of the semi-forbidden line CIII]1909 implies that the density is smaller than the critical collisional de-excitation density, i.e. $\lesssim$ a few $10^9 \mathrm{cm}^{-3}$. We will see, however, that there are indications for a larger density, which means that the BLR is made of different media. The presence of intense lines of low ionization species, such as MgII and FeII, indicates a weakly ionized medium and a high column density of neutral hydrogen, i.e. a high optical thickness at the Lyman limit, $\tau(912\ \text{Å}) \gtrsim 10^4$. The large ratio FeII(optical)/FeII(UV) implies even a higher value, $\tau(912$ Å$) \gtrsim 10^6$ (optical FeII photons are produced by the degradation of FeII UV photons after a great number of diffusions, and this process requires a large column density of Fe^+ ions, and consequently of H° atoms).

As for the NLR, an estimation of the mass of the emitting gas can be deduced simply. Here the Hβ line is of no use, since the Balmer line spectrum is not formed by recombination (see next section). But Lyα is given by recombination case B to good accuracy (better than a factor 2), and a useful equation is then:

$$L(\mathrm{Ly}\alpha) \sim 3\ 10^{-24}\ V_{\mathrm{eff}}\ n_e^2 \quad \mathrm{ergs}\ \mathrm{cm}^{-3}\ \mathrm{s}^{-1} \qquad (1\mathrm{bis})$$

Equation 1bis gives the ionized mass, once the density is known. It is of the order of 1 M$_\odot$. Note that such a small mass gives the most important spectral informations on AGN. However it is possible that a large fraction of the BLR is not emitting (it could be cold and neutral for instance), and that we are seeing only the "tip of the iceberg".

Another important information one can get from the value of $L(\mathrm{Ly}\alpha)$ is the proportion of ionizing photons absorbed by the BLR. If the optical thickness of the BLR at the Lyman limit is larger than unity (and this condition is fulfilled, as we shall see in the next section), all ionizing photons in the Lyman continuum which hit the BLR are absorbed and lead by subsequent recombination to a Lyα photon: measuring the number of Lyα photons is then equivalent to measure the number of ionizing photons which hit the BLR. On the other hand, it is possible to get the total number of ionizing photons in the continuum by extrapolating the spectral distribution in the UV range (this extrapolation is, however, hazardous, owing to the UV bump and to the soft X-ray excess). **Assuming that the continuum source is emitting isotropically**, the ratio of these numbers is equal to the *coverage factor* of the BLR (i.e. the proportion of the sky seen from the continuum source to be covered by the BLR). This is found of the order of 0.01 to 0.1. The assumption of isotropic central radiation is disputed now, as we shall see in the last section, but this does not strongly modify the result.

Contrary to the NLR, it is not possible to determine the reddening of the Broad Line spectrum. In the NLR, a powerful tool is indeed the Balmer decrement, whose theoretical value is known. In the BLR where Balmer lines are not formed by recombination, this determination is not possible. However we shall see in the next section that the theoretical

Hα/Hβ ratio is found to be of the order 3–5, while the observed ratio is generally small (about 3.5). This excludes the possibility of significant reddening, and it is likely that the Broad Line spectrum is modified only by interstellar reddening in our own Galaxy. This result is in agreement with the idea that we can see the BLR only when the conditions are favorable, i.e. when the line of sight is clear of absorbing matter (cf. the last section).

The observed Broad Line intensities are summarized in Table 1. For reasons which will become clear in the next section, they are divided into two classes, the High and the Low Ionization Lines (HILs and LILs). HILs, which are UV lines, are referred to Lyα, and LILs to Hβ, to avoid the uncertainty in line ratios due to reddening. The average observed Lyα/Hβ ratio is equal to 5. Taking into account interstellar reddening leads to a ratio of the order of 10, which is still much smaller than the value of 30–40 observed in planetary nebulae. This important result was established by Baldwin in 1977, and it opened a modern era for the study of the BLR.

Table 1. Observed average values of Broad Line intensities. Values for HILs are taken from Bechtold et al. (1987), and values for LILs are taken from Joly (1987)

HIL Spectrum		LIL Spectrum	
O VI	0.17 (0.07-0.28)	C II] 2326	0.3 (0.03-0.54)
Lα 1215	1	Mg II 2800	1.5 (0.2-4)
N V 1241	0.30 (0.5-0.58)	Fe II UV*	7 (3.5-18)
C IV 1549	0.42 (0.13-0.80)	Fe II opt**	7 (0.4-26)
He II 1640	0.07 (0.007-0.44)	Fe II 4570	0.7 (0.1-6.2)
C III] 1909	0.35 (0.025-0.37)	Hβ 4861	1
		Hα 6562	3.5 (2-10)
		BAL. CONT.	(0-10)
		PASC. CONT.	?

*multiplets with λ < 3000 Å
**multiplets with λ > 3000 Å

Finally, as for the NLR, it is possible to deduce directly the temperature of the emission medium from line ratios of the same ion: for instance the ratio CIII1909/977, when it can be measured, indicates a temperature smaller than 3×10^4 K, which implies that the BLR (or at least a part of it) is *not collisionally ionized*. Therefore photoionization models should be built for the BLR, but they are quite different from those of planetary nebulae and of the NLR (see next section). As for the NLR, they lead to an estimation of the overall dimension:

$$R(\text{BLR}) \sim 0.03 \sqrt{L_{44}} \text{ pc} \qquad (2\text{bis})$$

The dimension of the BLR is much smaller than that of the NLR. This is confirmed by the existence of line variations which have been intensively studied. The filling factor of the BLR can be determined from (3), and is found to be very small, $f(\text{BLR}) \sim 10^{-6} - 10^{-8}$.

A question is then immediately raised: what is the structure and the kinematics of such a "clumpy" and high velocity medium? Studies of line profiles and line variations can help to answer this question.

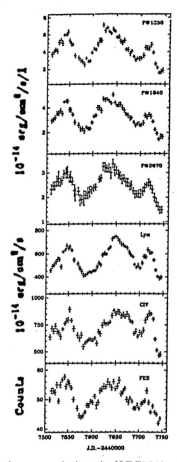

Fig. 5. Line and continuum variations in NGC5548, after Clavel et al. (1991)

Broad Line and continuum variability has been observed in AGN for 20 years. Recently several intense monitoring campaigns (in particular on NGC4151 and NGC5548, cf. Clavel et al. 1990, Peterson et al. 1991, Krolik et al. 1991, Maoz et al. 1991) have allowed to determine the relationship between line and continuum flux variations. Cross correlation studies of low luminosity AGN show that the time scale for large line flux variations is about equal to that of the UV continuum, which is a few weeks in general, and that *lines respond to continuum variations with a time lag which is smaller than the variation time scale*, typically a few days. This is illustrated in Fig. 5 which is a display for NGC5548 of fluxes of several lines and continuum bands versus time. This result confirms the existence of a strong link between the line spectrum and the UV continuum (and the optical continuum as well, since we recall that there is almost no time lag between the variations of the two bands), but it is not a direct proof of the validity of photionization models, as we shall see.

In quasars or luminous AGN, the line variation time scale is larger, typically a few months. In particular it is larger than the continuum variation time scale, and there is no direct link between UV continuum and line flux variations. It strongly suggests that the size of the BLR is larger than that of the continuum source (recall that the size is smaller than ct_{var}).

An interesting result is the discovery of a strong inverse correlation between the equivalent widths of several UV lines and the underlying continuum flux, in several well monitored low luminosity AGN. Apparently it has nothing to do with the "Baldwin effect" (which is averaged over time variations), and it shows that line fluxes are linked in a complex manner to the continuum flux.

Contrary to line intensities, line profiles are strongly variable from one object to the other. This is illustrated in Fig. 6, which displays broad Hα profiles in a few Seyfert 1 nuclei.

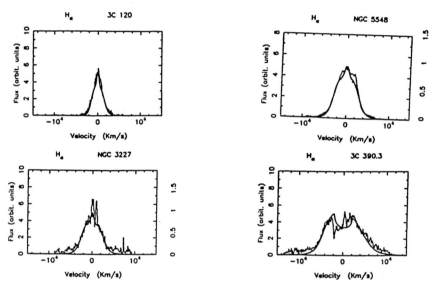

Fig. 6. Hα profiles in the Seyfert 1 galaxies 3C120, NGC3227, NGC5548, 3C390.3. Narrow Line components have been substracted. Solid lines correspond to theoretical fits obtained by Rokaki, Boisson and Collin-Souffrin (1992) and discussed in Sect. 3

The profiles can be very broad with a "flat" top, or narrow and "peaked". However these profiles show a few general trends:

- they seem to be more complex in Seyfert galaxies than in quasars
- in a majority of objects, Hβ has an extended red wing
- CIV and Lyα are blueshifted with respect to MgII and Hα by about 1000 km s^{-1} and often display extended blue wings
- all lines do not display the same profile. For instance, the ratio Hβ/Hα decreases systematically in the wings, from values close to 5 in the line center, to values as small as 2 in the wings
- FWHM(Hβ) shows a tendency to be small when the core to lobe radio flux density ratio is large (Wills and Browne 1986)

How profiles do vary with time is not clear. In some cases (NGC4151, M305 for instance) line wings display larger amplitude variations than cores, in other cases (3C273) line cores vary more rapidly than wings. More generally, many correlations between line widths and line profiles on the one hand and continuum and line intensities on the other hand, have been observed with various degrees of confidence. Many of these correlations can be attributed to observational bias (in particular the fact that only luminous objects are observed at high redshifts). It is difficult to extract from this huge amount of data those which are really significant and important for constraining the models. We shall limit ourselves to the above-mentioned results; they all argue for a complex structure and kinematics of the BLR.

Let us then summarize the picture which emerges already from this "first level study":

- at the smallest scale ($10^{-4} - 10^{-3}$ pc), probably located at the very center of the nucleus, lies a UV and X-ray source of continuum radiation, which is highly variable and displaying a complex spectral distribution

- further out is the BLR. It is consists of small, dense, optically thick clouds of high velocity, photoionized by the central source of UV-X continuum. These clouds fill a very small proportion of the whole volume of the BLR, whose dimension is itself smaller than a parsec. Another possibility is that the BLR is made of a very thin continuous shell with a large differential velocity. At the same scale, one finds sometimes compact radio sources with bulk relativistic motions

- on scales of tens or hundreds of parsecs, lies the NLR. It is made of dilute clouds, also photoionized by the UV-X continuum. The overall (often elongated) structure and the kinematics of the NLR is probably linked with the nuclear radio source.

Since these lectures aim at studying the central parsec of AGN, we shall not discuss the NLR any more, nor the more external part of the host galaxy. An important question is, however, what kind of relation exists between the BLR and the NLR? In this context we have already obtained some interesting results. From (3) and (3bis), $R(\text{NLR})$ = $10^3 R(\text{BLR})$. On the other hand the observed velocities are related by: $v(\text{NLR})$ = $v(\text{BLR})/10$. This corresponds to a variation of the velocity proportional to $R^{-1/3}$ and such a variation is less rapid than that of a gravitationally bound system with a given central mass (proportional to $R^{-1/2}$). It means that:

- either the NLR is gravitationally bound by a mass 10 times larger than the mass at the center of the BLR

- or it is not gravitationally bound.

As we shall see in the last section, values of the mass located inside the BLR have been quite accurately determined. They vary from 10^8 M$_\odot$ in Seyfert galaxies to 10^{10} M$_\odot$ in luminous quasars. The first hypothesis implies therefore the presence of a large mass - likely a star cluster - inside the NLR. According to (3) and (3bis), one obtains 10^9 M$_\odot$ inside a radius of 30 pc, in a low luminosity Seyfert galaxy, and 10^{11} M$_\odot$ inside a radius of 1 kpc in a quasar. In the other hypothesis, a likely solution for the NLR would be an outflowing accelerated flow which for instance may correspond to a nuclear wind.

For both models it is also interesting to compare the total mass flow of the BLR and of the NLR, assuming a radial motion. The mass flow dM/dt is then equal to $f\rho Rv$, where ρ is the density of the medium. According to the values obtained for the densities and for the filling factors of the BLR and NLR, one gets surprisingly similar values of dM/dt in the BLR and in the NLR. Although this result may not have any particular meaning (since the densities and filling factors correspond only to the *visible* part of the NLR and BLR), it can also imply that the NLR and the BLR both participate in the same flow. They can be two parts of an accretion flow, or the BLR is an accretion flow, and the NLR corresponds to the outflowing medium induced by the accretion activity.

2 The Broad Line Region

In the previous section it was shown that the BLR and the UV-X source of continuum are intimately linked and have small dimensions. In this section we shall concentrate on the BLR, and try to answer the following questions: what can the line intensities tell us about the physical state of the BLR? What can the line profiles and line variations tell us about the overall structure of the BLR? An answer to the first question can be obtained through the computation of photoionization models. The second problem involves a dynamical study which is the main topic of J. Perry's lectures: so the discussion will be restricted here to a few kinematical and phenomenological aspects.

The BLR has been the subject of many recent reviews (Ferland and Shields 1985, Mathews and Capriotti 1985, Osterbrock and Mathews 1986, Collin-Souffrin 1986a, 1987, 1989, Netzer 1987, 1989 and 1990, Collin-Souffrin and Lasota 1989).

2.1 Photoionization Models: Computations

In photoionization models, one solves the ionization balance and thermal equilibrium equations at each point of a slab of gas illuminated on one side by an ionizing continuum, and one computes the emerging line spectrum as a function of a set of given parameters, in order to compare it to the observed spectrum. The slab can adequately represent either an assembly of clouds or a continuous medium. The model assumes that the medium has reached a *state of equilibrium*. It is easy to verify that all important time scales are indeed smaller than the lifetime of a cloud, which is itself larger than the dynamical time scale t_{dyn} = cloud thickness/sound speed = 10^6 s (actually the lifetime of a cloud may be as large as $R(\mathrm{BLR})/\, v(\mathrm{BLR}) = 10^8$ s). We have to compare this time to:
  - the ionization time = 1/(ionizing photon flux × absorption cross section) = 10^{-2} s
  - the recombination time = $1/(n_e \alpha) = 10^4$ s, where α is the recombination coefficient
  - the cooling time = $(n_e kT)$ / cooling rate = 10^2 s

It is obviously not possible to describe all computations performed by sophisticated modern photoionization codes; only guide lines will be indicated. For more details one can read the above mentioned papers, or refer to the seminal paper of Kwan and Krolik (1981).

Ionization Equilibrium.

In a photoionized medium (where collisional ionizations are negligible) the fractional abundance ratio of two adjacent ionic species of the same element, $n(i)$ and $n(i+1)$, are given by the equation:

$$\frac{n(i+1)}{n(i)} = \frac{1}{4\pi R^2} \frac{1}{\alpha_{i+1} n_e} \int_{\nu_i} \frac{L_\nu\, \sigma_\nu(i)}{h\nu}\, d\nu \tag{4}$$

where R is the distance between the ionizing source and the cloud, α_{i+1} is the recombination coefficient from $i+1$ to i, L_ν is the differential luminosity, $\sigma_\nu(i)$ is the absorption cross section of ion i and ν_i is the threshold frequency for the ionization of ion i.

When applied to hydrogen, this equation gives the ionization degree of the medium. It makes evident the importance of a quantity called the *ionization parameter*, U:

$$U = \frac{1}{4\pi R^2} \frac{1}{c\, n_e} \int_{\nu_{\text{Lyman}}} \frac{L_\nu}{h\nu}\, d\nu \tag{5}$$

which is equal to the ionizing photon density to gas density ratio.

CAUTION! Several definitions exist for the ionization parameter and they do not necessarily have the same dimension or the same physical meaning. For instance in the first photoionization papers, U included the HI cross section. The following definition was introduced by Krolik, McKee and Tarter (1981):

$$\Xi = \frac{1}{4\pi R^2} \frac{1}{c\, k\, T\, n_e} \int_{\nu_{\text{Lyman}}} L_\nu\, d\nu \tag{6}$$

and is equal to 2.3 times the ionizing photon pressure to gas pressure ratio (the factor 2.3 takes into account ions and electrons due to heavy elements with cosmical abundances).

U and Ξ are both non-dimensional numbers, but one should notice that their ratio depends on the spectral distribution of the ionizing continuum. For instance, if the Lyman continuum is a power law with a spectral index equal to unity, Ξ/U is roughly equal to $100\, T_4$ (T_4 is the temperature expressed in 10^4 K). If an important UV-bump dominates the Lyman continuum, Ξ/U is of the order of $10\, T_4$.

U or Ξ determine several fundamental quantities:
- the thickness $H(\text{HII})$ of the ionized zone is given by equating the number of recombinations and the flux of ionizing photons:

$$n\, H(\text{HII}) = \frac{1}{\alpha_H\, n_e} \frac{1}{4\pi R^2} \int_{\nu_{\text{Lyman}}} \frac{L_\nu}{h\nu}\, d\nu = 10^{23}\, U \quad \text{cm}^{-2} \tag{7}$$

- Equation 5 gives the ionization degree of the gas: assuming a power law continuum with slope equal to unity, one gets $n(\text{H}^+)/n(\text{H}^\circ) \sim 10^5 U$: there is *no ionized zone if U is smaller than 10^{-5}*
- if U is larger than 10^{-5} and if the column density of the clouds (or of the shell) is larger than $10^{23}U$, the cloud is divided into two zones: an HII zone, ionized by the UV and soft X-ray continuum, and a neutral HI zone
- if the thickness of the clouds is larger than $10^{23}U$, and since hard X-ray photons are less absorbed than Lyman and soft X-ray photons, they can penetrate into the cloud much further (the absorption coefficient decreasing roughly as ν^{-3}), and the neutral zone is itself divided into two zones: a weakly ionized zone, heated and ionized by the hard X-ray flux, called the "excited HI zone (HI*)", and a cold neutral zone. In the HI* zone the ionization degree varies from 10^{-1} to 10^{-3}. Obviously **the extension of the HI* zone and the importance of its emission depends on the amount of hard X-ray photons**. Figure 7 depicts the structure of a BLR cloud.

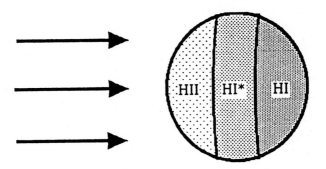

Fig. 7. Structure of a BLR cloud, whose left side is illuminated

NOTES ON THE IONIZATION EQUILIBRIUM
HII zone:

When τ(Lyman) is larger than unity, the diffuse radiation field J_ν should be taken into account, and the ionization equation becomes:

$$\frac{n(i+1)}{n(i)} = \frac{1}{\alpha_{i+1}n_e} \int_{\nu_i} \frac{\left\{ \frac{L_\nu \exp(-\tau_\nu)}{4\pi R^2} + 4\pi J_\nu \right\} \sigma_\nu(i)}{h\nu} d\nu \tag{8}$$

Then to compute the diffuse radiation field it is **necessary to solve the transfer of the ionizing continuum**.

HI* zone:

Several other ionization processes should be taken into account, as the medium is ionized by hard X-ray photons, and as it is only weakly ionized:

- charge exchanges
- inner shell photoionizations of heavy elements, including multiple Auger processes
- collisional ionizations from excited levels of H-atoms by thermal electrons (the excited level populations are considerably increased by Lyα trapping)
- collisional ionizations by non-thermal electrons created by inner shell photoionizations
- photoionizations from excited levels of H-atoms, either by the primary visible and infrared radiation, or by the diffuse radiation (continuum + **lines**).

Thus it is **necessary to solve the line transfer**.

Thermal Equilibrium.

There are two ways (obviously giving the same result) to compute the temperature of the gas: either write the equation corresponding to thermal balance of *the whole gas*, or write the equation corresponding to thermal balance of *the electron gas*. Let us describe the first one. In this case heating by absorption of radiation is equal to line and continuum emission.

In the HII zone heating is equal to:

$$H = \Sigma_{i,j}\ n(i,j) \int_{\nu_{i,j}} \left\{ \frac{L_\nu\ \exp(-\tau_\nu)}{4\pi R^2} + 4\pi J_\nu \right\}\ \sigma_\nu(i,j)\ d\nu \quad \text{ergs cm}^{-3}\ \text{s}^{-1} \quad (9)$$

where the summation is performed on all ions i and elements j. These processes include not only the usual absorption of UV and X-ray continuum by the photoelectric effect, but also free-free and bound-free absorptions of infrared and visible continuum, absorption by H^- ions, and Compton scatterings of hard X-ray and gamma-ray photons. Cooling is equal to $C = C(\text{lines}) + C(\text{free-free}) + C(\text{bound-free})$.

Since the HII zone is optically thin to all transitions (including resonance transitions, whose *effective optical thicknesses*, taking into account diffusions, are also smaller than unity), line cooling reduces to:

$$C(\text{lines}) = \Sigma_{\text{lines}}\ n_u\ h\nu_{ul}\ A_{ul} \quad (10)$$

where u and l correspond to the upper and lower levels of the transition, n_u is the upper level population, and A_{ul} is the Einstein probability coefficient. To establish the degree of line cooling it is therefore necessary to solve **the statistical equilibrium equations** for all ions. These equations include collisional excitations, spontaneous and collisional de-excitations, but **not radiative excitations**.

Free-free and bound-free coolings are given by:

$$\begin{aligned} C(\text{free}-\text{free}) &= 1.6 \times 10^{-27}\ \sqrt{T}\ n_e^2 & \text{ergs cm}^{-3}\ \text{s}^{-1} \\ C(\text{bound}-\text{free}) &= \Sigma\ (kT + X_{i,j})\ \alpha_{i,j}\ n_e\ n(i,j) & \text{ergs cm}^{-3}\ \text{s}^{-1} \end{aligned} \quad (11)$$

where $X_{i,j}$ is the ionization potential of the ion i,j.

In the HI* case, and contrary to the HII zone, the HI* zone is optically thick in many lines, and sometimes in the bound-free and free-free continuum. The problem is therefore not simple, since it requires us to solve **the coupled transfer equations for line and continuum radiation**. It is similar to a stellar atmosphere problem, but is in a sense more complex, since the BLR gas is more dilute than a stellar atmosphere, and heating is due to hard X-ray photons.

The complete solution of such a problem is presently out of the scope of the biggest existing computers, and it has required the development of approximation methods, which unfortunately are still not completely satisfactory. These methods are based on *the escape probablity approximation*. To summarize the question, we shall consider a simple two-level atom.

The state of equilibrium of the population levels implies:

$$n_u \left(B_{ul} \int J_\nu \psi_\nu d\nu + n_e C_{ul} + A_{ul} \right) = n_l \left(B_{lu} \int J_\nu \phi_\nu d\nu + n_e C_{lu} \right) \quad (12)$$

where ϕ_ν and ψ_ν are the absorption and emisison line profiles, B_{ul} and B_{lu}, C_{ul} and C_{lu} are the usual radiative (Einstein) and collisional excitation and de-excitation coefficients. Here J_ν is the diffuse line radiation.

Let us now define the **divergence flux**, ρ:

$$\rho_{ul} = \frac{1}{n_u A_{ul}} \left\{ n_u \left(B_{ul} \int J_\nu \psi_\nu d\nu + A_{ul} \right) - n_l \left(B_{lu} \int J_\nu \phi_\nu d\nu \right) \right\} \quad (13)$$

With this definition (12) becomes simply:

$$n_u \left(n_e C_{ul} + \rho_{ul} A_{ul} \right) = n_l \, n_e C_{lu} \tag{14}$$

which is identical to the classical statistical equation in HII regions or planetary nebulae, provided:

$$n_e \ll \frac{\rho_{ul} A_{ul}}{C_{ul}} \tag{15}$$

This inequality defines the critical density for an optically thick line. It also explains why resonant diffusions can be neglected in the HII zone. Indeed the diffuse radiation field in the line is small and ρ_{ul} is dominated by spontaneous transitions and is therefore close to unity.

Now let us make the **approximation that the emission and the absorption profiles are equal**. Then the source function is constant in the line and is equal to:

$$S = \frac{n_u A_{ul}}{n_l B_{lu}} \frac{1}{1 - \frac{n_u B_{ul}}{n_l B_{lu}}} \tag{16}$$

and ρ_{ul} can be written:

$$\rho_{ul} = 1 - \frac{\int J_\nu \phi_\nu d\nu}{S} \tag{17}$$

This fundamental equation allows us to relate the emergent flux in a line to the divergence flux. The transfer equation, assuming that the line is the only contributor to the emissivity and absorption coefficient at frequency ν is:

$$\cos\theta \, \frac{dI_\nu}{d\tau} = I_\nu - S_\nu \tag{18}$$

which implies, integrating over frequencies and angles:

$$\frac{1}{4\pi} \frac{dF}{d\tau} = \int J_\nu \phi_\nu d\nu - S = \rho_{ul} \tag{19}$$

where F is the local flux integrated on the line profile. Integrating this equation over the optical thickness, one gets the emerging line flux:

$$F = 4\pi \int \rho_{ul} \, S \, d\tau = \int \rho_{ul} \, \varepsilon \, dh \tag{20}$$

Here τ is the optical thickness at the line centre, ε is the usual emissivity and dh is the differential geometrical thickness. Again this equation is similar to that defining the emerging flux in HII regions or planetary nebulae, providing the divergence flux is close to unity. According to this last equation **the divergence flux can therefore be identified with the photon escape probability** Pe, i.e. the probability that, once emitted at a point τ (optical thickness from the surface), a photon can escape from the medium.

This escape probability is easily integrated over the line profile:

$$Pe = \frac{1}{2} \frac{\int_0^1 \int \frac{h\nu}{4\pi} n_u \, A_{ul} \, \phi_\nu \, \exp(-\tau\phi_\nu/\mu) d\nu \, d\mu}{\int \frac{h\nu}{4\pi} n_u \, A_{ul} \, \phi_\nu \, d\nu} \tag{21}$$

where $\mu = \cos\theta$.

According to the shape of the line profile, Doppler or Voigt, the functional dependence of Pe on τ is different:

$$
\begin{aligned}
Pe &= \frac{1}{\sqrt{\pi}}\frac{1}{\tau\sqrt{(\ln\tau)}} \qquad \text{for a Doppler profile} \\
Pe &= \frac{2\sqrt{a}}{\pi\sqrt{\pi}\sqrt{\tau}} \qquad\qquad \text{for a Voigt profile}
\end{aligned}
\tag{22}
$$

where a is the usual damping constant $A_{ul}/(4\pi\,\Delta\nu_{\mathrm{D}})$

So one can get the impression that we have succeeded in replacing frequency dependent transfer equations with frequency integrated ones. However the equality of the escape probability and the divergence flux was assumed under three basic conditions:

- the emission and the absorption profiles are identical. This is achieved if there is a *complete redistribution* of frequencies after a diffusion process, or on the contrary in a completely coherent diffusion. It is not valid for instance in the case of Lyα or other resonance lines which have a large opacity in the wings (so core photons are shifted in wings)

- the atom is restricted to two levels. This assumption is not valid for Balmer or Paschen lines which are strongly interlocked with Lyman lines

- at line frequency there is no other contribution to the absorption coefficient. This approximation fails if the opacity in the continuum is not negligible, as it is the case in BLR clouds.

Finally the computation of a photoionization model requires us to assume that the expression of the escape probability computed above for the whole medium is valid for each small slab. It amounts to replacing a *mean* value by a *local* value, and this implies that **the absorption and the emission of a photon takes place locally**, or equivalently that the medium is homogeneous. This assumption fails in particular in the transition region between the HII and the HI* zone, where physical conditions vary on a length scale which is smaller than the mean free path for a majority of photons. The divergence flux takes then negative values, while the escape probability is always positive.

So in general **THE CONDITIONS FOR THE ESCAPE PROBABILITY APPROXIMATIONS ARE NOT FULFILLED**. Several ways to circumvent this problem have been proposed (Elitzur and Netzer 1984; Netzer, Elitzur and Ferland 1985) :

- partial frequency redistribution is approximated by using expressions such as:

$$
Pe = \frac{1}{1 + g(\tau)\tau}
\tag{23}
$$

where the function $g(\tau)$ depends on the transition.

- the continuum opacity (or the overlapping of two lines) is taken into account by:

$$
Pe_{\mathrm{eff}} = \frac{\kappa_{\mathrm{cont}} + \kappa_{\mathrm{line}} Pe(\tau)}{\kappa_{\mathrm{cont}} + \kappa_{\mathrm{line}}}
\tag{24}
$$

where κ_{cont} and κ_{line} are the opacities in the continuum and at the line center. However this expression **is not valid if τ_{cont} is of the order of unity**.

Avrett and Loeser (1988) and Collin-Souffrin and Dumont (1986) have compared the results found for the same model by using an exact transfer treatment in the lines and an escape probability approximation. A general effect of the escape probability approximation is to decrease (by a factor which can be as large as two) the ratio of a lower to a higher transition, such as $H\alpha/H\beta$ or $Ly\alpha/H\alpha$.

2.2 Photoionization Models: Results

To compute a photoionization model one first chooses a set of *input parameters* for the cloud or shell:
   - spectral distribution of the incident ionizing continuum
   - ionization parameter
   - abundances
   - pressure P_0 or density n_0 at the illuminated surface (generally one assumes that the gas pressure is constant inside the cloud or shell; some authors deal with a constant density, which is less physical, or with different pressure laws such as hydrostatic equilibrium)
   - geometrical thickness H of the cloud, or column density N.
   The best procedure is the following. The shell is divided into about a hundred slabs and one chooses initial values for the total optical thicknesses at the centre of the lines. The computation progresses from the illuminated side towards the back, solving in each slab the transfer of continuum, the ionization, the line escape probabilities (taking into account escape towards **both** the illuminated side and the back of the shell). Upon reaching the back, one gets new values for the line optical thicknesses. The procedure is iterated until convergence. Then the emergent line spectrum can be compared to the observed one.

2.2.1 The Standard Model

The so-called "**standard model**", computed by Kwan and Krolik (1981), gives a relatively good agreement with the observed spectrum and was used in many studies of the BLR. Its input parameters are: $n_0 = 4 \times 10^9$ cm^{-3} , $N = 10^{23}$ cm^{-2}, $U = 0.03$. The abundances are normal, and the ionizing spectrum does not include an EUV excess. The spectral index in the X-ray range and the X-ray flux are similar to that of 3C273. Compared to a standard AGN, this spectral distribution overestimates the X-ray/UV flux ratio.
   A striking result is that the lines are almost completely produced, either in the HII zone (Lyα, CIV, CIII], HeII, NV, OVI...), or in the HI* zone (Balmer lines, FeII, CII..., and Balmer and Paschen continua). It is why we have separated in the previous section the emission lines into High Ionization Lines (HIL) and Low Ionization Lines (LIL). Note the paradox that HILs are emitted by the HII zone which draws its energy from UV and soft X-ray photons, and LILs are emitted by the HI* zone which draws its energy from hard X-ray photons! Since Lyα is emitted exclusively by the HII zone, and Balmer lines are emitted mainly by the HI* zone, the Lyα/Hβ ratio is much lower than in planetary nebulae: in the HI* zone, trapping of Lyα photons (corresponding to a very small value of Pe) brings the population of the second level close to its thermodynamical value ($n_2/n_1=$ the Boltzmann ratio), so Balmer lines are enhanced by collisional excitations from the second level, while Lyα is collisionally de-excited. Finally, since lines are produced either in the HII or in the HI* zones which are optically thick, they emerge preferentially from

the illuminated face of the clouds (as Lyα) or from the back (as Hα). This result is important from a kinematical point of view, as we shall see later.

Although the computed line intensities account roughly for the observed spectrum, some important discrepancies remain:

- the computed Hα/Hβ ratio is too large (5 instead of 3.5); note that this discrepancy would be increased by an exact line transfer treatment, as explained above
- the computed OVI/Lyα ratio is too small by a factor 5
- the computed CIII]/CIV ratio is too small by a factor 3 (it is linked with the "ionization parameter problem")
- the FeII(total)/Hβ ratio is too small by a factor 5.

2.2.2 Influence of Parameters and Other Models

Many photoionization models have been computed since the standard model. Let us summarize briefly the influence of the input parameters.

- Changing the incident continuum by the introduction of an EUV excess increases the OVI/Lyα and HIL/LIL ratios (the He^{++} Strömgren zone is larger); a break at a higher X-ray energy (in the standard model the continuum breaks at 10 keV) increases the LIL/HIL ratio, provided N is large enough (the shell should be able to absorb hard X-ray photons). The addition of a millimeter and infrared component increases the OVI/Lyα ratio (the temperature of the HII zone is larger), and the LIL/HIL and Hβ/Hα ratios (the HI* zone is more excited).

- An increase of the ionization parameter changes all HIL intensity ratios (it increases the CIV/CIII] ratio for instance); it also increases the Hβ/Hα ratio and the Balmer and Paschen continuum intensities (the HI* zone is more excited).

- An increase of the density, assuming no change of the ionization parameter, leads to an increase of the CIII]/CIV ratio due to the influence of CIII] collisional de-excitations, and to an increase of the Hβ/Hα ratio due to the increased collisional excitation rate.

- An increase of the column density increases the LIL/HIL ratio (the HI* zone is larger).

- When hydrostatic equilibrium instead of pressure equilibrium is assumed, the LIL/HIL ratio is increased (the density of the HI* zone is increased with respect to that of the HII zone.

- Finally, no model with strong abundance anomalies (except a few Fe overabundant models) have been computed.

The "Energy Budget" Problem.

Netzer (1985) stressed the fact that the whole computed Broad Line intensities are about 4 times smaller than observed with respect to Lyα, whatever the incident ionizing spectrum. This is actually due to the fact that the computed LIL/HIL ratio is smaller than the observed ratio by about one order of magnitude (Collin-Souffrin, 1986a). A way to solve the problem is to assume an X-ray continuum extending up to the hard X-ray range (100 keV at least) and a medium with a very large column density (in Netzer's computations the column density is limited to 10^{23} cm^{-2}) that is able to absorb these photons in an extended HI* zone heated mainly by Compton diffusions (Collin-Souffrin et al. 1986, 1988).

The Ionization Parameter Problem.

Another long-standing problem is the remarkable uniformity of the ionizing parameter deduced from photoionization models in objects displaying luminosities ranging over 5 orders of magnitude. As shown by Davidson already in 1977, this parameter is determined mainly by two HIL ratios, namely CIII]/CIV/Lyα. Since these ratios are almost constant in AGN spectra the ionization parameter is found to be constant also: $0.01 \lesssim U \lesssim 0.1$, corresponding to $0.1 \lesssim \Xi \lesssim 1$. This result has several implications. First it means that the radiative pressure equals the gas pressure and explaining this fine tuning has been a challenge. An explanation has been given in terms of a thermal instability and a "two-phase medium" by Krolik et al (1981) but it is not entirely satisfying (cf. later). Second we know that the density in the HII zone is restricted to a small range by the CIII] intensity, $10^9 \lesssim n \lesssim 10^{10}$. According to (6), this leads to the value of the BLR radius (or to the distance between the clouds and the continuum source) as given by (2bis). This radius is about one order of magnitude larger than the "reverberation radius" deduced from the time lag between continuum and line fluxes: the equality between the two radii would require a density of the order 10^{12}!

An elegant solution to both problems has been found by Ferland and Persson (1989). They have shown that the ionization parameter is not limited to the above range, **in the presence of an ionizing continuum with a strong EUV** excess: a value of the ionization parameter about 10–100 times larger also gives a correct fit to the CIII]/CIV/Lyα ratios. For large value of U, a large He^{++} zone is created, where carbon is present as C^{+4} ions, and not as C^{+3} ions. On the other hand a He^o Strömgren zone is also created (owing to the large number of He$^+$ ions present which absorb energetic photons), where carbon is present as C^{+2}. As a result, the CIII] /CIV ratio remains high in spite of the large value of the ionization parameter. The reverberation radius and the radius given by the ionization parameter (for $n = 10^9$) are reconciled, and the radiation pressure can be larger than the gas pressure.

Finally one should mention that the ionization parameter may not be completely constant among AGN; Mushotzky and Ferland (1984) suggested that the Baldwin effect can be explained by a decrease of the ionization parameter when the luminosity increases (U being roughly proportional to $L^{-1/4}$), inducing a decrease of the CIV/Lyα and CIV/CIII] ratios. This is not sufficient to account for the Baldwin effect, and they also propose that the covering factor decreases with increasing luminosity, from values close to unity for low luminosity Seyfert to values close to 0.01 for luminous quasars. Other explanations for the effect are now invoked. Netzer (1990) suggested that the presence of a geometrically thin accretion disc leads to an equivalent width (EW) luminosity dependence. If lines originate in an isotropically emitting system (as is probably the case for HILs, cf. below), their observed fluxes do not depend on the observer's viewing angle, contrary to that of the underlying UV continuum, if the latter is emitted by an accretion disc: it is more intense in the direction perpendicular to the disc. An observational bias will therefore emerge, namely that more luminous object are more likely to be seen perpendicular to the disc, and they have at the same time smaller line to continuum ratio, i.e. smaller EW. The effect will be stronger at frequencies where the contribution of the disc to the continuum emission is larger.

Homogeneous Models.

A severe problem with "homogeneous" models where all clouds are similar, is that a better fit than the standard model would require us to change the input parameters in two opposite directions. For instance to increase the $H\beta/H\alpha$ ratio requires an increase in the density, while in order to increase the CIII]/CIV ratio it is necessary to decrease it. In fact the most important parameter is the spectral distribution of the incident continuum and this is largely unknown, as we have seen!

Assuming a continuum extending from 1 mm to 100 keV, and consisting of two separate power laws in the radio to visible range and in the X-ray range, on which is superimposed an ultraviolet component peaking at 60 eV, as proposed by Mathews and Ferland (1987), Ferland and Persson (1989) found a satisfactory fit to the Broad Line intensities with a low density medium ($n = 3 \times 10^9$ cm^{-3}), made of clouds of very high column density ($N = 10^{25}$ cm^{-2}) with a large ionization parameter ($U \sim 1$). Such a continuum provides a strong source for heating the deepest layers by both the hard X-rays and the infrared continuum (through H- and free-free absorption), and therefore produces an extended hot HI* region. It intensifies on the one hand lines from highly ionized species (OVI), since the HII zone is hotter and more ionized, and on the other hand LILs, since the HI* zone is hot and thick. A problem with this model is however that it corresponds to a very large thickness of the emitting clouds ($H = 10^{16}$ cm), of the order of the size of the whole BLR in low luminosity AGN. It seems impossible to reconcile the geometry of the BLR and its small filling factor with such large clouds. On the other hand we have seen in the first section that the far infrared continuum is probably produced in a region which is much larger than the BLR. Finally this model still gives too large CIV/Lyα and Lyα/Hβ ratios.

Heterogeneous Models.

Since homogeneous models are not able to account for the Broad Line intensities, one has to look for multi-component or heterogeneous models. As we shall see below, such models are also more satisfactory from the kinematical and physical points of view.

In one model (Netzer 1987b), the BLR is made of similar clouds (i.e. having the same range of density), distributed in two systems, a flat disc-like structure and a spherical one. The inhomogeneity is due to a non-isotropic UV continuum source: the EUV flux is preferentially emitted in the axial direction (as it would be the case if the EUV continuum is due to an accretion disc, cf. Sect. 3), and therefore the spherical system "sees" the UV source, while the flat system receives a lower EUV flux. On the contrary, the X-ray source is assumed to emit isotropically, so both systems receive the same amount of X-rays. In the flat system, HILs are therefore weak, and they are produced mainly in the spherical system. Such a model partly solves the energy budget problem, but does not give a low enough Hα/Hβ ratio.

A suggestion of Collin-Souffrin (1986a) is that the BLR is made of two media of different density and column density. One medium has a high density and a high column density ($n \sim 10^{11-12}$ cm^{-3}, $N = 10^{24-25}$ cm^{-2}) and emits mainly LILs, the second medium is dilute with a small column density, ($n = 10^9$ cm^{-3} and $N = 10^{21-22}$ cm^{-2}) and it emits mainly HILs. This model can account satisfactorily for all the Broad Line intensities, except the FeII ones.

Rees, Netzer and Ferland (1989) considered a distribution of clouds ranging from very small radii and very high densities (10^{13} cm^{-3}) up to large radii and small densities, linked by different simple pressure laws and the assumption of cloud matter conservation. They integrated the cloud emission over the radial distribution to obtain the entire line spectrum of the BLR, as well as the line profiles. This model improves but does not solve the Hα/Hβ nor the Lyα problem.

At least one result concerning the ionizing continuum seems to have been definitively established: all studies (including those taking into account line variations, cf. below), agree that the ionizing continuum includes a UV bump with a maximum in the range 10–50 keV. This continuum is not necessarily "seen" by the whole BLR: it may be absorbed in a part of the BLR itself, and the other part is illuminated by harder photons (Ferland, Korista and Peterson 1990), or it may be emitted in a preferential direction and illuminate only a part of the BLR, as explained above. This result is very important for the continuum emission mechanism.

The FeII Problem.

No solution has yet been found for the FeII problem, although a great deal of work has been performed on it and several thousands of FeII lines have been included in photoionization codes (Wills, Netzer and Wills 1985). Neither an increase of the iron abundance, nor of the density or column density are able to give at best an overall FeII intensity larger than about a third of what is observed in many AGN. Possibly some unknown atomic process populate the upper levels of the ion. A more likely solution is that FeII lines are emitted by a dense **non-photoionized but mechanically heated medium**, whose existence may be linked with the radio emission (Joly 1987a, 1991).

2.3 Overall Structure of the BLR

Line intensities have provided information on the physical conditions prevailing in the Broad Line medium, although many ambiguities remain on the best solution, multicomponent or continuous model, high density or low density, clouds or continuous medium... But they were not able to give an overall picture of the BLR, its structure and kinematics. Only line and continuum variability studies, coupled with the study of line profiles, can provide insight to the geometrical and kinematical structure of the BLR. Moreover they are probably the best way to discriminate between physical models. However at the present time they are still chiefly observational, and have not led to definitive conclusions. Another potential tool is the study of X-ray absorption and emission features (fluorescent lines and continuum edges), which can also help us to understand the structure of the inner regions of AGN.

2.3.1 Line Profiles and BLR Kinematics

We have mentioned in the first section several characteristics of broad line profiles. One of the most noticeable is the decrease of the Hα/Hβ ratio in the line wings, i.e. with increasing velocity. Photoionization models have shown that a small Hα/Hβ ratio implies a large density. A natural idea is that the density increases towards the centre (this is obviously the case for the NLR and the BLR). One is then led to conclude that the velocity **and** the density both increase towards the center. This result strongly suggests a

gravitationally bound medium, where the velocity - Keplerian or free-fall - is proportional to $R^{-1/2}$. Moreover the extended red Hβ wing observed in a majority of objects is easily explained by gravitational redshift in the vicinity of a massive object. The correlation between FWHM(Hβ) and the core-to-lobe flux density ratio is also an argument in favour of rotational broadening. The core-to-lobe ratio is indeed assumed to be increased by relativistic boosting effects, so a large value for the ratio means that the jet is pointing towards us. On the other hand a small FWHM is observed if lines are emitted by a disc structure perpendicular to the jet axis.

HILs profiles have different characteristics. Instead of an extended **red** wing, these lines often display extended **blue** wings. Such profiles could be produced by an outward flow whose remote part is hidden. Two observations corroborate this interpretation. First the existence of blue-shifted absorption components in HIL profiles is naturally explained by outflowing absorbing clouds. Second the fact that HILs are blue-shifted with respect to LILs is easily interpreted if HILs are formed in an outflowing medium and LILs by a flat structure (a disc?) partially obscuring the HIL emitting region. Such a picture also agrees with the absence of a strong asymmetry in the Lyα profile.

The Asymmetry Problem

We have seen that some lines, such as Lyα, should be emitted exclusively by the illuminated face of the clouds (in the standard model at least). The anisotropy of the emergent lines appears also as a strong dependence of the intensity on the angle between the light ray and the cloud surface. As shown for the first time by Ferland, Netzer and Shields (1979), these effects have important consequences on line profiles: if the clouds are in radial motion, Lyα should display an asymmetric profile: the line should be blue-shifted with an extended blue wing for an inflow, and red-shifted with an extended red wing for an outflow. Other intense lines, such as CIV1549, which are not re-absorbed by the HI* zone, should not exhibit such an asymmetry, and LILs should be affected in the opposite direction. These effects are not observed. There are of course some shifts and some asymmetries, but not so strong as predicted by the cloud model in the case of a radial motion.

This result does strongly argue in favour of rotational or non-organized turbulent motions, but there are other solutions to the problem. Kallmann and Krolik (1986) suggested that Broad Line clouds are embedded in a hot medium of large Thomson optical thickness, so that the line emission is redistributed isotropically by the diffusion process. The drawback of such a model is that it does not allow one to observe rapid line and continuum variations, since they should be smoothed by the diffusion process. With the recent variation studies, it is clear that the optical thickness of a hot medium surrounding the BLR (and the continuum source as well) must be smaller than unity.

Collin-Souffrin (1986a) suggested another solution in the framework of the two component model: HILs are emitted by optically thin clouds i.e. with a column density smaller than N(HII). In other words the HIL emitting region is *"matter bounded" and not "ionization bounded"*. The emergent HIL profiles should then be symmetrical in the case of symmetrical radial motion.

2.3.2 Line Variations and BLR Structure

In the photoionized picture of the BLR, lines should respond to a variation of the ionizing flux, after a time lag corresponding to the distance between the continuum source and the BLR. So studies of correlations between line and continuum fluxes can lead to a real *"mapping of the BLR"*. Several methods have been developed for these studies (for a review, see Netzer 1990).

- The cross correlation function method:

The maximum of the cross correlation coefficients between line and continuum fluxes as a function of the time lag corresponds to the radius of the BLR. This method requires a very good sampling of the light curve.

- The response function method:

One assumes that L_{line} is proportional to a power n of $L_{continuum}$. It is then possible to get the Fourier transform of the BLR radius through the equation:

$$L_{line}(t) = \int L_{continuum}^n(t)\, R(t-\tau)\, d\tau \qquad (25)$$

and then:

$$F[R(\omega)] = \frac{F[L_{line}(\omega)]}{F[L_{continuum}(\omega)]} \qquad (26)$$

which leads to $R(t)$, and therefore to the space distribution of the emissivity. Unfortunately $R(t)$ does not provide directly the geometry of the BLR: this would require us to know the emissivity of the lines as a function of the continuum flux, and therefore to build grids of photoionization models as functions of **all** the physical parameters, which is clearly impossible.

There are several other difficulties which make this "inversion method" difficult to handle.

First it can be used only for low luminosity AGN, not because the dimension of the BLR in high luminosity objects would require monitoring it over several years, but because the radius of the BLR is larger than the radius of the UV and X-ray source, so the variations of the ionizing continuum are smoothed in the BLR. The only possibility for studying the BLR in high luminosity objects would be during "quiescent phases", when the continuum flux varies in a time which is comparable to $R(BLR)/c$. We do not know if such phases exist.

The second problem is that the monitored continuum is not the ionizing continuum: it is the optical or UV continuum, and a hazardous extrapolation is required to get the variations of the ionizing EUV flux. Moreover, as far as LILs and not HILs are concerned, the photoionizing continuum is the hard X-ray continuum, which is a priori not linked with the EUV continuum, as we have seen in the previous section. So one can legitimately ask why the variations of the optical continuum are so well correlated with LIL variations, since it is not predicted by photoionization models! The answer is that **not only the line variations, but also the optical and UV continuum variations, are induced by the same primary phenomenon, namely variations of the hard X-ray flux.** This result will be very important in the discussion concerning the origin of the optical-UV continuum (cf. next section).

Finally a difficulty lies in the fact that lines do not respond proportionally to the continuum flux, so any real "mapping" would require us to compute a photoionization model for the whole BLR. It is why (in my opinion) the best method to get the structure of the BLR is to have in mind a given model, and to compute in a detailed way for this model the correlation between line and continuum fluxes, to see if it matches the observations.

Some qualitative results have already been obtained through the use of the cross correlation or the response functions:

- the size of the BLR is small, which implies a large value of the ionizing parameter, as already discussed

- the emission region is strongly heterogeneous, with high and low density regions

- the observed bumps and shoulders observed in line profiles can, however, result from a homogeneous spherical distribution

- there are still no conclusive results concerning the geometry of the BLR, spherical or flat.

2.3.3 Links with the Dynamics

A possible model which seems satisfactory from the physical and kinematical points of view has therefore emerged from the discussion. The BLR is probably made at least of two different media. One medium is a spherical or simply an axisymmetric structure (two cones, for instance) in outward motion, made of optically thin and dilute clouds. This medium sees both the EUV and the X-ray continuum, and emits mainly HILs. The second medium is a flat, rotating, dense and optically thick structure, it sees the X-ray continuum, but not necessarily the EUV continuum, and it emits mainly LILs. This medium partly obscures the HIL emitting region.

Such a model also agrees with the X-ray observations mentioned in the first section. It explains why the column density of the region absorbing the soft X-ray continuum is low: it does not correspond to the column density of the whole BLR, but only to that of the HIL emitting region (recall that the BLR, or more precisely the HIL emitting region, covers about 10% of the central source). Indeed if the LIL emitting region is a flat rotating structure, and is **geometrically thin** (its thickness can be as small as the thickness of a unique cloud, i.e. 10^{14} cm), it will not absorb the continuum source, except when the line of sight of the continuum sources crosses the disc. Since it is illuminated by the central X-ray source, the flat structure will also be a reprocessor of X-ray photons, and it will emit reprocessed continuum and line radiation. Reprocessing in the optical UV continuum will be discussed in Section 3. The disc can also account for the reprocessing in the X-ray continuum (fluorescent K-line and emission hump). And finally it can account for the line emission (cf. below).

Very little is known about the cloud dynamics itself.

The "radiatively driven" cloud model (cf. Mathews and Cappriotti 1985, for a review) has been intensely studied. In order to be accelerated efficiently the clouds should be mainly made of fully ionized gas, so the model is valid only for the HIL emitting region. This model leads to logarithmic profiles in the wings, comparable to the observed profiles, and for this reason it was the most popular model used for more than a decade. It has been shown that the radiatively accelerated clouds, which rapidly develop into "pancake" shapes, are Rayleigh-Taylor unstable. Orbiting clouds have also been envisaged, but they are also subject to severe instabilities (due to shear flows in particular) which should destroy them. More generally if the clouds are not confined, for instance by a gravitational field, they will expand and dilute rapidly in a dynamical time scale and become very hot under the action of the ionizing continuum. So one needs a confinement mechanism.

The Cloud Confinement Problem and the Two-Phase Theory.

Krolik, McKee and Tarter (1981) have proposed a very interesting model, based on thermal confinement by a hot medium, which could explain the uniformity of the ionization parameter among AGN. However a careful analysis of this model shows some weak points.

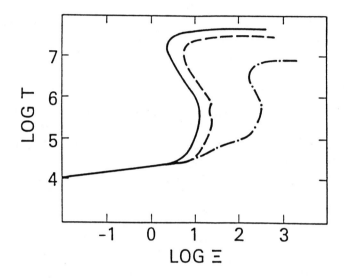

Fig. 8. The two-phase medium, after Krolik and Kallmann (1988). The solid line corresponds to a "bare power law" continuum, with a slope of 1.2 under 2 keV and 0.7 above 2 keV; the dashed curve corresponds to the addition of a small "10 eV-bump", and the dot-dashed curve to the addition of a large "80 eV-bump"

The equation of state of a photoionized gas in thermal equilibrium, given by the equality between gains and losses, defines a relation $\Xi(T)$, shown on Fig. 8 for different spectral distributions of the ionizing continuum. Note that the $\Xi(T)$ curves display an "S" shape, which allows the existence of three solutions between the values Ξ_h and Ξ_c. The three solutions correspond to the same gas pressure, for a given radiative flux.

The solutions located on the portion of the curve having a negative slope, between Ξ_h and Ξ_c, cannot exist because they are thermally unstable (a perturbation in the pressure or in the density is amplified: for instance if the temperature increases, the representative point moves up, heating becomes larger than cooling, and T goes on increasing). The solutions located on the hot and cold branch between Ξ_h and Ξ_c are thermally stable and correspond to the same gas pressure. A hot medium could therefore confine the cold medium. As one can see on Fig. 8, the shape of the $\Xi(T)$ curve varies with the incident continuum (it can be understood easily by studying the physical processes of losses and gains in the different portions of the curve, cf. Collin-Souffrin and Lasota 1989). The observations of the UV bump and of the soft X-ray excess in the ionizing continuum led to the discovery that the range $\Xi_h - \Xi_c$ is considerably reduced (Fabian et al. 1986; Krolik and Kallmann 1988). So the value of Ξ deduced from the study of the HIL emitting region, which lies at the lower end of the interval Ξ_h and Ξ_c if the continuum does not include a UV-bump, is completely outside this interval if it includes a large UV-bump. For the LIL emitting region, which has a larger gas pressure, it in any case lies outside this interval.

Moreover in the case of luminous AGN and quasars the hot medium should be Compton thick, as shown by Perry and Dyson (1985) and Kallmann and Mushotsky (1985). This is in contradiction with the observations of rapid variations of the X-ray flux.

The existence of the double solution as well as the remarkable coincidence of the values found for Ξ when the UV-bump was not discovered, were the origin of the great success of this "two phase" theory. The physical picture emerging for the BLR was a hot medium (a wind or an accretion flow) where clouds were formed by thermal instability at a given gas pressure, corresponding to a range of distances from the central source, and then confined by the hot gas. This picture should now be abandoned.

Several other models have then been proposed to circumvent this problem of confinement. First, clouds can be transient and continuously reforming from the hot medium. Perry and Dyson (1985) have proposed a very interesting model which explains **dynamically** the constancy of the ionization parameter.They assume that AGN contains a star cluster. A high-velocity nuclear wind (which is naturally created as a result of the heating of the upper layers of an accretion disc by the X-ray continuum, for instance, cf. Begelman and McKee 1983) hits the star envelopes. As a consequence of this, shocks are created and the post-shock gas cools almost isobarically. The pressure of the cooled gas is therefore close to the stagnation pressure of the wind, or:

$$nkT(\text{cooling gas}) \;=\; \rho_{\text{wind}} V_{\text{wind}}^2 \tag{27}$$

Using (6), the ionization parameter becomes:

$$\Xi \;=\; \frac{1}{4\pi R^2}\; \frac{L_{\text{ion}}}{c\rho_{\text{wind}} V_{\text{wind}}^2} \tag{28}$$

Let us call $\dot{M}_{\text{wind}} \;=\; 4\pi R^2\, \rho_{\text{wind}} V_{\text{wind}}$ the mass rate flow of the wind (assuming spherical symmetry) and $\dot{M}_{\text{acc}}$ the accretion rate giving rise to the bolometric luminosity with a mass-energy efficiency coefficient η (of the order of 0.1 for an accreting black hole, cf. next section). If $L_{\text{ion}} = f_{\text{bol}} L_{\text{bol}}$ (f_{bol} is of the order unity), one gets finally:

$$\Xi \;=\; \frac{\eta}{f_{\text{bol}}}\; \frac{\dot{M}_{\text{acc}}}{\dot{M}_{\text{wind}}}\; \frac{c}{V_{\text{wind}}} \tag{29}$$

which shows that Ξ is of the order of $\dot{M}_{\text{acc}} / \dot{M}_{\text{wind}}$, which is itself likely to be of the order of unity or larger.

Second, clouds can be confined by another mechanism. A possibility is magnetic confinement (Rees 1987). Another natural mechanism is gravitation confinment, which is realized if the BLR consists of stellar atmospheres, or of the surface of an accretion disc.

The suggestion that the BLR can be made from the atmospheres of giant and supergiant stars was first proposed by Edwards (1980). The density in these atmospheres is indeed not very different from that of the BLR clouds deduced from photoionization models. The problem with this idea is the very large number of stars required to account for the line emission, about 10^9 for quasars. This would require that a large proportion, if not all, of the central star cluster is in the form of giant stars. It has been suggested that they could be post-main-sequence stars due to the evolution of a co-eval dense starburst (Norman and Scoville 1988). They could also be normal stars "bloated" by the intense UV radiation field in which they are immersed (Puetter 1988, Penston 1988). But it is not clear if the radiation field is intense enough to produce an important effect, since the radiation pressure is strongly reduced by opacity. The interesting point with this model is that such giant stars would not only account for the BLR, but could also feed the black hole at a high accretion rate through mass-loss from their envelopes (Scoville and Norman 1988).

The suggestion that the accretion disc can contribute to the Broad Line emission was first made by Jones and Raine (1980): the surface of the accretion disc is heated and ionized by the radiation, backscattered by a hot medium located above the disc, and consequently produces an emission spectrum. Dumont and Collin-Souffrin (1989a, b and c) have shown that if the disc is illuminated directly by a central non-thermal source, or indirectely by the backscattering of this radiation by a hot medium (this same hot medium which gives the "reflection hump" in the X-ray continuum) the disc emits an LIL spectrum, quite comparable in intensities and profiles to the observed one. In particular, the advantage of this model, compared to a spherical model, is that it leads to a broad range of profiles, owing to the influence of the inclination parameter. On the contrary, the intensities are about constant, whatever the luminosity, because the illuminated surface of the disc is a dense photoionized shell, whose state is very close to thermodynamical equilibrium (Boltzman level populations).

So finally we see that from the dynamical point of view, as well as from the previous physical and kinematical study, the best solution for the BLR is also a two-component model. A fraction of the BLR is likely to be gravitationally confined in a disc and, as a reprocessor of the central continuum source, it will emit mainly LILs. The other fraction could be composed of transient more dilute clouds, formed in the cooling post-shock medium created in the nuclear wind (Collin-Souffrin et al. 1988). This fraction will emit mainly HILs.

3 The Accretion Disc

3.1 The Black Hole Model

Since the discovery of quasars, black holes have been suspected to power the central engine of these objects. But it was only during the seventies that black hole models were developed intensively in the AGN context, and it was only about a decade ago that these models were almost universely accepted as the only ones able to account both for the very small size of the energy source and for the high rate of energy emission (cf. Rees 1984).

It is out of the scope of these lectures to discuss black hole models, but it is necessary to recall the values of a few basic parameters:
- the gravitational radius R_G:

$$R_G = \frac{2GM}{c^2} = 3 \times 10^{13} M_8 \text{ cm} \tag{30}$$

where G is the gravitational constant, M is the black hole mass and M_8 the mass expressed in 10^8 solar masses
- the Eddington luminosity L_{Edd}, which is the maximum luminosity of an object fuelled by accretion; it corresponds to the equality of radiation pressure and gravitational force:

$$L_{Edd} = \frac{4\pi cGm_p M}{\sigma_T} = 1.3 \; 10^{46} \; M_8 \text{ erg s}^{-1} \tag{31}$$

where σ_T is the Thomson diffusion cross section, and m_p the proton mass. Although this expression assumes that radiation is emitted isotropically and that the gas is completely ionized, it is valid to within about an order of magnitude for any type of accretion, except for short transient phenomena
- the mass efficiency conversion factor η:

$$\eta = \frac{L_{bol}}{\dot{M}_{acc}c^2} \tag{32}$$

which can take values from 0.05 for a non-rotating black hole up to 0.4 for a strongly rotating black hole. Assuming $\eta = 0.1$, one gets:

$$\dot{M}_{acc} = 1.75 \; L_{46} \; M_\odot \text{ y}^{-1} \tag{33}$$

where L_{bol} is expressed in 10^{46} ergs s^{-1}.

An average quasar with a bolometric luminosity equal to 10^{46} ergs s^{-1}, radiating close to its Eddington luminosity, should therefore harbor a black hole of about 10^8 M$_\odot$. It is interesting to note that the dimensions given by the X-ray variability is then equal to about 10 R_G for this black hole. We shall see below that the bulk of the soft X-ray luminosity is indeed emitted close to this radius.

3.2 Spherical or Disc Accretion?

The question to ask now is whether accretion is taking place with or without angular momentum (i.e. as spherical accretion or as an accretion disc).

We have seen that there are some hints for accretion with angular momentum: the existence of a privileged axis, most probably linked with rotation (recall the correlation between FWHM(Hβ) and core-to-lobe ratio); the "dust molecular torus" in Seyfert 2 galaxies, whose presence is suspected in all AGN (cf. the "unified scheme" in the next section). Above all, the accretion disc is an appealing model which easily accounts for the optical and UV continuum of AGN, and is a reprocessor of radiation explaining many observed features, including broad optical lines and X-ray lines.

However, observational tests for accretion discs based on orientation effects are not clear and their interpretation is generally model-dependent. Several orientation effects on line-continuum or continuum-continuum correlations have been predicted, as part of the radiation is expected to be dependent on the direction. The optical-UV continuum, if mainly emitted by the accretion disc, is expected to be inclination dependent. We have seen for instance in the previous section that the Baldwin effect can be interpretated within this model. But a fraction of lines may be emitted in a disc structure and be inclination dependent, and another fraction may be emitted more isotropically. So line-continuum correlations are not easy to interpret. Predictions concerning the polarization of the continuum are also questionable, since the same dependence on wavelength or on orientation can be obtained with a non-thermal and with a disc continuum. Perhaps the most promising effect is predicted by Laor and Netzer (1990): it is a change in polarization at the Lyman edge, due to a change in the opacity regime for disc emission (atomic versus Thomson).

Emission line profiles are also a topic of controversy. It is generally believed that double peaked lines are the best signature of disc emission. This is not definite, since several other effects can lead to this kind of profiles, such as binary black holes, or simply two emitting regions in motion with respect to each other (such as symmetrical jets). Moreover discs do not inevitably lead to double peaked profiles (cf. Dumont and Collin-Souffrin 1990b and c). So finally, although there are *many suggestive facts in favour of accretion discs, none constitute real evidence at the moment.*

Accretion discs were not always as fashionable as they are at present for explaining AGN fuelling. In the seventies spherical accretion was more often envisaged. And it was only after Shields suggested in 1978 that the "big blue bump", observed in the optical continuum of 3C273, might be the black-body-like emission of an accretion disc, and after the study of similar features in the spectrum of other quasars and AGN by Malkan and Sargent (1982) and Malkan (1983), that accretion discs became very popular, although they have always been considered by theoreticians as an interesting possibility for AGN. However it became rapidly evident that there are severe problems with the so-called "thin standard discs" in AGN. The most conspicuous one is that at the high level of accretion rate required to fuel quasars, these discs should be supported by radiation pressure in their central regions and self-gravitating in their external regions, such a situation encourages strong instabilities, probably leading to the destruction of the disc.

It is nevertheless possible to conclude through the study of the UV-continuum that the only possible accretion mechanism is a disc.

3.3 Emission of the UV-Bump
3.3.1 Radiation Regimes

There are two extreme possibilities, according to the value of the ratio t_{cool}/t_{acc}, where t_{cool} is the characteristic time for gas cooling, and t_{acc} the characteristic accretion time.

1) $t_{acc}/t_{cool} \ll 1$: the "Virial regime".

In this case the accretion energy is given directly to radiation and the temperature is equal to:

$$T_{Vir} = \frac{GM m_p}{kR} \approx 5 \times 10^{11} (\frac{R}{10 R_G})^{-1} \text{ K} \tag{34}$$

This upper limit is generally not reached, but the temperature can stay quite high. For instance, in a strong radial shock, the temperature is equal to 0.2 T_{Vir}. Radiation can be produced through a variety of mechanisms, ie non-thermal emission, bremsstrahlung or comptonized free-free emission, or, if a magnetic field is present, infrared cyclotron radiation transferred into the X-ray range through Compton interactions. So in any case the bulk of emission will take place in the hard X-ray or gamma-ray range.

2) $t_{\mathrm{acc}}/t_{\mathrm{cool}} \gg 1$: the "optically thick regime".

In this case the medium cools to a low temperature. The minimum temperature is that of a dense and thick medium radiating as a black-body:

$$T_{\mathrm{eff}} \approx \left(\frac{GM\dot{M}}{4\pi R^3 \sigma}\right)^{1/4} = 10^5 \left(\frac{\eta}{0.1}\right)^{-1/4} \left(\frac{L}{L_{\mathrm{Edd}}}\right)^{1/2} \left(\frac{R}{10R_G}\right)^{-3/4} L_{47}^{-1/4} \mathrm{K} \qquad (35)$$

Radiation is then emitted in the UV and soft X-ray range.

Spherical accretion, where the accretion takes place roughly in a free-fall time, leads most naturally to the Virial regime. However a fraction of the hard radiation can be *reprocessed* into a softer range (UV) by dense and thick blobs, if they condensate in the hot flow, as proposed by Guilbert and Rees (1988) and Lightman and White (1988). Accretion with angular momentum leads most naturally to the optically thick regime, since the gas has ample time to cool and a dense and thick disc can be formed. However in this case too, hard X-rays can be emitted, either as Compton *reprocessed* radiation, through a hot corona or through a hot part of the disc (a "two-phase disc", cf. Wandel and Liang 1990), or they can (for instance) even be emitted directly through hydromagnetic disc winds.

A relevant observation would then be to know in which energy band is the bulk of radiation emitted, in the hard X-ray range or in the EUV band. Unfortunately none of these bands are well observed, as we have seen. In order to make progress with this problem, we have to discuss in more detail the shape of the UV continuum emitted in both cases.

3.3.2 Basic Concepts for the UV-Continuum Emission

A first concept is that the "UV bump" should be emitted by a gas whose temperature is close to 10^5 K. At a higher temperature the bulk of the emission would stay in the X-ray range, and at a lower temperature, it would stay in the optical or infrared range.

The following basic concept is that a *smooth* UV continuum, not displaying strong Balmer or Lyman discontinuities, should necessarily be emitted by an *optically thick* medium. In particular the effective optical thickness at *the red side of the Lyman edge* should be much larger than unity. Since the temperature of the gas is of the order of 10^5 K, the gas should be dense to enable a large number of neutral excited hydrogen atoms to absorb this radiation, and it should have a large column density.

The Spherical Model.

In this model, dense clouds located in a small volume R $<$ $100R_G$, are illuminated by the non-thermal X-ray continuum produced in a more central region. The emitted spectrum is composed of a mixture of directly seen non-thermal radiation and reprocessed thermal radiation. Ferland and Rees (1988) computations of UV continuum due to reprocessed non-thermal continuum by dense clouds show that, even at high densities, a column density of about 10^{25} cm^{-2} is required to get a small Lyman discontinuity in agreement with the absence of any observed discontinuity.

First, we estimate the density of the region emitting the UV continuum. We have seen in the previous section that a gas exposed to a non-thermal spectrum extending into the hard X-ray range displays a bimodial behaviour according to the value of Ξ. For Ξ larger than a few tens, the equilibrium gas temperature is close to the "Compton temperature" (equilibrium between Compton heating and inverse Compton cooling), equal to a few 10^7 K for a typical AGN spectral distribution; for Ξ smaller than unity, the equilibrium temperature is governed by photoionization processes and collisional cooling, and is close to 10^4 K. A temperature of the order 10^5 K is obtained only in a small range of values of Ξ, with $\Xi \sim 10$, and this gives:

$$n \sim 5\,10^{15}\,(\frac{\eta}{0.1})\,(\frac{L}{L_{\mathrm{Edd}}})^3\,(\frac{R}{100R_{\mathrm{G}}})^{-2}\,L_{44}^2\,T_5^{-1}\,\mathrm{cm}^{-3} \qquad (36)$$

Second, in order to reprocess efficiently the non-thermal continuum the shell of clouds should have a large covering factor, $\Omega/4\pi \geq 0.5$. We have seen that at the same time it should have a large column density, $> 10^{24}$ cm^{-2}, in order not to display strong spectral discontinuities. In this case a strong absorption feature should be present in the X-ray range, up to 10 keV or more, at least in half of the objects. This is in contradiction with observations, which show that the continuum is never heavily absorbed in the hard X-ray range.

A way to circumvent the problem is to assume that the *clouds are embedded in the hot medium* producing the X-ray luminosity. But in this case, not only the density should be very high, of the order 10^{18} cm^{-3}, but also the UV continuum should display variations in the same time scale as the X-ray continuum, and this is not observed.

In summary it seems that there is no way to accomodate spherical accretion with the emission of the UV-bump.

The Disc Model.

On the contrary, an accretion disc offers ideal conditions for the emission of a UV-bump, because it is made of a very dense and thick medium. Its mean density at a radius R is larger than that of a spherical accretion flow by a factor $\sim (Rv_{\mathrm{ff}})/(Hv_{\mathrm{r}}) \gg 1$ where H and v_{r} are the thickness and radial velocity of the disc, and its column density is larger by a factor $v_{\mathrm{ff}}/v_{\mathrm{r}} \gg 1$ (assuming that the velocity of the spherical flow is close to the free fall velocity v_{ff}). We shall see below that an accretion disc, either geometrically thin or thick, emits a smooth continuum which peaks in the EUV range. This UV bump is produced directly by gravitational release in regions of the disc located between $10R_{\mathrm{G}}$ and $1000R_{\mathrm{G}}$, and the hard X-ray continuum comes either from a more central region, or from a region located above the disc, as proposed in several models. In both cases the disc will act as a reprocessor of X-ray photons, since it can absorb this radiation and re-emit in another wavelength range.

3.4 Generalities on the Structure of Accretion Discs

A considerable literature has been written on accretion discs ((cf. Pringle 1981; Frank, King and Raine 1985; Begelman 1985, for a review of the subject).

3.4.1 The Different Regimes

If accretion onto the black hole is taking place with angular momentum, the accreted matter will settle in an accretion disc and lose its angular momentum through the action of viscous stresses, which transport matter inward and angular momentum outward. Gravitational energy is converted into heat which is radiated away by the surface of the disc. Matter spirals inward, until it reaches the last stable orbit, which is equal to $1.24R_G$ in the case of a rotating black hole, and to $6R_G$ for a non-rotating ("Schwarzschild") black hole. This difference explains why rotating black holes are more efficient than non-rotating ones in converting mass into energy.

According to the mass accretion rate, there are several possible regimes for the accretion disc. They have been explored in a number of papers. These regimes are determined mainly by the geometrical thickness of the disc, or more exactly by the thickness to radius ratio, H/R. Let us describe very schematically these different regimes. The accretion rate will be referred to as the *"critical accretion rate"*, defined as $\dot{M}_{\text{crit}} = L_{\text{Edd}}/\eta c^2$. (Caution, in many papers the critical accretion rate is defined as L_{Edd}/c^2)

- If $\dot{M}/\dot{M}_{\text{crit}} \lesssim 10^{-2}$: the disc has a small density, and interactions are not rapid enough to provide an efficient energy transport between protons and electrons. Consequently ions stay very hot, the disc is supported by the ionic pressure and is thick. This kind of disc is called "ion torus" (cf. for instance Rees et al. 1982). Since the luminosity is small, these authors have suggested that the model can apply to radio galaxies. Recent developments have however shown that ionic tori may become cold under the action of collective phenomena coupling ions and electrons.

- If $\dot{M}/\dot{M}_{\text{crit}} \gtrsim 1$: the disc has a high accretion rate. Such discs have a large optical thickness and therefore a large radiation pressure, are consequently supported by radiation pressure and are also geometrically thick. Thick discs have been extensively studied by Abramovicz et al. (1980 and subsequent works), who have shown that particle acceleration can take place in the "funnel" created at the centre of the disc near the black hole. The accelerated particles provide the radio jets and the non-thermal continuum. This model could apply to luminous quasars. There are, however, a few severe problems. First these discs are suspected to be dynamically unstable. Second they imply a very large accretion rate since they are inefficient in the emission of radiation (at high accretion rate the luminosity is no longer proportional to $\dot{M}$ but to $\log \dot{M}$). It is difficult to imagine that an accretion rate as large as 100 or 1000 M$_\odot$ y^{-1} can be sustained for long periods of time.

- If $0.01 < \dfrac{\dot{M}}{\dot{M}_{\text{crit}}} < 1$: the accretion rate is appropriate to *"geometrically thin discs"*. These kinds of discs have been mostly studied, since they are thought to provide the mass transport between the companion star and the compact object in X-ray binaries.

As we shall see in Sect. 4, the $\dot{M}/\dot{M}_{\text{crit}}$ ratio - that we can identify with $L_{\text{bol}}/L_{\text{Edd}}$, except for very high accretion rates - is of the order of unity for luminous quasars, and of the order of 0.01 to 0.1 for Seyfert nuclei. Therefore it is likely that thin disc models are valid for a majority of AGN.

Although these lectures aim only at giving a phenomenological description of AGN, and although the broad characteristics of the emitted radiation do not depend on the disc structure, we shall show in a very rough way how one can get the structure of thin discs, since it is relevant for our understanding of the central engine.

3.4.2 Geometrically Thin Accretion Discs

These discs are characterized by a value of the H/R ratio much smaller than unity.

We shall use the cylindrical coordinates R, z, and ϕ. Let us denote by Σ the disc surface density, Ω the angular velocity, V_R and V_ϕ the radial and azimuthal velocities, V_K the keplerian velocity, c_s the sound velocity, and ν the kinematic viscosity. We consider **stationary** discs, in which nothing depends explicity on time. This implies in particular that the mass transfer rate takes place at large radii and is constant within time scales of the order of the viscous time (cf. below). The vertical component of the velocity is assumed to be negligible, and the heat transport is assumed to take place only vertically. Apart from the conservation equations in a ring, one assumes that vertical hydrostatic equilibrium is realized. The equations giving the structure of the disc are then:

- the vertical hydrostatic equilibrium equation:

$$\frac{1}{\rho}\frac{\delta P}{\delta z} = -\frac{GM}{R^3}\,z = \frac{V_K^2}{R} \tag{37}$$

where P is the total pressure (gas plus radiative pressure). Identifying z with H, one gets:

$$\frac{c_s}{V_K} \sim \frac{H}{R} \ll 1 \tag{38}$$

which shows that **the keplerian velocity is strongly supersonic**. This equation gives the value of the scale height of the disc, or the disc thickness, H, once c_s is known

- the mass conservation equation:

$$\frac{\delta}{\delta R}(R\Sigma V_R) = 0 \;\Rightarrow\; \dot{M} = 2\pi(R\Sigma V_R) = \text{constant} \tag{39}$$

- the angular momentum conservation equation:

$$\frac{\delta}{\delta R}(R\Sigma V_R R^2\Omega) = \frac{\delta}{\delta R}(R\nu\Sigma R^2\frac{\delta\Omega}{\delta R}) \tag{40}$$

Identifying $\delta\Omega/\delta R$ with Ω/R, (40) gives an approximate value of the radial velocity:

$$V_R = \frac{\nu}{R} \tag{41}$$

We reach here the crucial step of the computation, i.e. we have to guess what is the viscosity. In a most famous paper, Shakura and Sunayev (1973) have proposed that the main dissipation occurs through turbulent viscosity, namely $\nu = V_{turb}\,H_{turb}$ where H_{turb} is the typical dimension of turbulent eddies. Assuming that H_{turb} is equal to the disc thickness, H, and that the turbulent velocity, V_{turb}, is of the order of the sound velocity or is subsonic, i.e. $V_{turb} = \alpha c_s$ with $\alpha \lesssim 1$, one gets:

$$V_{\mathrm{R}} \; = \; \alpha \, c_{\mathrm{s}} \, \frac{H}{R} \; \ll \; c_{\mathrm{s}} \tag{42}$$

which shows that **the radial velocity is largely subsonic**.

In these so-called "α-discs", α is assumed to be a constant. With the additional assumptions that the opacity is dominated by electron scattering, and that the disc can be described by vertically averaged quantities, it is a "standard α-disc". This modelization has often been called a "parametrization of our ignorance". It neglects other possible energy transport mechanisms, such as magnetic, which may be important in AGN. However standard discs have proved to be very successful for accretion discs in binary stars, and therefore they are often used in the AGN context

- the equation of conservation of the impulsion:

$$V_{\mathrm{R}} \frac{\delta V_{\mathrm{R}}}{\delta R} \; - \; \frac{V_\phi^2}{R} \; + \; \frac{1}{\rho} \frac{\delta P}{\delta R} \; + \; \frac{GM}{R^2} \; = \; 0 \tag{43}$$

According to the above discussion, $V_{\mathrm{R}} \frac{\delta V_{\mathrm{R}}}{\delta R}$ and $\frac{1}{\rho} \frac{\delta P}{\delta R}$ are negligible, so one gets:

$$V_\phi \; = \; \sqrt{\frac{GM}{R}} \; = \; V_{\mathrm{K}} \tag{44}$$

which shows that the **azimuthal velocity is equal to the keplerian velocity**.

Finally one should add the equation for the vertical transport of energy. Assuming that it is purely radiative (no convective or magnetic transport for instance), the solution of the diffusion equation, if the disc is optically thick, can be approximated by:

$$F_{\mathrm{rad}} \; = \; \frac{\sigma T^4}{\tau} \tag{45}$$

where τ is a frequency vertically averaged optical thickness and T a vertically averaged temperature (note that this temperature is larger by a factor $\tau^{1/4}$ than the effective temperature T_{eff}). F_{rad} is the flux emitted by each face of the disc. This equation shows the **fundamental importance of the opacity** in this problem. Twice F_{rad} is equal to the gravitational dissipation rate:

$$\nu \Sigma \, (R \frac{\delta \Omega}{\delta R})^2 \; = \; \frac{3GM\dot{M}}{4\pi R^3} \, (1 - \sqrt{\frac{\beta R_{\mathrm{G}}}{R}}) \; \mathrm{ergs \; cm}^{-2} \, \mathrm{s}^{-1} \tag{46}$$

(for more details, see one of the review papers referred to above). The numerical factor β is of the order of unity, depending on the angular momentum of the black hole (β=6 for a non-rotating black hole).

It is then possible to solve the previous set of equations at each radius, by dividing the disc into concentric rings, and assuming that each ring behaves like an independent plane-parallel radiating slab. For **a given opacity regime**, one gets the disc structure, i.e $H(R)$, $c_{\mathrm{s}}(R)$ and $\Sigma(R)$. It is necessary to check a posteriori that the assumed opacity is dominant in the physical conditions prevailing at each point. A "thin disc solution" is not always found, if the cooling mechanism is not efficient and the condition $c_{\mathrm{s}} \ll V_{\mathrm{K}}$ not fulfilled. In this case the disc will become hot and thick, and the model does not apply. Note that c_{s} includes not only the gas pressure, but also the radiation pressure, which is dominant in the inner regions of the disc.

Characteristic Times.

A most important question is to know if the disc is stable or not. There are many possible instabilities: gravitational (cf. below), dynamical, thermal, viscous and so on. Piran (1978) has shown that the criterium for viscous stability is:

$$\frac{d \ln \dot{M}}{d \ln \Sigma} > 0 \tag{47}$$

and that for standard discs it is equivalent to the thermal stability criterium:

$$\left(\frac{\delta \ln Q^-}{\delta \ln H}\right)_\Sigma > \left(\frac{\delta \ln Q^+}{\delta \ln H}\right)_\Sigma \tag{48}$$

where Q^- and Q^+ are respectively the radiative flux and the heat generation due to gravitational release given by (46).

Corresponding time scales can be compared to the observations:

- the dynamical time t_{dyn}, or the time for a dynamical perturbation to propagate vertically through the disc:

$$t_{\text{dyn}} = \frac{H}{c_s} = \frac{R}{V_K} \tag{49}$$

- the viscous time t_{visc}, or the time for a perturbation of the mass rate to propagate radially through the disc:

$$t_{\text{visc}} = \frac{R}{V_R} = \frac{t_{\text{dyn}}}{\alpha} \left(\frac{R}{H}\right)^2 \tag{50}$$

- the thermal time t_{therm}, or the time for the disc to cool:

$$t_{\text{therm}} = \frac{\Sigma c_s^2}{F_{\text{rad}}} \tag{51}$$

According to (37), (42), (44) and (46), one then gets:

$$t_{\text{therm}} = \frac{t_{\text{dyn}}}{\alpha} \tag{52}$$

and finally $t_{\text{dyn}} \lesssim t_{\text{therm}} \ll t_{\text{visc}}$.

It is clear that any variation affecting the whole disc structure requires a time of the order of t_{visc} to be established, while a perturbation at a given radius is settled in a dynamical time. The values of characteristic times can be computed once the disc structure is known.

Application.

Let us assume an AGN with the following parameters: $L_{\text{bol}} = 10^{44}$ ergs s^{-1}, $L_{\text{bol}}/L_{\text{Edd}} = 0.1$ and consequently $R_G = 2 \times 10^{12}$ cm. NGC4151 is probably well represented by these parameters. One gets, for $\eta = 0.1$ and $\alpha = 1$, after solving the disc structure:

- for $R = 100 R_G$, $t_{\text{dyn}} = 0.5$ days, $t_{\text{visc}} = 1$ year
- for $R = 1000 R_G$, $t_{\text{dyn}} = 15$ days, $t_{\text{visc}} = 100$ years

We shall see below that these radii correspond approximatively to the emission of the UV ($10 R_G - 100 R_G$) and of the visible continuum ($100 R_G - 1000 R_G$). These regions can therefore not exchange information through a dynamical process in a time smaller than the viscous time, i.e. a few years. Since the time lag between the UV and the optical flux is smaller than a few days, it implies that the variations of the UV and optical continuum fluxes cannot be induced by dynamical changes of the disc.

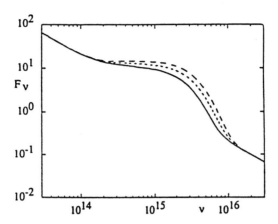

Fig. 9. Response of the accretion disc to variations of a central spherical X-ray source, as described in the text (Collin-Souffrin 1991)

Now let us assume that the disc is irradiated by a variable source of hard X-rays. These photons penetrate under the photosphere of the disc, and the surface temperature is increased with respect to pure gravitational release. The disc can adjust to a local perturbation in a time smaller than the dynamical time (since only the upper layers are affected), therefore this is a time much smaller than a few days. So finally the disc emission will respond to the X-ray variations after a time lag corresponding to the light travel time between the source and the UV and optical emission regions, which are about one light day apart. Such a time lag corresponds indeed to the absence of detection of any time lag between the UV and optical flux variations. This is a supplementary proof of the existence of a central X-ray source partially reprocessed by the accretion disc in the optical and UV range. Besides, the strong correlation observed between the lines and the underlying continuum leads to the conclusion that the hard X-rays (largely responsible for the Balmer line emission, as we have seen), the soft X-rays, the UV and the optical continuum, are all tightly linked through complex emission and reprocessing mechanisms. As an illustration Fig. 9 shows the possible link between the optical-UV continuum and the X-ray source, for the same object as considered previously. The central X-ray source is spherical and has a radius of $30R_G$. An X-ray luminosity corresponding to 0%, 40% and 80% of the bolometric luminosity is considered (it takes into account the continuum up to the MeV range). A ν^{-1} power law continuum corresponding to a contribution equal to 50% of the total flux at 1 μm is added to the disk emission: this does not take into account the starlight which might dominate in the red range. This figure shows that the reprocessed radiation can represent a non-negligible part of the total flux in the optical-UV range.

3.5 Results: Structure and Emission of Geometrically Thin Discs

We know the value of the inner radius of the disc (the radius of marginal stability), but not that of the outer radius: it depends on the origin of the fuelling matter. We shall assume that the disc extends up to a large distance, say $10^6 R_G$, which is of the order of one parsec for an object like NGC4151.

A general result is that the disc has a so-called *"flaring structure"*, which means that the H/R ratio increases with R. This result is very important, since it implies that the outer regions of the disc can be irradiated by the emission from the inner regions, or by a central source lying near the centre, even if this source has a small dimension. Owing to the decrease of the gravitational release with increasing radius, the temperature of the disc drops to very low values, of the order of 100 K. On the other hand the surface density decreases with increasing radius and the disc becomes optically thin at large radii. At large radii the disc is no longer geometrically thin, because the H/R ratio becomes larger than unity. The temperature increases again, owing to external irradiation. A possible picture of the accretion disc is shown in Fig. 10.

3.5.1 The Inner Region, a Few $R_G \lesssim R \lesssim 10^3 R_G$

This is the most studied part of the disc, because it emits the UV bump and the soft X-ray excess. In this region the opacity is dominated by Thomson scattering and free-free absorption, and the radiative pressure dominates the gas pressure. It is not clear if the disc is thermally and viscously unstable in this region (Pringle 1981).

The emitted spectrum is computed using different approximations, ordered by increasing complexity:

- the black-body approximation. The disc is assumed to emit at each radius a black-body spectrum at the effective temperature given by the viscous flux. If there is an external irradiating flux due to a central X-ray source, this equation becomes:

$$\sigma T_{\text{eff}}^4 = \frac{3GM\dot{M}}{8\pi R^3}\left(1 - \sqrt{\frac{\beta R_G}{R}}\right) + F_x(R) \tag{53}$$

where F_x is the irradiation flux. The superposition of black bodies gives the continuum displayed in Fig. 11, where T_{in} and T_{out} are the temperature at the inner and outer radii (for $F_x = 0$).

The maximum of the black-body at a radius R corresponds to a wavelength:

$$\lambda \sim 200\left(\frac{\eta}{0.1}\right)^{1/4}\left(\frac{L}{L_{\text{Edd}}}\right)^{-1/2}\left(\frac{R}{10R_G}\right)^{3/4} L_{44}^{1/4} \text{ Å} \tag{54}$$

This equation shows that for a given radius expressed in R_G, the emitted spectrum is harder for a lower luminosity. It can partly explain the Baldwin effect, since the continuum underlying Lyα and CIV would be less intense for low-luminosity objects. For an object like NGC4151, it leads to the results mentioned above for the characteristic radii corresponding to the UV and optical emission

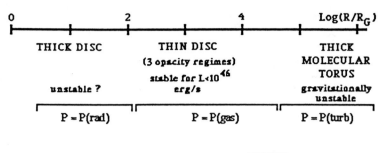

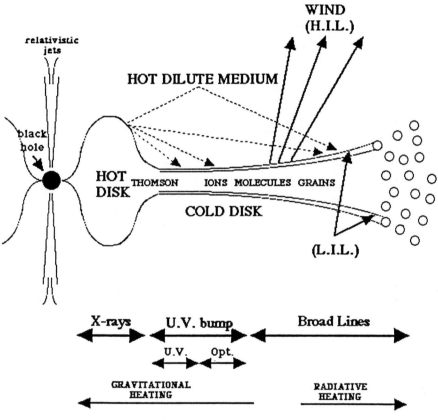

Fig. 10. Possible picture of an accretion disc in AGN

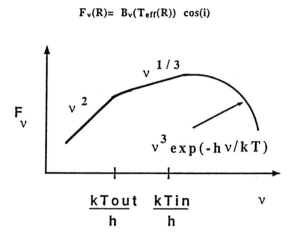

Fig. 11. Emission of the inner region in the black body approximation

- a superposition of black bodies with relativistic effects taken into account. The introduction of relativistic effects is important, especially for rotating black holes seen at a large inclination. The spectrum is harder, due to light bending.

Note that **the spectrum does not depend on the viscosity in the two previous approximations** (but it depends on the thin disc assumption)

- black bodies on which is superposed a stellar atmosphere with the same gravity (Sun and Malkan 1988). This model gives a Lyman edge in absorption

- reprocessing of the black-body spectrum through Compton diffusions in the disc or above the disc (Czerny and Elvis 1987)

- and finally complete solution of the vertical structure of the accretion disc, taking into account the radiation transfer and the thermal and hydrostatic equilibrium (Laor and Netzer 1989). Unfortunately this treatment implies the knowledge of the vertical heat deposition, which is completely unknown. According to the assumptions concerning this point, the vertical distribution of temperature gives a Lyman edge in emission or in absorption, or no Lyman edge at all.

It is therefore clear that only broad characteristics of the spectrum emitted by the disc are yet known. We have seen that it is not even clear if the disc is geometrically thin or thick near the centre. The spectrum emitted by a geometrically thin disc would actually not be very different from the spectrum emitted by a geometrically thick disc, except that the dependence on the viewing angle is opposite for a thin and a thick disc. In a thin disc, the spectrum is harder at large inclinations, while in a thick disc, where the soft X-ray emission takes place in a central funnel, the spectrum is harder at small inclinations (Madau 1988). One could hope to be able to test this prediction in the future, where more observations concerning the soft X-ray spectrum and the core to lobe ratio will be obtained. Predictions concerning the polarization have also been made: Netzer

(1990) stressed that thin discs should be characterized by a change of the polarization degree at the Lyman edge, due to a change in the opacity regime.

Finally one should mention several variants of the standard disc model. A recently proposed promising model is the so-called "two phase" accretion disc, in which the disc is divided into a hot optically thin and a cold optically thick phase, owing to a stabilizing mechanism induced by the presence of the soft photons (Wandel and Liang 1990). The hot medium is located near the center at $R \lesssim 30R_G$, and the cold medium is located at larger radii. It emits the UV and soft X-ray bump while the hot medium emits hard X-rays. The two media interact through Compton processes, a part of the UV photons being upscattered to the soft X-ray range, the hot medium being stabilized by these inverse Compton scatterings.

3.5.2 The Intermediate Region, $10^3 R_G \lesssim R \lesssim 10^5 R_G$

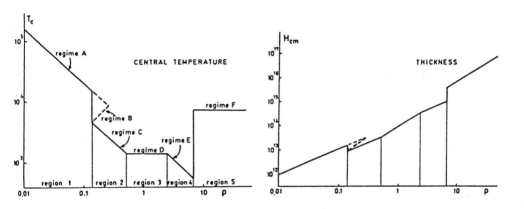

Fig. 12. Temperature (equatorial) and thickness, as functions of the radius expressed in $10^4 R_G$ (equal to $2\ 10^{16}$ cm), in the thin disc model, after Collin-Souffrin and Dumont (1990). Values of the parameters are given in the text, and correspond to an object like NGC4151. The different opacity regimes are indicated, the broken lines being for non-physical regimes

This part of the disc is cold, the opacity being dominated by atomic and molecular processes, and by dust. It is viscously and thermally stable, sustained by gas pressure, and contrary to the inner part where the disc is puffed up by radiation pressure, **it stays geometrically thin even at a high accretion rate**. But it is auto-gravitating (i.e. the disc gravity is larger than the gravity of the central object) for relatively modest accretion rates, $\dot{M} \gtrsim 1\ M_\odot y^{-1}$, and it probably fragments into small clouds above this rate. A schematic of the disc is displayed in Fig. 12, for the same values of the parameters as before. This figure shows the flaring of the disc, which becomes geometrically thick for $R \gtrsim 10^5 R_G$, and the different opacity regimes: Regime A correponds to electron scattering, Regime C to atomic and molecular opacity, Regimes D and E to dust opacity, and Regime F is the optically thin part of the disc where the heating is provided by the central X-ray source.

In this region the illumination of the disc by the central X-ray source induces several important effects:

- the creation above the disc of an irradiated "chromosphere" which gives rise to line emission. This process was mainly invoked to account for line emission with double peaked profiles, as in Arp102 (Chen, Halpern and Fillipenko 1989) or in 3C390.3 (Peres et al. 1990). But Dumont and Collin-Souffrin (1990) have shown that it is certainly a more general process occuring in all AGN. The lines cannot be produced for $R <$ a few $10^2 R_G$, because the line excitation temperature is smaller than the disc photospheric temperature, and moreover a major part of the irradiating flux is absorbed under the photosphere and reprocessed in the continuum as explained above. For $R \geq 10^4 R_G$ the flux from the central source is too small to heat the disk. This gives *a natural explanation for the line widths*, as they should be about equal to the keplerian velocity at $R = 10^3 R_G$, 10 000 km s^{-1}. On the other hand the disc emission matches quite well the observations concerning line intensities and line profiles. Figure 13 displays several kinds of Hα profiles obtained with this disc model, for different inclinations and different disc radii. It illustrates the fact that all types of profiles, including double or simple peaked profiles, can be obtained according to the radius of the disc (regions distant from the centre emits the cores of the lines) and to the inclination. For small inclinations, an extended red wing due to the Doppler gravitational effect is produced. In Fig. 6 the fits to the observed profiles are given, obtained with an accretion disc model which also accounts for the UV continuum (Rokaki, Boisson and Collin-Souffrin 1992).

- the creation of a hot corona due to heating of the upper layers of the disc at the Compton temperature, according to the discussion of Sect. 2. At large radii the thermal velocity of the corona is larger than the keplerian velocity, so the corona is no longer bound, and gives rise to an hydrodynamic wind (Begelman and McKee 1983), which can decrease appreciably the accretion rate, and produce shocks responsible for the emission of HILs (cf. Sect. 2).

3.5.3 The External Region, $R \gtrsim 10^5 R_G$

This part of the disc is not a continuous medium. The disc is geometrically thick and optically thin, it is gravitationally unstable and certainly fragmented into clouds with large turbulent velocities (Krolik and Begelman 1986). Most probably this part of the disc constitutes the "dust-molecular torus" present in a few Seyfert 2 galaxies, and perhaps in all AGN, which hides the central continuum and the BLR (Antonnucci and Miller 1985, cf. next section). This torus is heated by the central X-ray source up to a temperature of about 1000 K (Krolik and Lepp 1989), and it may be the origin of the strong molecular H$_2$ lines observed in many Seyfert nuclei.

3.6 Masses and Accretion Rates

We have now at our disposal two methods for determining the mass and the accretion rate of a given object.

The "Accretion Disc Method".

This method amounts to fitting the UV-bump with the emission of an accretion disc, combining this emission with the other components (Malkan 1983, Wandel and Petrosian

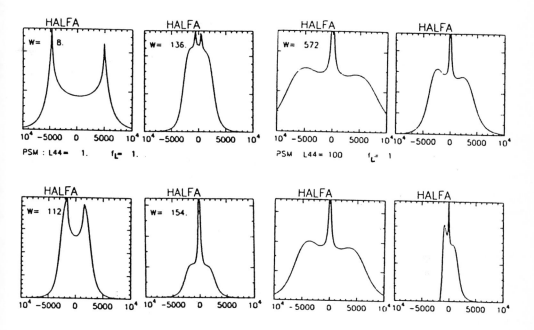

Fig. 13. Profiles produced by the disc for different values of the radius (left panel, $R = 10^3 R_G$ to $10^6 R_G$) and of the inclination (right panel, $i = 80$ to $i = 10$ degrees), after Dumont and Collin-Souffrin (1990)

1988, Sun and Malkan 1989, Laor 1990). The method can be summarized in the following equations:

$$L_{\text{disc}} = \frac{3}{2} \frac{GM\dot{M}}{R_{\text{in}}} \qquad (55)$$

where L_{disc} is the frequency-integrated disc luminosity and R_{in} the inner radius,

$$T(\text{UV})^4 = \frac{3}{8\pi\sigma} \frac{GM\dot{M}}{R_{\text{in}}^3} \qquad (56)$$

where $T(\text{UV})$ is the temperature measured on **the high frequency side** of the blue bump. Combined with $L_{\text{bol}} = \eta\dot{M}c^2$, and assuming $\eta=0.1$, we have a system of three equations which leads to the values of the three parameters M, $\dot{M}$ and R_{in}, once the three "observables", $T(\text{UV})$, L_{disc} and L_{bol} are known.

This method, once given the working model (thick or thin disc, rotating or non-rotating black hole) suffers from the uncertainties discussed above concerning the disc emission (cf. Laor 1990). In particular the inclination i is an almost free parameter: the mass determined from the fits is roughly inversely proportional to cos i, and varies by more than an order of magnitude from a face-on to an edge-on disc (Sun and Malkan 1989). Observationally L_{disc} and L_{bol} are very difficult to determine, as the disc is emitting most of its energy in the far UV range.

The "Dynamical Method".

This method amounts to accounting for the Broad Line velocities by a photoionized gravitationally bound region (Joly et al. 1985, Wandel and Yahil 1985, Reshetnikov 1987, Joly 1987a, Padovani and Rafanelli 1988, Padovani 1989).The method can be summarized in the following equations:

$$V(\text{H}\beta) \sim \frac{\text{FWHM}(\text{H}\beta)}{2} \sim \sqrt{\frac{GM}{R_{\text{line}}}} \tag{57}$$

where R_{line} is a characteristic distance of the line emission region,

$$U \propto \frac{L_{\text{bol}}}{n R_{\text{line}}^2} \sim 0.1 \tag{58}$$

These equations, coupled with the equation giving the bolometric luminosity, give the three parameters M, $\dot{M}$ and R_{line}, once the two observables, FWHM($\text{H}\beta$) and L_{bol} and the two parameters U and n are known.

This method suffers from strong uncertainties concerning the value of the ionization parameter and of the density in the line emission region, which are in a sense arbitrarily chosen (average values deduced from photoionization models).

Another method has also been used for the determination of the black hole mass. It is based on the measure of the variation time scale of the X-ray flux, and assumes that the X-ray continuum is emitted within a region of a few Schwarzschild radii from the black hole (Wandel and Mushotzky 1986). This method gives only an upper limit of the size of the X-ray source, and consequently of the black hole, and sampling problems may lead to an overestimation of the time scales.

The best method is certainly to combine the accretion disc and the dynamical method through the computation of a complete self-consistent photoionization and disc model, with the obvious drawback that the results are model dependent (Rokaki et al. 1992).

In spite of their large uncertainties, the two first methods reach roughly the same conclusion (from a statistical point of view): **the Eddington ratio** (i.e. the bolometric to Eddington luminosity ratio) **is close to unity for luminous quasars, and of the order 0.01 to 0.1 for Seyfert galaxies.** They lead also to the value of the black hole mass. One finds that the black hole mass of a quasar with a bolometric luminosity of the order of 10^{48} ergs s^{-1} - such as 3C273 - is of the order of 10^{10} M$_{\odot}$, and the mass of a black hole AGN with a bolometric luminosity of 10^{44-45} ergs s^{-1} - respectively NGC4151 and NGC5548 - is of the order of 10^{7-8} M$_{\odot}$.

Cosmological Consequences.

The previous results are of considerable importance for our understanding of the physical and cosmological evolution of AGN. Since this is not the topic of these lectures, we will only briefly mention the problem.

Evidence has been growing in recent years that quasars evolve in luminosity more than in density, since a pure density evolution would lead to a much larger local number of "dead quasars" than the number of normal galaxies (for a review, see for instance Weedman 1985). It is also possible to determine the total mass accumulated in quasar black holes, assuming that they are fueled by accretion and using quasar counts (Soltan 1982). The mass per **covolume** unit is:

$$< \rho > = \frac{4\pi}{c} \frac{(1-\eta)}{\eta c^2} \int (1+z) \, dz \int F_{\mathrm{bol}} N(F_{\mathrm{bol}}, z) \, dF_{\mathrm{bol}} \qquad (59)$$

where F_{bol} is the observed monochromatic flux corrected to the bolometric flux (obviously with a large uncertainty) and N the quasar luminosity function. The value found using the most recent quasar surveys is very large, about 10^{14} $M_\odot$ Gpc^{-3}. The problem is to find out how this mass is **now** distributed among galaxies.

High redshift quasars display a luminosity function similar to that of Seyfert galaxies. A very natural idea would then be that quasars have dimmed monotonically, giving rise to a population of less luminous AGN, namely Seyfert galaxies. In this case Seyfert galaxies should harbor the very massive black holes necessary to account for the high quasar luminosity in the past. However this explanation does not hold, as we have seen above that quasar masses are larger than the masses of black holes in local less luminous AGN. More quantitatively Cavaliere and Padovani (1989) have shown that a pure luminosity evolution, with quasar lifetime of the order 1/3 of the Hubble time, would lead to very low Eddington ratios for Seyfert galaxies, of the order of 10^{-4}, much smaller than the real Eddington ratios (this is due to the very large masses - 10^{11} $M_\odot$ - reached by black holes during their long luminous phase).

Moreover, Padovani, Burg and Edelson (1990) have estimated the mass function of Seyfert nuclei from Eddington ratios and shown that their integrated mass per covolume unit is at least two orders of magnitude smaller than the integrated quasar mass. They conclude that the accreted mass cannot reside only in Seyfert nuclei, and should on the contrary be dispersed among a much larger number of galaxies. Precisely, **massive black holes ($M \gtrsim 10^6$ $M_\odot$) should be present in the nuclei of all galaxies comparable in size to our own Galaxy**, and the mass function of these holes should extend up to 10^{10} $M_\odot$. This conclusion rules out a long-lived activity phenomenon, and implies a short life-time for quasars, of the order 1/100 of the Hubble time, either in one unique event or in recurrent episodes of activity. This time is actually even smaller than the "*Eddington time*", i.e. the time it takes for a black hole to double its mass, assuming it accretes at the critical rate.

That central massive black holes are present in many galaxies, which do not exhibit an obvious Seyfert activity, agrees with the fact that a weak Seyfert activity exists in a large fraction of galaxies, the so-called LINERS in particular (cf. below), and could be due to a black hole accreting at a very low rate (cf. for instance Filipenko 1988). Also recent dynamical studies have clearly shown the presence of large non-luminous mass concentrations of 10^7 - 10^8 $M_\odot$ in the nuclei of otherwise inactive galaxies such as M31, M32, or even of larger masses of 10^9 $M_\odot$ as in M104 . As we know, the Galactic centre may also harbor a black hole of 1 - 3 10^6 $M_\odot$.

Cavaliere and Padovani (1989) have estimated the mass distribution of black holes harboured by inactive galactic nuclei from the above-mentioned central mass of a few galaxies, extended over the whole luminosity function of galaxies. From their mass distribution one deduces that the total mass per covolume unit is very close to the mass stored in quasar black holes and that the average mass of a black hole is $3 \ 10^7 \ M_{\odot}$, a value almost equal to the average mass of Seyfert nuclei. This agreement indicates that **Seyfert activity is probably indeed a recurrent phenomenon affecting all currently inactive galaxies**.

4 A Few Concluding Remarks...

The previous discussion has shown that the presence of a massive black hole should be quite common among nuclei of galaxies, and that the AGN phenomenon should represent a short-lived and probably recurrent phenomenon during the life of these otherwise inactive galaxies. The AGN phase might be triggered by a variety of mechanisms able to bring new fuel to the nucleus, like interactions with other galaxies and mergers. We are therefore led to conclude that the AGN phenomenon should take different aspects according to the degree of activity of the nucleus, i.e. the accretion rate of the black hole.

We have also seen that the inner part of a quasar or of a Seyfert 1 nucleus includes several components which should be linked in the accretion process: an accreting black hole, a rapidly variable X-ray source (whose nature was not discussed in this chapter), an accretion disc, and a high-velocity gas, both in a hot and in a cold phase. Since there are lots of other objects displaying various kinds of "activity", we are led to ask if they do have the same central engine, and differ only in one aspect, namely the accretion rate.

Finally, since AGN imply necessarily an accretion disc, we expect to observe several important effects of the **viewing angle** of the object.

In fact, it is possible to explain all the types of nuclear activity, by invoking only the variation of the mass of the black hole, the accretion rate and the viewing angle.

The "Zoo" of AGN.

It is difficult to give a logical and exhaustive description of all classes of active objects. We shall try to order them by "increasing activity", although it is a somewhat subjective classification.

- *Radio galaxies*. They all have a nucleus with emission lines. We have seen that they could be explained by massive black holes with a very low accretion rate.

- *Low Ionization Emission Line Regions (LINERS)*. These nuclei, which display a line spectrum completely different, either from an HII spectrum or from a Seyfert Narrow Line spectrum, can well be explained by photoionization models with a hard, non-stellar, ionizing continuum. Moreover, in several of these nuclei (whose representative objects are M81 and NGC1052) a weak broad line component underlying the narrow Hα profile has been discovered.

- *Seyfert 2 galaxies*. Their nuclei show a Narrow Line spectrum quite similar to that of Seyfert 1 galaxies, but they do not have broad lines and their X-ray continuum is very weak. In NGC1068, Antonnucci and Miller (1985) have found the presence of polarized broad lines which they attributed to a Seyfert 1 nucleus hidden inside an absorbing dust torus, whose light is diffused by a hot diffuse medium surrounding the central engine. This discovery has given impulse to the idea of a "unified scheme", although this phenomenon does not seem to be very common among Seyfert 2, since only a few new Seyfert 2 displaying polarized broad lines have been discovered since this epoch. Moreover, Seyfert 2 galaxies seem to be tightly connected with an intense starburst activity (cf. J. Perry's lectures).

- *Narrow Line Radio Galaxies (NLRG)*. These nuclei are identical to Seyfert 2 nuclei, except that they lie in radiogalaxies (i.e. elliptical galaxies) and not in spiral galaxies.

- *Narrow Line X-ray Galaxies (NLXG)*. These strong X-ray emitters are in other aspects comparable to Seyfert 2 nuclei. They, however, show broad line components in the infrared. They are therefore suspected to be highly reddened Seyfert 1 nuclei (X-rays are less severely absorbed than visible photons).

- *Seyfert 1 galaxies...*

- *Broad Line Radio Galaxies (BLRG)*. These nuclei are identical to Seyfert 1 nuclei, except that they lie in radiogalaxies.

- *Ultra-luminous Far Infrared galaxies (FIR galaxies)*. They radiate about 10^{45} ergs s^{-1} in the far infrared range, and they show an infrared line emission spectrum with "broad" components (actually not as broad as in Seyfert 1) and a highly reddened narrow line spectrum. They are suspected to be quasars enshrouded in a cocoon of dust, or in a dust-molecular torus.

- *B Lac objects*. This complex class is characterized by its strong radio and X-ray emission, the absence of emission lines, a strongly variable and highly polarized continuum. For a long time these objects have been suspected to be relativistically boosted radio galaxies or quasars, seen almost in the direction of the jet axis. The non-thermal continuum is therefore amplified, while the thermal emission (UV-bump and emission lines) is not.

- *Radio quiet and radio loud quasars...*

- *Blazars and Optically Violent Variables (OVV)*. They are similar to B Lac, except that they do have an emission line spectrum. They could be boosted luminous quasars.

The "Unified Scheme".

It is clear that the viewing angle is important for B Lac and for Blazars. The question, which arises then, is: are there other manifestations of the viewing angle? Another important question is: what makes the difference between radio loud and radio quiet objects?

Concerning the second question, which is still not answered, it appears that radio loud objects are located systematically in elliptical galaxies, which are known to contain less gas than spiral galaxies. The idea is that the jet is blocked by a denser environment in spiral galaxies, and cannot fuel the radio lobes. So the radio phenomenon would be confined in the very nucleus, where self-absorption inhibits the synchrotron emission. In other aspects the radio loud and radio quiet objects could be identical.

Now regarding the first question, the discovery of Antonnucci and Miller (1985) has perhaps opened a new era for the understanding of AGN. If Seyfert 2 are indeed Seyfert 1 hidden by a dust-molecular torus, what could be the "Seyfert 2 of quasars"? Barthel (1989) made the suggestion that strong radio galaxies could be the counterparts of radio loud quasars, when viewed "edge on", i.e. in the direction of the torus. Since the torus has an opening angle of about 45 degrees in NGC1068, it would mean that about half of the quasars would have their central engine - the BLR and partly the NLR, the central source of continuum - hidden from our view. Going one step further, Barthel also suggested that FIR galaxies would be the counterparts of radio quiet quasars.

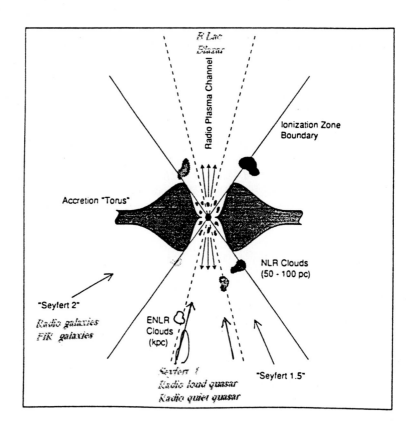

Fig. 14. The unified scheme of AGN, after Pogge (1988). Small changes (in italics) have been introduced by the author.

Several tests should be performed in order to be able to conclude definitively on this point. For instance, it would be necessary to observe FIR galaxies in the hard X-ray and gamma-ray range: these photons should not be absorbed by the torus, except if it has an average column density larger than a few g cm^{-2}. Statistical studies, determination of luminosity functions in particular, should also help.

There are already strong hints that the model could be valid, at least partly. We have mentioned in the first section the existence of the Extended Narrow Line Regions (ENLRs) which are aligned with the radio axis, as in NGC1068, or have pure conical structure, as in NGC5252. These structures might prove that the ejection of matter from the nucleus is preferential in the direction perpendicular to the disc, owing to the presence of a blocking torus. ENLRs correspond also sometimes to a level of ionization higher than computed, assuming that the central continuum is emitted isotropically (cf. Penston et al. 1990, for NGC4151). This could be explained if we are seeing only a fraction of the continuum source.

To conclude, let us have a look at Fig. 14 which could be the picture of a "Unified AGN", according to Pogge (1988).

A CAVEAT. Although I have tried to be as objective as possible, I recall that several ideas developed here are not necessarily shared by all specialists working on the subject.

References

Abramovicz, M. , Calvani, M., Nobili, L. (1980): *Astrophys. J.*, **242**, 772

Antonucci, R.R.J., Miller J.S. (1985): *Astrophys. J.*, **297**, 621

Antonnucci, R.R., Kinney A.L., Ford H.C. (1989): *Astrophys. J.*, **342**, 64

Avrett, E.H., Loeser, R. (1988): *Astrophys. J.*, **331**, 211

Baldwin, J.A. (1977): *Mon. Not. R. astr. Soc.*, **178**, 67

Barthel, P.D. (1989): *Astrophys. J.*, **336**, 606

Bechtold, J., Weyman, R.J., Lin, A., Malkan, M.A. (1987): *Astrophys. J.*, **315**, 180

Begelman, M.C. (1985): in the Proceedings of the Santa Cruz Workshop on *Active Galactic Nuclei and QSO*, ed Miller

Begelman, M.C., McKee C.F. (1983): *Astrophys. J.*, **271**, 89

Bregman, J.N. (1990): *Astron. Astrophys. Rev.*, **2**, 125

Cavaliere, A., Padovani P. (1988): *Astrophys. J. (Lett.)*, **333**, L33

Cavaliere, A., Padovani P. (1989): *Astrophys. J. (Lett.)*, **340**, L5

Chen, K., Halpern J.P., Filippenko A.V. (1989): *Astrophys. J.*, **339**, 742

Clavel, J. (1990): in the Proceedings of the Workshop on *Variability of Active Galaxies*, eds. Wagner and Camenzind, Springer-Verlag

Clavel, J., Wamsteker, W., Glass, I. (1989): *Astrophys. J.*, **337**, 236

Clavel, J. et al (1991): *Astrophys. J.*, **366**, 64

Collin-Souffrin, S. (1986a): in the Proceedings of the IAU Symposium 119, *Quasars*, eds Swarup and Kapahi, Springer-Verlag

Collin-Souffrin, S. (1986b): *Astron. Astrophys.*, **166**, 115

Collin-Souffrin, S. (1987): in the Proceedings of the 6th Workshop of Astronomy and Astrophysics, ed Gondakher, Rutherford Appleton Laboratory, Chilton, England

Collin-Souffrin, S. (1989): in the Proceedings of the 11th European Astronomy Meeting, Cambridge University Press

Collin-Souffrin, S. (1991): *Astron. Astrophys.*, **249**, 344

Collin-Souffrin, S., Dumont, S. (1986): *Astron. Astrophys.*, **166**, 13

Collin-Souffrin, S., Dumont, A.M. (1990): *Astron. Astrophys.*, **229**, 292

Collin-Souffrin, S., Hameury, J.M., Joly, M. (1988): *Astron. Astrophys.*, **205**, 19

Collin-Souffrin, S., Joly M., Pequignot D., Dumont S. (1986): *Astron. Astrophys.*, **166**, 27

Collin-Souffrin, S., Lasota, J-P. (1989): *Publ. Astr. Soc. Pac.*, **100**, 1041

Collin-Souffrin, S., Perry J.J., Dyson J.E., McDowell J.C. (1988): *Mon. Not. R. astr. Soc.*, **232**, 539

Czerny, B., Elvis, M. (1987): *Astrophys. J.*, **321**, 243

Davidson, K. (1977): *Astrophys. J.*, **218**, 20

Dumont, A.M., Collin-Souffrin S. (1990a): *Astron. Astrophys.*, **229**, 302

Dumont, A.M., Collin-Souffrin S. (1990b): *Astron. Astrophys.*, **229**, 313,

Dumont, A.M., Collin-Souffrin S. (1990c): *Astron. Astrophys. Suppl. Ser.*, **83**, 71

Edwards, A.C. (1980): *Mon. Not. R. astr. Soc.*, **190**, 757

Elitzur, M., Netzer; H. (1984): *Astrophys. J.*, **291**, 464

Fabian, A.C., Guilbert, P.W., Arnaud, K.A., Shafer, R.A., Tennant, A.F., Ward M.J. (1986): *Mon. Not. R. astr. Soc.*, **218**, 457

Ferland, G.J., Korista K.T., Peterson B.M. (1990): *Astrophys. J. (Lett.)*, **363**, L21

Ferland, G.J., Netzer, H., Shields, G.A. (1979): *Astrophys. J.*, **232**, 382

Ferland, G.J., Persson, S.E. (1989): *Astrophys. J.*, **347**, 656

Ferland, G.J., Rees M.J. (1988): *Astrophys. J.*, **332**, 141

Ferland, G.J., Shields, G.A. (1985): in the proceedings of the Santa Cruz Workshop on *Active Galactic Nuclei and QSO*, ed Miller

Filippenko, A.V. (1988): in *Supermassive Black Holes*, ed. Kafatos

Frank, J., King, A.R., Raine, D.J. (1985): *Accretion Power in Astrophysics*, Cambridge University Press

Halpern, J.P., Chen K. (1989): in the Proceedings of IAU symposium no. 134, *Active Galactic Nuclei*, eds. Miller and Osterbrock, Kluwer Dordrecht, p. 245

Guilbert, P.W., Rees M.J. (1988): *Mon. Not. R. astr. Soc.*, **233**, 475

Joly, M. (1987a): in the Proceedings of the 6th Workshop of Astronomy and Astrophysics, ed. Gondalekhar, Rutherford Appleton Laboratory, Chilton, England

Joly, M., (1987): *Astron. Astrophys.*, **184**, 33

Joly, M., (1991): *Astron. Astrophys.*, **242**, 49

Joly, M., Collin-Souffrin S., Masnou J.L., Nottale L. (1985): *Astron. Astrophys.*, **152**, 282

Jones, B.C., Raine D.J. (1980): *Astron. Astrophys.*, **81**, 128

Kallman, T., Krolik, J. (1986): *Astrophys. J.*, **308**, 805

Kallman, T.R., Mushotzky R.F. (1985): *Astrophys. J.*, **292**, 89

Krolik, J.H. (1988): *Astrophys. J.*, **325**, 182

Krolik, J.H., Begelman, M.C. (1986): *Astrophys. J. (Lett.)*, **308**, L55

Krolik, J.H, Horne K., Kallman T.R., Malkan M.A., Edelson R.A., Kriss G.A. (1991): *Astrophys. J.*, **371**, 541

Krolik, J.H., Kallman, T.R. (1988): *Astrophys. J.*, **324**, 714

Krolik, J.H., Lepp, S. (1989): *Astrophys. J.*, **347**, 179

Krolik, J.H., McKee C.F., Tarter C.B. (1981):*Astron. J.*, **249**, 422

Kwan, J., Krolik, J.H. (1981): *Astrophys. J.*, **250**, 478

Laor, A. (1990): *Mon. Not. R. astr. Soc.*, **246**, 369

Lightman, A.L., White, T. (1988): *Astrophys. J.*, **335**, 57

Madau, P. (1988): *Astrophys. J.*, **327**, 116

Malkan, M.A. (1983): *Astrophys. J.*, **268**, 582

Malkan, M.A., Sargent, W.C.W. (1982): *Astrophys. J.*, **254**, 122

Maraschi, L., Molendi, S. (1990): *Astrophys. J.*, **353**, 452

Maoz D., Netzer H., Mazeh T., Beck S., Almoznino E., Leibovitz E., Brosch N., Mendelson H., Laor A., (1991): *Astrophys. J.*, **367**, 493

Mathews, W.G., Cappriotti, E.R. (1985): in the Proceedings of the Santa Cruz workshop on *Active Galactic Nuclei and QSO*, eds Osterbrock and Miller

Mathews, W.G., Ferland G.J. (1987): *Astrophys. J.*, **323**, 456

Mushotzky, R.F. and Ferland, G.J. (1984): *Astrophys. J.*, **278**, 558

Netzer, H. (1985): *Astrophys. J.*, **289**, 451

Netzer, H. (1987a): in the Proceedings of the 6th Workshop of Astronomy and Astrophysics, ed Gondalekhar, Rutherford Appleton Laboratory, Chilton, England

Netzer, H. (1987b): *Mon. Not. R. astr. Soc.*, **216**, 63

Netzer, H. (1990): in *Santa-Fee Advanced Course 20*, eds. Courvoisier and Mayor, Springer-Verlag

Netzer, H. (1991): in Proceedings of IAU Colloquium 129 *Structure and Emission processes in Accretion Disks*, ed. Frontieres

Netzer, H., Elitzur, M., Ferland, G.J (1985): *Astrophys. J.*, **299**, 752

Netzer, H. et al (1990): *Astrophys. J.*, **353**, 108

Norman, C., Scoville N. (1988): *Astrophys. J.*, **332**, 124

Osterbrock, D.E. (1989): *Astrophysics of Gaseous Nebulae and Active Galactic Nuclei*, University Science Book

Osterbrock, D.E. (1991): *Rep. Prog. Phys.*, **54**, 579

Osterbrock, D.A., Mathews, W.G. (1986): *Ann. Rev. Astron. Astrophys.*, **24**, 171

Padovani, P. (1988) *Astron. Astrophys.*, **192**, 9

Padovani, P., Burg R., Edelson R.A. (1990): *Astrophys. J.*, **353**, 438

Padovani, P., Rafanelli P. (1988): *Astron. Astrophys.*, **205**, 53

Penston, M.V. (1988): *Mon. Not. R. astr. Soc.*, **233**, 601

Penston, M.V. et al. (1990): *Astron. Astrophys.*, **236**, 53

Perez, E., Penston M.V., Tadhunter C., Mediavilla E., Moles M. (1988): *Mon. Not. R. astr. Soc.*, **230**, 353

Perry, J.J., Dyson, J.E. (1985): *Mon. Not. R. astr. Soc.*, **213**, 665

Peterson, et al (1991): *Astrophys. J.*, **368**, 119

Piran, T. (1978): *Astrophys. J.*, **221**, 652

Pogge, R.W. (1988) in *Extranuclear Activity in Galaxies* eds. E.J.A. Meurs and R.A.E. Fosbury, p. 411

Pounds, K.A., Nandra K. Stewart G.C., George I.M., Fabian A.C. (1990): *Nature*, **344**, 132

Puetter, R.C. (1988): in Proceedings of the IAU Symposium 134 *Active Galactic Nuclei*, eds. Miller and Osterbrock, Kluwer, Dordrecht

Pringle, J.E. (1981): *Ann. Rev. Astron. Astrophys.*, **19**, 137

Rees, M.J. (1984): Ann. Rev. Ast. Ap. 22, 471

Rees, M.J. (1987): *Mon. Not. R. astr. Soc.*, **228**, 47

Rees, M.J., Begelman, M.C., Blandford, R.D., Phinney, E.S. (1982): *Nature*, **295**, 17

Rees, M.J., Netzer H., Ferland G.J. (1989): *Astrophys. J.*, **347**, 640

Reimers, D. et al (1989): *Astron. Astrophys.*, **218**, 71

Reshetnikov, V.P. (1987): *Astrofisica*, **27**, 283

Rokaki, E., Boisson C., Collin-Souffrin S. (1992): *Astron. Astrophys.*, **253**, 57

Sanders, D.B., Phinney, E.S., Neugebauer, G., Soifer, B.T., Matthews, K. (1989): *Astrophys. J.*, **347**, 29

Scoville, N.Z., Norman C. (1988): *Astrophys. J.*, **332**, 163

Shakura, N.I., Sunayev R.A. (1973): *Astron. Astrophys.*, **24**, 337

Shields, G.A. (1978): *Nature*, **272**, 423

Soltan, A. (1982): *Mon. Not. R. astr. Soc.*, **200**, 115

Sun, W-H., Malkan, M.A. (1989): *Astrophys. J.*, **346**, 68

Turnshek, D.A., Foltz, C.B., Crillmair, C.J., Weymann, R.J., (1988): *Astrophys. J.*, **325**, 651

Wandel, A., Liang, (1990): *Astrophys. J.*, preprint

Wandel, A., Mushotzky R.F. (1986): *Astrophys. J. (Lett.)*, **306**, L61

Wandel, A., Petrosian V. (1988): *Astrophys. J. (Lett.)*, **319**, L11

Wandel, A., Yahil, A. (1985): *Astrophys. J. (Lett.)*, **295**, L1

Weedman, D.W. (1985): *Quasar Astronomy*, Cambridge University Press

Wills, B., Browne I.W.A. (1986): *Astrophys. J.*, **302**, 56

Wills, B.J., Netzer, H. and Wills, D. (1985): *Astrophys. J.*, **288**, 94

Wilkes, B., Elvis, M. (1987): *Astrophys. J.*, **323**, 243

Infrared Activity of Galaxies

T. de Jong

Astronomical Institute 'Anton Pannekoek', University of Amsterdam, Kruislaan
403, 1098 SJ Amsterdam, The Netherlands
and
SRON Laboratory for Space Research Zernike Gebouw, P.O. Box 800, 9700 AV
Groningen, The Netherlands

Abstract: A summary of the main topics covered in the four lectures that I gave during the Dublin Summer School on "Central Activity in Galaxies" follows. The school was organized in September 1990 under the auspices of the European Astrophysics Doctoral Network.

Lecture 1: An Introduction

Most galaxies emit the bulk of their energy at infrared wavelengths (longwards of 10 μm). Before the launch of the InfraRed Astronomical Satellite (IRAS) in 1983 only a handful of galaxies had been studied at infrared wavelengths (see review by Rieke and Lebofsky 1979). Right now IRAS data for over 20000 galaxies are available. The main limitations of the IRAS survey are its poor spatial resolution (several arc minutes) and its limited wavelength coverage (no data longward of 100 μm). Roughly 30 galaxies are resolved by the IRAS survey instrument (Rice et al. 1988) and about twice as many by the 1.5 arcminute beam of the less sensitive Chopped Photometric Channel (van Driel et al. 1992).

Infrared activity of galaxies is one of the most powerful tracers of recent star formation. The progress made in our understanding of star formation in galaxies based on IRAS data has been summarized in recent reviews by Soifer et al. (1987) and Telesco (1988). Early studies of the infrared characteristics of optical and infrared complete samples of normal galaxies were reviewed by de Jong (1986). Infrared activity turns out to increase from early to late morphological type and is enhanced in galaxies experiencing interaction. The infrared "excess" is correlated with far-infrared colour indicating the existence of a warm (about 60 K) dust component associated with recent star formation and a cool (about 15-20 K) dust component associated with a more quiescent/extended interstellar cloud component.

The IRAS-infrared (10-100 μm) spectral energy distribution of active galaxies is usually also dominated by thermal dust emission. Only in a few quasars does the non-thermal synchrotron emission of the central radio source dominate (Soifer et al. 1987). Seyfert galaxies often show a warm mid-infrared (25 μm) component, probably associated with the active nucleus (Miley et al. 1985). This has been used by de Grijp et al. (1985) as a criterion to search for Seyferts in the infrared.

The far-infrared and the radio continuum emission of spiral galaxies were found to be tightly correlated by de Jong et al. (1985) and Helou et al. (1985). Both are tracers of recent massive star formation. The infrared is emitted by dust particles directly heated by photons of recently formed massive main sequence stars. The radio continuum is generated when relativistic electrons produced in supernova explosions of massive stars interact with the interstellar magnetic field. Typical diffusion times of relativistic electrons are about 100 million years so that the observed correlation implies that star formation typically does not change on shorter timescales. Since some galaxies in the sample studied are suspected of experiencing a recent starburst, the observed correlation may constrain the duration of the starburst. The tightness of the correlation has implications for the equipartition of the magnetic field energy density (see Hummel 1986).

Using published data for the so-called mini-survey galaxies van den Broek et al. (1991) found a correlation between the near- and far-infrared emission. They interpret this in terms of star formation. The far-infrared traces recently formed massive main sequence stars while the near-infrared is due to (somewhat older) giants and supergiant stars. Thus the period of enhanced star formation in the mini-survey galaxies must last longer than the typical lifetime of supergiant stars (about 30 million years).

Finally, I discussed the correlation between the Hα luminosity (corrected for extinction using the observed Hα/Hβ ratio) and the far-infrared luminosity of galaxies (van den Broek 1990, 1992a). The data seem to indicate that extinction increases with increasing star formation rate, possibly due to geometrical effects (more centrally condensed star formation?).

Lecture 2: Interstellar Dust

Since in almost all galaxies the infrared is emitted by dust particles heated by stellar photons our present knowledge of the absorption and emission properties of interstellar dust was reviewed. Some of the relevant physics can be found in a paper of Hildebrand (1983).

Lecture 3: Simple Two-Component Models for the Infrared Emission of Galaxies

To be able to consistently interpret near-infrared, far-infrared and optical data of galaxies, simple two-component models of galaxies were discussed based on the approach of de Jong and Brink (1987) and the simple starburst evolution models of van den Broek et al. (1991).

Lecture 4: Infrared-Bright Galaxies

Using infrared flux densities and redshifts of over 300 galaxies in the infrared-complete Bright Galaxy Sample (BGS), Soifer et al. (1986, 1987) have determined the infrared luminosity function of galaxies. The results were discussed and compared to other luminosity functions. The infrared luminosity function contains a population of infrared luminous galaxies that is not found in optical surveys.

Infrared activity in galaxies increases from the Bright Galaxy Sample (Soifer et al. 1987), via "Extreme" IRAS Galaxies (van den Broek 1990, 1992b, van den Broek et al. 1991, van Driel et al. 1991) to the so-called "Ultra-luminous" Bright Galaxies (Sanders et al. 1988). The latter class consists of the 10 most luminous galaxies in the BGS (infrared luminosities larger than $10^{12}\,L_\odot$). They show evidence for extremely high star formation rates (about 1000 solar masses per year), although a substantial non-thermal contribution to the infrared luminosity cannot be excluded. A large fraction of the interstellar medium of ultra-luminous galaxies consists of high-density molecular clouds (Solomon et al. 1990). All show distorted morphology often due to close interaction or merging with another galaxy. The speculation that these ultraluminous galaxies are related to quasars was discussed.

Based on the large amount of observational material recently gathered the prototype ultraluminous galaxy Arp 220 (IC 4553) was discussed in detail. Some of the relevant references can be found in Scoville et al. (1986) and Solomon et al. (1990).

References

de Grijp, M.H.K., Miley, G.K., Lub, J., de Jong, T. (1985): *Nature*, **314**, 240

de Jong, T. (1986): in *Spectral Evolution of Galaxies*, eds. C. Chiosi and A. Renzini, p. 111

de Jong, T. Brink, K. (1987): in *Star Formation in Galaxies*, ed. C.J. Lonsdale, NASA CP-2466, p. 323

de Jong, T., Klein, U., Wielebinsky, R., Wunderlich, E. (1985): *Astron. Astrophys. (Lett.)*, **147**, L6

Helou, G., Soifer, B.T., Rowan-Robinson, M. (1985): *Astrophys. J. (Lett.)*, **298**, L7

Hildebrand, R. (1983): *Quart. J. R. astr. Soc.*, **24**, 267

Hummel, E. (1987): *Astron. Astrophys. (Lett.)*, **160**, L4

Miley, G.K., Neugebauer, G., Soifer, B.T. (1985): *Astrophys. J. (Lett.)*, **293**, L11

Rice, W., Lonsdale, C.J., Soifer, B.T., Neugebauer, G., Kopan, E.L., Lloyd, L.A., de Jong, T., Habing, H.J. (1988): *Astrophys. J. Suppl.*, **68**, 91

Rieke, G.H., Lebofsky, M.J. (1979): *Ann. Rev. Astron. Astrophys.*, **17**, 477

Sanders, D.B., Soifer, B.T., Elias, J.H., Madore, B.F., Matthews, K., Neugebauer, G., Scoville, N.Z. (1988): *Astrophys. J.*, **325**, 74

Scoville, N.Z., Sanders, D.B., Sargent, A.I., Soifer, B.T., Scott, S.L., Lo, K.Y. (1986): *Astrophys. J. (Lett.)*, **311**, L47

Soifer, B.T., Houck, J.R., Neugebauer, G. (1987): *Ann. Rev. Astron. Astrophys.*, **25**, 187

Soifer, B.T., Sanders, D.B., Madore, B.F., Neugebauer, G., Danielson, G.E., Elias, J.H., Lonsdale, C.J., Rice, W.L. (1987): *Astrophys. J.*, **320**, 238

Soifer, B.T., Sanders, D.B., Neugebauer, G., Danielson, G.E., Lonsdale, C.J., Madore, B.F., Persson, S.E. (1986): *Astrophys. J. (Lett.)*, **303**, L41

Solomon, P.M., Radford, S.J.E., Downes, D. (1990): *Astrophys. J. (Lett.)*, **348**, L53

Telesco, C.M. (1988): *Ann. Rev. Astron. Astrophys.*, **26**,343

van den Broek, A.C. (1990): *A Study of Extreme IRAS Galaxies*, Ph.D. Thesis, University of Amsterdam

van den Broek, A.C. (1992a): *Astron. Astrophys. (Lett.)*, in press

van den Broek, A.C. (1992b): *Astron. Astrophys.*, submitted

van den Broek, A.C., de Jong, T. Brink, K. (1991): *Astron. Astrophys.*, **246**, 313

van den Broek, A.C., van Driel, W., de Jong, T., Lub, J., de Grijp, M.H.K., Goudfrooij, P. (1991): *Astron. Astrophys. Suppl. Ser.*, **91**, 61

van Driel, W., de Graauw, Th., de Jong, T., Wesselius, P.R. (1992): *Astron. Astrophys. Suppl. Ser.*, submitted

van Driel, W., van den Broek, A.C., de Jong, T. (1991): *Astron. Astrophys. Suppl. Ser.*, **90**, 55

Radio Galaxies and Jets

Peter L. Biermann

Max-Planck-Institut für Radioastronomie
Auf dem Hügel 69
D-5300 Bonn 1 Germany

Abstract: We review the physics of radio galaxies and their jets. We emphasize the physics of gas flow and high energy particle interactions as well as shock acceleration

1 Radio Galaxies and Their Jets: Observations

1.1 Surveys

There are various radio astronomical sky surveys, the most important ones are the 3rd Cambridge catalogue 3C at 159 MHz, then the twice revised 3rd Cambridge catalogue 3CRR at 178 MHz, the more sensitive 4th Cambridge catalogue 4C at 178 MHz, then the Parkes survey PKS at 408 MHz, the 2nd Bologna catalogue B2 also at 408 MHz, then the higher frequency Parkes survey PKSF at 2700 MHz, and finally the survey at the highest frequency, the MPIfR/NRAO survey at 5 GHz. In addition there are many surveys that cover only part of the sky, done at Cambridge, Molonglo, Westerbork, NRAO and the MPIfR. A compilation of such surveys has been given by Fricke and Witzel (1982). The high frequency radio spectra of the sources found in these surveys ranges in $\alpha(11 - 6)$, the spectral index between 11 cm or 2.7 GHz and 6 cm or 5 GHz, from about -1 to $+1$; here we define the spectral index such that the flux density $S_\nu \propto \nu^\alpha$.

1.1.1 Spectral Index Distributions

The observations show that those sources with $\alpha(11 - 6)$ flatter than about -0.5 are generally variable and show compact radio structure; such spectral indices we call "flat". These sources are dominated by their compact structure, compact in this context referring to milliarcseconds, i.e. observable with intercontinental radio interferometry, or very long baseline interferometry, VLBI. Such flat spectrum radio sources show a variety of interesting phenomena, such as evidence for bulk relativistic motion. In fact, as we will see below, a small but complete survey of 13 radio sources with flat radio spectra from the MPIfR/NRAO radio survey at 5 GHz showed evidence for bulk relativistic motion for the entire sample (Witzel et al. 1988). On the other hand, radio sources with "steep" spectra, i.e. α typically in the range -0.8 to -1.0, are normally dominated by extended radio structure and are not variable; there are a few noteworthy exceptions to such simple rules. Compact steep-spectrum sources have aroused considerable interest (see, e.g., the review by Fanti et al. 1988). In a compilation of the strongest radio sources at 5 GHz (Kühr et al. 1981a, b) the proportion of flat radio spectrum sources is quite high at around 50%; among the first five sources in order of decreasing flux density, there are, e.g., three sources with flat or inverted spectra and two with steep spectra. Those sources which are variable, show a variety of characteristics in their variability (Quirrenbach et al. 1991). One mode of variability can be described as synchrotron emission from an expanding homogeneous sphere filled with a population of relativistic electrons (van der Laan 1966). In this case the source shows a peaked spectrum, where the peak frequency moves to lower frequencies with time, and the peak intensity decreases with time; the explanation for the spectrum is synchrotron self-absorption, where the frequency at which the optical depth is unity decreases with time. Such a simple behaviour is seen, for example, in 3C120. Many other sources demonstrate a much more erratic mode of variability, sometimes correlated over a very large band in frequency. The sources detected in all these sources have been subject of a large number of structural investigations with radio interferometers (see Fricke and Witzel 1982). Identifications for sources from the MPIfR/NRAO survey are about 50% galaxies, which here means a red image on the Palomar Sky Survey, and about 50% quasars, with an increasing proportion of undetected sources on the Palomar Sky Survey, as one looks at sources with lower and lower flux density at 5 GHz. The radio spectral index distributions peak near -0.9 for the sources selected by low frequency surveys, such as the 3CRR, and show a double-peaked distribution for the sources selected at high frequency, such as the MPIfR/NRAO survey. In the latter case the two peaks are near -0.9 again, and near 0.0, with a long tail to $+1.0$. Note that all these sources in the second peak of the spectral index distribution are generally variable, and so no unique spectral index can actually be associated with them, since the emission varies differently at different frequencies. The spectra are often straight, and then usually steep, but also often convex or concave, simply peaked, or because of variability so complex that no simple shape can be used to describe them. It is quite common that a source which shows a simple steep spectrum at low frequency, becomes dominated by a compact core at high frequency, which is variable and shows a concave spectrum. This is usually due to an extended structure which dominates at low frequencies, and a compact structure which dominates at high frequencies. On the other hand, those sources which show a steepening of their spectrum with increasing frequency, can be understood as "aged" synchrotron sources, in which the injection of energetic electrons has been terminated some time in the past, and now those electrons have already lost their energy, and the electron energy distribution itself is a power law with a gradual cut-off.

1.2 Morphology

Radio morphology has been classified by Miley (1980) and he distinguishes four basic structures.

1.2.1 Narrow Edge-Brightened Sources

The narrow edge-brightened double sources such as 3C452 are the famous Fanaroff-Riley class II sources (this is usually abbreviated as FR II). These show narrow weak channels (often embedded in a region of broad emission) normally referred to as tails, bridges and jets, where the energy is transported to the outer lobes, which usually have one or more hot spots. These hot spots are typically 1 kpc in size, and there is evidence from detailed observations of a small number of such sources (Meisenheimer et al. 1989) that these hot spots are parallel shock fronts, which limit a flow with a bulk speed of order 0.1 to 0.4c. Most radio galaxies with a power of 10^{25} W Hz^{-1} at 178 MHz or more show such a structure. The central object is often an elliptical galaxy or a quasar with usually weak compact radio emission. The outer two hot spots and the central compact source are collinear and usually symmetrical with respect to a rotation by 180 degrees. If VLBI detects a jet, it is one-sided and lies on that side which shows a stronger jet and a stronger hot spot. A word of caution is in order here: if the source happens to have a structure nearly parallel to the line of sight, then any bending in the elongated structure is greatly amplified and so will not fit into the simple scheme presented here. Recent work (Barthel 1989) has demonstrated that radio galaxies and radio quasars can be understood in a common framework if it is assumed that there is a torus of obscuring material close to the active nucleus which obscures a large part of the sky as seen from the active nucleus, and which makes it impossible for an observer within about 45 degrees of the plane of this torus to actually see the broad emission lines characteristic of many active galactic nuclei. This, of course, refers to a particular sample, for which the 3CRR has similar numbers of radio galaxies in redshift and luminosity range. Such an argument introduces an important selection effect in the classification of sources according to their electromagnetic emission. Later we will see that another important selection effect is the Lorentz boosting which occurs when we view emission from a source component which moves at bulk relativistic speeds close to the line of sight; in that case the emission is boosted relative to a comoving frame.

1.2.2 Narrow Centre-Brightened Sources

Narrow centre-brightened sources such as 3C449 (Perley et al. 1979) represent the weaker sources generally, and are referred to as Fanaroff-Riley class I sources. Here we also find central compact components, often rather strong jets, and often no outer clear boundary. With increasing sensitivity one often finds more and more extended emission at weaker and weaker levels which seems to be self-similar (see Cen A or NGC5128, Burns et al. 1983), i.e. appears to have the same shape at different length scales at different emission levels.

1.2.3 Narrow Tailed Sources

Narrow tailed sources such as NGC1265 can be seen as an extreme form of the second class, where the two jets are turned around into a common direction and form a common tail (Miley et al. 1973, 1975). One reasonable explanation for this morphology is the ram pressure exerted by the differential motion of hot cluster gas and the radio galaxy (Begelman et al. 1979). Other well-known names for this class are "head-tail" sources or "narrow-angle-tail" sources.

1.2.4 Cluster Sources

Finally there are the cluster halos where no particular structure is seen. Such sources have very steep radio spectra. All of these sources can also be seen as a sequence in bending, in edge-brightening, and in rotational symmetry.

1.2.5 Luminosities and Scales

The luminosity function of these sources has a break at 10^{26} W Hz^{-1} at 178 MHz, just where the radio galaxies divide into Fanaroff-Riley class I and II. Newer data show this break only for the blue radio galaxies (Windhorst 1984). The scales for the size of these radio galaxies range up to 4 Mpc, for 3C286 (using $H_o = 75$ km s^{-1} Mpc^{-1}). Most sources have linear sizes in the range from 100 to about 300 kpc on the sky, with only a few % of all sources extending over 1 Mpc. The large sources are almost all of the Fanaroff-Riley class II. Some FR I sources are also about 1 Mpc in size. The luminosities range up to 10^{48} erg s^{-1} for flat spectrum radio sources, assuming isotropic emission, and integrating over the entire known electromagnetic spectrum. There is, however, an important caveat: our common physical interpretation suggests that just for these sources the emission is, in fact, highly anisotropic. There is no part of the electromagnetic spectrum, except the radio radiation of the very extended structure, for which we can be entirely certain that the emission is really isotropic at cosmic distances although it is likely to be a fair approximation for dust emission at infrared wavelengths. But the very interpretation of infrared radiation as emission from dust is contested (Chini et al. 1989a, Sanders et al. 1989, de Kool and Begelman 1989, Schlickeiser et al. 1991), however it is very likely to be correct for many sources such as Seyfert galaxies, radio galaxies and radio weak quasars. Here the luminosities are, e.g. 10^{43} erg s^{-1} for Cen A, a FR I radio galaxy, and can be even higher for radio quasars by about two orders of magnitude (see Chini et al. 1989b).

1.3 Jets
1.3.1 Jet Morphology

From an early stage, supply channels were thought to exist to transport the energy continuously to the very large radio lobes (themselves with luminosities up to 10^{45} erg s^{-1}), especially in the FR II radio galaxies such as Cyg A. Bridle and Perley (1984) give long lists of the now well-established jets, detailing their properties. It appears that a very large proportion of radio galaxies have detectable jets, which is to say for all practical purposes, visible for the Very Large Array (VLA) in New Mexico. In the second Bologna list, B2, 9 out of 11 radio galaxies within $\delta \geq 10^o$, $b \geq 10^o$, and $z \leq 0.05$ show jets, and in the 3CRR list 12 out of 22. The visibility of the jet is increased for sources with fairly low overall luminosities. Some very nice pictures have been published by Perley (1989). It is remarkable, how some of the well-observed jets break up into patterns of regularly spaced knots, rather reminiscent of the regularly spaced biconical shocks and Mach disks seen in the exhausts from jet aircraft. An important difference is, of course, that a) aircraft have well-defined nozzles, while b) astrophysical jets have both stationary and moving knots. Some jets are not straight but we must remember that their angle to our line of sight can be small so that a minor wiggle might appear as a great spiral or in the shape of a question mark (e.g. 3C418).

1.3.2 Jet Asymmetry

The distribution of jet prominence, i.e. the ratio of jet luminosity over total extended luminosity plotted versus the total lobe power, clearly shows an envelope which decreases with increasing lobe power in FR I sources to reach values near 0.1 and less for FR II sources. Plotting such a graph for the counterjet only shows an even more restricted envelope of the distribution, suggesting that the counterjet-to-jet ratio also decreases with total lobe power (Bridle 1989). Most jets are one-sided near the core; in weak radio galaxies they become two-sided after a few kpc, the side with the one-sided section corresponding to the brighter jet on the larger scale. Most jets with a total power at 1.4 GHz $< 3 \times 10^{24}$ W Hz^{-1} are two-sided, while most jets in stronger sources are totally one-sided. This also corresponds well to the separation between FR I and FR II sources.

1.3.3 Jet Magnetic Configuration

The radiation from jets is usually strongly polarized with levels up to $\sim 40\%$ and in some locations it may be even up to 50%. Low polarization is rare and it follows that the magnetic field has to be well-ordered. We note that usually observations with at least four different radio frequencies are required to derive the magnetic field configurations, because the Faraday rotation has to be determined first and at fairly high spatial resolutions. There are several typical magnetic field configurations B_a, the magnetic field structure weighted with the synchrotron emission (Bridle and Perley 1984). The configuration referred to as $B_\parallel$, shows a magnetic field which is mostly parallel to the jet axis over the entire cross-section; second, $B_\perp$, where the magnetic field is mostly perpendicular to the jet cross section; third, $B_{\perp-\parallel}$, where the magnetic field is perpendicular to the jet axis close to the centre line of the jet, and mostly parallel to the jet axis along the sides. Most two-sided regions of jets have either the $B_\perp$ or the $B_{\perp-\parallel}$ configuration. For low-luminosity sources, i.e. the power at 1.4 GHz is $< 3 \; 10^{24}$ W Hz^{-1}, the configuration B_a turns from $B_\parallel$ to $B_\perp$ or $B_{\perp-\parallel}$ within the first 10% of the distance from the nucleus along the jet. In the very strong sources, the jet is dominated over its entire length by $B_\parallel$. It follows that we have the simple correlation: FR II sources are edge-brightened, have strong hot spots, and the $B_\parallel$ configuration, while the FR I sources are edge-darkened, have no hot spots, and have the $B_\perp$ or $B_{\perp-\parallel}$ configuration. There are some common variations on the theme. For example along curved sections of a jet one often finds the $B_{\perp-\parallel}$ configuration with the $B_\parallel$ edge stronger on the outside. One also finds that some knots in one-sided jets have the $B_\perp$ configuration, although the weaker emission close by may be $B_\parallel$ dominated.

All this can be readily interpreted as due to magnetic flux conservation in a plasma flow along the jet:

- since $B_\parallel \sim r^{-2}$ and $B_\perp \sim r^{-1}$, it follows that further outside $B_\perp$ readily dominates
- in a shock wave $B_\perp$ is increased by the density jump, while $B_\parallel$ remains constant, and knots, interpreted as shocks, should have some preference for the $B_\perp$ configuration
- shear across the wall of a jet with momentum transfer strengthens $B_\parallel$.

1.3.4 Length, Bending and Non-Linear Structure

Jets are short (i.e. < 40 kpc) in weak sources and in core-dominated sources (perhaps because of projection effect). But jets are often strongly bent in the same two cases, because ram pressure affects weak jets more strongly, and because a direction close to the line of sight (small aspect angles) increases projection effects and also increases, in the common interpretation, the core flux density by Lorentz boosting. That the latter effects are important, becomes obvious from the correlation between non-linearity of compact with extended structures and core-dominance. Again, projection effects, coupled with differential Lorentz boosting, offer an obvious explanation.

1.3.5 Collimation

In some jets the lateral structure can be resolved and generally the emission is stronger along the central axis of the jet. At very high spatial resolution (e.g. the M87 jet), however, there appears to be a dark channel along the very central axis. Since $B_\perp$ has to go to zero on the axis, because $\nabla B = 0$, we expect less emission; on the other hand, what we observe is an integration along the line of sight through the cross-section of the jet, where the very centre does not contribute all that much. The physical interpretation of this observation is therefore not clear. If we define Φ as the full-width half-maximum jet diameter after deconvolution, and consider this width Φ as a function of distance θ from the core along the jet, then $d\Phi/d\theta$ ought to be constant for a freely expanding jet, i.e. when the jet pressure $P_j \gg$ external pressure P_{ext}. The observations demonstrate that only a few jets are totally free in this sense. In fact $d\Phi/d\theta$ is lower for more luminous radio jets, and often oscillates from values near zero to some maximum number (see Bridle and Perley 1984), which also strongly decreases with increasing jet-power. Again, it is rather likely that bulk relativistic motion explains these trends, with the hypothesis that bulk relativistic motion is prevalent in all radio galaxies on the compact scales, and is more common even on fairly extended scales in powerful radio galaxies. In extreme cases it may well extend to the hot spots at speeds near to $0.3c$.

1.3.6 Spectra and Intensity Patterns

Some 40% of all jets have radio spectra in the range $\nu^{-0.6}$ to $\nu^{-0.7}$, but over 90% have spectra in the range $\nu^{-0.5}$ to $\nu^{-0.9}$. In those cases where the spectral index has been measured as a function of position along the jet, the spectrum either stays the same along the jet or steepens towards the outside (Perez-Fournon et al. 1988). A number of jets have been detected in the optical (e.g. Stocke et al. 1981, Röser and Meisenheimer 1986, 1987) including 3C31, 3C66B, M87, 3C273, 3C277.3, and possibly Cen A. This optical emission is also highly polarized and so is likely to be synchrotron emission as is the case at radio wavelengths. In X-rays only a few jets have been reliably detected notably M87, Cen A and 3C273. This situation will almost certainly improve as ROSAT data become available. It appears that the structure of jets is the same at all wavelengths where observed. This may be interpreted as synchrotron emission at all wavelengths, with the electrons causing the X-ray emission arising as secondaries from $p - p$ and $p - \gamma$ collisions (Klemens 1987a, b). Further explorations of this model have shown some interesting and fruitful insights for compact jets (Mannheim et al. 1991). In particular, the emission caused by proton initiated hadronic cascades (PIC), i.e. relativistic protons, can easily dominate over synchrotron-self-Compton in the observed emission at X-ray wavelengths. One tell-tale sign of hadronic cascades is that they easily produce strong emission far beyond the MeV peak typical for electron-positron plasma emission (e.g., Zdziarski et al. 1987).

1.4 Physical Parameters

1.4.1 Energies and Magnetic Fields

Let us assume that the electron energy distribution can be written as a power law:

$$N(\gamma)d\gamma = C\,\gamma^{-p}d\gamma \tag{1}$$

then the synchrotron emissivity is

$$\epsilon_\nu^S = \text{const } C\, B^{(p+1)/2}\, \nu^{-(p-1)/2}. \tag{2}$$

The energy content in electrons can be written as

$$u_e = C\, m_e\, c^2 \int_{\gamma_1}^{\gamma_2} \gamma^{-p+1} d\gamma = C\, \frac{m_e\, c^2}{2-p}\left(\gamma_2^{2-p} - \gamma_1^{2-p}\right) \text{ for } p \neq 2$$

$$= C\, m_e\, c^2\, \ln(\gamma_2/\gamma_1) \text{ for } p = 2. \tag{3}$$

The total emission is then given by

$$S_\nu = \frac{\text{Vol}}{4\,\pi\,D^2}\,\epsilon_\nu = \frac{\text{Vol}}{4\,\pi\,D^2}\text{const } C\, B^{(p+1)/2}\, \nu^{-(p-1)/2} = \frac{\text{Vol}}{4\,\pi\,D^2}\text{const } C\, B^{1-\alpha}\, \nu^{\alpha} \tag{4}$$

where $\alpha = -(p-1)/2$, and so $(p+1)/2 = 1 - \alpha$. Hence

$$C = S_\nu\, \nu^{-\alpha}\, \frac{\text{Vol}}{4\,\pi\,D^2}\, B^{\alpha-1}\, \text{const} \tag{5}$$

and

$$u_e = S_\nu\, \nu^{-\alpha}\, \frac{\text{Vol}}{4\,\pi\,D^2}\, B^{\alpha-1}\, \text{const}\, \left(\gamma_2^{2-p} - \gamma_1^{2-p}\right) \tag{6}$$

if we consider only relativistic electrons. Protons add an additional contribution to the energy density with $u_p = k\, u_e$. In our Galaxy $k = 100$ at energies above 1 GeV but it is usually argued that $k \approx 1$ on the basis of diffusive shock acceleration in relativistic shocks. Hence in our case we have no idea what value k might actually take, anything between 1 and 100 is plausible at this time. The observed synchrotron frequency is

$$\nu \sim \gamma^2\, B \implies \gamma \sim (\nu/B)^{1/2} \tag{7}$$

and

$$\left(\gamma_2^{2-p} - \gamma_1^{2-p}\right) = \left(\nu_2^{\frac{1}{2}+\alpha} - \nu_1^{\frac{1}{2}+\alpha}\right)\text{const } B^{\frac{1}{2}+\alpha}. \tag{8}$$

The magnetic field energy density is, obviously,

$$u_B = B^2/(8\,\pi). \tag{9}$$

Hence the energy content of the relativistic components is, in total, given by

$$u = \frac{B^2}{(8\,\pi)} + (1+k)\,S_\nu\,\nu^{-\alpha}\,\frac{4\,\pi\,D^2}{\mathrm{Vol}}\,B^{\alpha-1}\,\mathrm{const}\,B^{-1/2-\alpha}\,\left(\nu_2^{\frac{1}{2}+\alpha}\,-\,\nu_1^{\frac{1}{2}+\alpha}\right). \quad (10)$$

The magnetic field strength B is unknown, but we can derive a minimum for this sum by varying B so as to determine a minimal energy density as a function of B. Putting the derivative $\frac{du}{dB} = 0$ we derive an expression for the strength of the magnetic field at that point where the total energy density is a minimum:

$$B_{\mathrm{min,en}} = \left(4\,\pi\,(1+k)\,S_\nu\,\nu^{-\alpha}\,\frac{4\,\pi\,D^2}{\mathrm{Vol}}\,\mathrm{const}\,\left(\nu_2^{\frac{1}{2}+\alpha}\,-\,\nu_1^{\frac{1}{2}+\alpha}\right)\right)^{2/7}. \quad (11)$$

This leads then to values of the magnetic field of order $\sim 3\,10^{-7}$ G in cluster halos, $\sim 3\,10^{-6}$ G in diffuse radio lobes, $\sim 10^{-4}$ G in hot spots, and $\sim 10^{-3}$ G in core sources at low-frequency VLBI spatial resolution. These values depend almost linearly on the spatial scale used:

$$B_{\mathrm{min,en}} = \mathrm{scale}^{-6/7} \quad (12)$$

and so tend to increase for observations at higher spatial resolution. This has to borne in mind when consulting the literature. At this minimum energy density the magnetic field and the relativistic particles are in approximate equipartition, just as it is observed in the interstellar medium. The total energies derived, when multiplying the minimum energy density with the volume of the emission, can exceed 10^{60} erg.

Inverse Compton emission also produces X-ray emission by scattering off the microwave background. As, however, the relativistic electrons responsible for the inverse Compton process are the same ones that cause the synchrotron emission, we obtain a limit on the strength of the magnetic field:

$$S_{\nu,R} \sim C\,B^{1-\alpha}\,\nu_R^\alpha \quad (13a)$$

$$S_{\nu,X} \sim C\,u_{ph}\,\nu_X^\alpha \quad (13b)$$

Hence

$$S_{\nu,R}/S_{\nu,X} \sim \frac{B^{1-\alpha}}{u_{ph}}\,\frac{\nu_R^\alpha}{\nu_X^\alpha} \quad (14)$$

and so our estimate for the field is

$$B = \mathrm{const}\,\left(S_{\nu,R}\,\nu_R^{-\alpha}\,S_{\nu,X}^{-1}\,\nu_X^\alpha\right)^{\frac{1}{1-\alpha}} \quad (15)$$

Since the Inverse Compton emission is clearly only one of many possible contributions to any observed X-ray emission, the derived estimate for the magnetic field B is a lower limit. In a very small number of cases it is possible to use this argument to derive this lower limit for B in extended radio lobes, where one finds $B \geq 10^{-6}$G which is not far from the minimum energy condition.

1.4.2 Distribution of Polarization

The direction of linear polarization is turned by a magnetic field through an amount

$$\Delta\chi \; = \; 5.73 \; 10^{-3} \; R \; \lambda^2 \; \text{degrees} \tag{16}$$

where λ is measured in cm and R in rad m^{-2} by convention. The rotation measure R is then given by

$$R \; = \; 812 \int n_t(s) \; B_{\|}(s) \; ds \; \simeq \; 8.1 \times 10^8 \; n_t \; B_{\|} \; s \; \text{rad m}^{-2} \tag{17}$$

where n_t is measured in cm^{-3}, B in Gauss, and s in kpc, Such conventions with mixed units are not unusual in astronomy. Clearly, it is important to use short wavelengths to determine R; and as previously mentioned, normally four or more frequencies are used, with at least one pair of frequencies that are closely spaced, in order to make sure that the polarization plane does not turn by 2π between any two neighbouring frequencies.

Most sources have integral values of the rotation measure of $<$ 50 rad m^{-2}, but some have several hundred. There is a small but growing number of cases where high spatial resolution rotation measure maps show values of R of over 10^3 rad m^{-2}. These regions seem to show a wavy structure of sign reversals, with the crests of these "waves" nearly perpendicular to the symmetry axis of the source.

1.5 Physical Concepts of Radio Sources - No Jets

The basic ideas can be summarized as follows:
    - Diffuse emission usually has a steep spectrum, and can be interpreted as an "aging" of a population of relativistic electrons along the direction of flow for edge-darkened sources, with the opposite behaviour for edge-brightened ones. This suggests that particle acceleration is strong "inside" for the former and "outside" for the latter.
    - Buoyancy can lift clouds of hot gas with a population of relativistic particles (as they have practically no "weight") produced by old radio sources in a cluster of galaxies. Observation of this behaviour can be used to estimate the gravitational mass of clusters, leading to values $\sim 10^{15}$ M$_\odot$, in agreement with other arguments.
    - Hot spots can be understood as the final shock in supersonic flow where the bulk of the flow is decelerated to subsonic speeds, with B usually perpendicular to the jet axis, and a spectrum typically flatter than the surroundings, suggesting particle acceleration at the shock. Brightness and distances from the core for the two opposite heads in a FR II radio galaxy lead to an estimate of the net advance speed of the heads of $\leq$ 0.3c, which in turn leads from ram pressure arguments to a lower limit of the density of the surrounding medium of $n_{\text{ext}} = 6 \times 10^{-5}$ cm^{-3} corresponding to very large distances in the gaseous halos of clusters. This may imply that the observed heads of radio galaxies are never seen to reach the true intergalactic medium. This leads one to speculate that there may be no head left as soon as it is reached, thus changing the overall appearance of the radio galaxy.

1.6 Physical Concepts for Extended Jets

Today the usual supposition is the following: all jets are initially relativistic and are braked to subrelativistic, but supersonic flow, in edge-darkened sources which may turn to subsonic flow far outside. In edge-brightened sources they stay relativistic - albeit only marginally so - all the way to the hot spot. In this context relativistic flow means velocities such that $(v/c)^2 \simeq \frac{1}{3}$ to 1. This interpretation of the observations means the following:

- The apparent superluminal motion observed in many compact sources is a kinematic illusion.

- We get Doppler or Lorentz boosting of the radiation emitted by material moving along the jet with relativistic speeds.

- Extreme brightness temperatures are obtained. This is commonly referred to as the Compton dilemma, when the inverse Compton emission implied by the compactness of the source (as derived from its variability or direct VLBI imaging) is larger than the X-ray flux actually observed. The dilemma is resolved simply if the emitting region moves at relativistic speeds.

- Projection effects strongly increase the apparent bending of the jet at small aspect angles. The latter are required if one is to obtain strong Lorentz boosting.

- We can understand why stronger hot spots correlate with the stronger jet.

1.6.1 Breakout and Reconfinement of Jets

The initial formation of the jet flow has been investigated by, e.g., Camenzind (1986, 1987, 1990). Following Sanders (1983) we can use classical jet flow physics to discuss the radial dependence of the pressure inside a jet. Note that if $P_j \gg P_{ext}$ the jet freely expands with a constant opening angle. We consider transsonic flow and first discuss breakout and then reconfinement. The continuity equation then relates the cross-section F with the velocity v and the density ρ:

$$F\,v\,\rho \, = \, \text{const.} \tag{18}$$

The energy equation with pressure P_j, using γ for the adiabatic index of the gas, reads

$$F\,v\,\rho \left(\frac{1}{2}\,v^2 \,+\, \frac{\gamma}{\gamma-1}\,\frac{P_j}{\rho} \,+\, \phi \right) \,=\, \text{const} \tag{19}$$

where ϕ is the gravitational potential and where we have for strongly supersonic flow

$$v^2 \,\gg\, \frac{P_j}{\rho}. \tag{20}$$

If we introduce an asterisk ($\star$) to denote quantities at the point where the flow reaches and exceeds the local sound speed, we then have for the radius of the cone of the flow

$$(r/r_\star)^2 \,=\, \frac{\rho_\star\,v_\star}{\rho\,v} \,=\, \frac{\rho_\star\,c_{s\star}}{\rho\,v} \tag{21}$$

and using the equation of state it follows that

$$(r/r_\star)^2 = \frac{c_{s\star}}{v}\left(\frac{P_\star}{P}\right)^{\frac{1}{\gamma}}. \tag{22}$$

The energy equation can then be written as

$$\frac{1}{2}v^2 + \frac{\gamma}{\gamma-1}\frac{P}{\rho} + \phi = \frac{1}{2}\frac{\gamma+1}{\gamma-1}c_{s\star}^2 + \phi_\star. \tag{23}$$

Hence it follows that

$$\left(\frac{v}{c_{s\star}}\right)^2 = \frac{\gamma+1}{\gamma-1}\left(1 + 2\frac{\gamma-1}{\gamma+1}\frac{\phi_\star-\phi}{c_{s\star}^2} - \frac{2}{\gamma+1}\left(\frac{c_s}{c_{s\star}}\right)^2\right) \tag{24a}$$

and

$$\frac{r}{r_\star} = \left(\frac{\gamma-1}{\gamma+1}\right)^{\frac{1}{4}}\left(\frac{P_\star}{P}\right)^{\frac{1}{2\gamma}}\left(1 + 2\frac{\gamma-1}{\gamma+1}\frac{\phi_\star-\phi}{c_{s\star}^2} - \frac{2}{\gamma+1}\left(\frac{c_s}{c_{s\star}}\right)^2\right)^{-\frac{1}{4}}. \tag{24b}$$

For the Mach number it follows that

$$\frac{v}{c_s} = \left(\frac{\gamma-1}{\gamma+1}\right)^{\frac{1}{2}}\left(\frac{P_\star}{P}\right)^{\frac{\gamma-1}{2\gamma}}\left(1 + 2\frac{\gamma-1}{\gamma+1}\frac{\phi_\star-\phi}{c_{s\star}^2} - \frac{2}{\gamma+1}\left(\frac{c_s}{c_{s\star}}\right)^2\right)^{-\frac{1}{2}}. \tag{25}$$

If the jet is "hot" and so the jet temperature is much higher than the hydrostatic "temperature", we have

$$\left|\frac{\phi_\star-\phi}{c_{s\star}^2}\right| \ll 1. \tag{26}$$

Far from the point where the local sound speed is passed, we have then a continuous density decrease outwards and $\left(\frac{c_s}{c_{s\star}}\right)^2 \ll 1$ so

$$\frac{r}{r_\star} = \left(\frac{\gamma-1}{\gamma+1}\right)^{\frac{1}{4}}\left(\frac{P_\star}{P}\right)^{\frac{1}{2\gamma}} \tag{27}$$

and the Mach number

$$M = \frac{v}{c_s} = \left(\frac{\gamma+1}{\gamma-1}\right)^{\frac{1}{2}}\left(\frac{P_\star}{P}\right)^{\frac{\gamma-1}{2\gamma}} \tag{28}$$

assuming the jet is confined by outside pressure. If, on the other hand, the velocity which is implied by the lateral increase, exceeds the velocity of sound substantially, then the jet becomes free and independent of its surroundings. If we assume that in a time Δt the jet expands sideways by an amount Δr and radially by Δz, then breakout means

$$\Delta r > c_s \Delta t = \frac{\Delta z}{M}, \tag{29a}$$

$$\frac{dr}{dz} = -r_\star\left(\frac{\gamma-1}{\gamma+1}\right)^{\frac{1}{4}}\left(\frac{P_\star}{P}\right)^{\frac{1}{2\gamma}}\frac{1}{2\gamma}\frac{d\ln P}{dz}, \tag{29b}$$

$$M = \frac{v}{c_s} = \left(\frac{\gamma+1}{\gamma-1}\right)^{\frac{1}{2}} \left(\frac{P_\star}{P}\right)^{\frac{\gamma-1}{2\gamma}} > \frac{2\gamma}{r_\star} \left(\frac{\gamma-1}{\gamma+1}\right)^{\frac{-1}{4}} \left(\frac{P_\star}{P}\right)^{\frac{-1}{2\gamma}} \left(-\frac{d\ln P}{dz}\right)^{-1},$$

$$(29c)$$

$$\text{or} \quad -\frac{d\ln P}{dz} > 2\gamma \frac{1}{r_\star} \left(\frac{\gamma-1}{\gamma+1}\right)^{\frac{1}{4}} \left(\frac{P}{P_\star}\right)^{\frac{1}{2}}. \tag{29d}$$

Assuming then in this case that the jet pressure initially matches the external pressure and has a radial dependence of the form

$$P = P_\star \left(\frac{z}{z_\star}\right)^{-\alpha} \tag{30}$$

it follows then that

$$\frac{\alpha}{z} > 2\gamma \left(\frac{\gamma-1}{\gamma+1}\right)^{\frac{1}{4}} \left(\frac{z_\star}{z}\right)^{\frac{\alpha}{2}} \frac{1}{r_\star} \tag{31a}$$

or

$$\alpha > 2\gamma \left(\frac{\gamma-1}{\gamma+1}\right)^{\frac{1}{4}} \left(\frac{z_\star}{z}\right)^{\frac{\alpha}{2}-1} \frac{z_\star}{r_\star}. \tag{31b}$$

For $\alpha > 2$ this is always possible but if $\alpha \leq 2$ this is not in general the case, since we usually have $z_\star/r_\star \gg 1$. It follows that jets cannot break out in regions where $P_{\text{ext}} \sim r^{-3/2}$ or r^{-2}, both of which are possible standards. The first dependence arises from a simple isothermal cooling flow in an elliptical galaxy (e.g. Biermann et al. 1989), the second one from an isothermal outflow. However, if jets are free, then their internal pressure dependence (for a free supersonic jet) is

$$P_{\text{j}} \propto r^{-\frac{1}{2\gamma}} \tag{32}$$

much steeper than any expected outside pressure, and thus leading to reconfinement. Reconfinement causes the formation of biconical shock structures with sharply increasing pressures that bring the jet pressure to exceed the surrounding pressure again. After such a reconfinement the same process is likely to happen again, leading to a semiperiodic structure of biconical shocks, which can then be identified with the knots observed in jets. In the simplest of such pictures, all parts of the biconical shocks are highly oblique and so the overall supersonic flow remains supersonic throughout. It is also important to remember that the synchrotron emission samples the regions of highest pressure, i.e. the regions immediately behind the downstream cone of the biconical shock structure where the pressure strongly exceed the surrounding pressure. It is quite likely that both inside and outside the jet magnetic fields and a relativistic particle population strongly contribute to the overall energy density and pressure, just as in the interstellar medium. Hence it should come as no surprise that the minimum pressures derived for knots exceed the surrounding pressure derived from X-ray measurements bremsstrahlung emission of the hot gas.

1.6.2 Orientation with Respect to the Line Emitting Material

In one standard model (Orr and Browne 1982, Wills and Browne 1986) one assumes for the distribution of emission line clouds a random isotropic velocity component v_r, and a component perpendicular to the jet axis v_p in the plane of the proposed disk. The observed velocity width is then given by

$$\left(v_r^2 + v_p^2 \sin^2 \theta\right)^{1/2} \tag{33}$$

where θ is the angle between the line of sight and the radio axis. Let us define $R(\theta)$ as the ratio of the flux density of the beamed, i.e. Lorentz boosted, components to the flux density of the unbeamed components, then with $R_T = R(90 \text{ degrees})$, we obtain from the angle dependence of the Lorentz boosting the following relation

$$R/R_T = \frac{1}{2}\left(1 - \beta \cos\theta\right)^{-2} + \frac{1}{2}\left(1 + \beta \cos\theta\right)^{-2} \tag{34}$$

assuming the individual radio spectra are flat - although the generalization is easily made. Since sources out in the universe are randomly oriented with respect to us, we have for the probability of having a particular value for the angle θ the relation $P(R)\,dR = d(\cos\theta)$, and so

$$\cos\theta = \frac{1}{\beta}\left(\frac{1}{2R}\left(2R + R_T - \left(R_T\left(8R + R_T\right)\right)^{1/2}\right)\right)^{1/2} \tag{35}$$

where we have $R_{\min} = R_T$ and $R_{\max} = R_T\,\gamma^2\,(\gamma^2 - 1)$. Comparing then these relations with observations shows that there is an envelope in R versus the width of the emission line H_β, which leads to estimates for $v_r = 4,000$ km s^{-1}, and $v_p = 13,000$ km s^{-1}. This velocity in turn can be related to the distance from an assumed black hole with $v = \frac{1}{2} v_p$

$$G\,M_{\rm bh}/v^2 = r \rightarrow r = 3.14 \times 10^{16}\,M_{\rm bh,8}/(2\,v/v_p)^2 \text{ cm} \tag{36}$$

as $M_{bh,8}$ corresponds to an Eddington luminosity $L_{\rm edd}^*$ of 10^{46} erg s^{-1} then

$$L_{\rm edd} = \alpha_{\rm edd}\,L_{\rm edd}^* = 10^{46} \text{ erg s}^{-1}\,r_{16.5}\left(\frac{2\,v}{v_p}\right)^2. \tag{37}$$

$L_{\rm edd}^*$ in turn corresponds to an approximate UV flux of $4.6\ 10^{56}$ ph/sec. We should then compare these numbers with derived UV-luminosities for various active galactic nuclei, and their radial scales as derived from emission line variability. Gondhalekhar (1990) has prepared such a graph. Putting then the resulting relationship into this graph shows fair agreement for $\alpha_{\rm edd} \cong 0.1$ to 1.

1.7 Suggested Reading List

1) Wills and Browne (1986)
2) Perez-Fournon et al. (1988)
3) Chini et al. (1989a)
4) Chini et al. (1989b)
5) Witzel et al. (1988)
6) Barthel (1989)

2 Physics of Jets - Compact Jets

2.1 Hydrodynamical Modelling of Jet Flow

One of the best radio pictures of the M87 jet has been presented by Muxlow et al. (priv.comm.): it shows clearly various knots, and then a rather sharp edge facing the core at, what is commonly called knot A, and an apparent helical structure beyond knot A. Norman et al. (1982) have presented many calculations of gaseous jet flow and tried to compare and interpret the observations with their modelling. The questions which they have tried to answer in these calculations are the following:

- Can the observed jets be modelled as the propagation of a supersonic beam of gas? Here the tentative answer is yes, and so it is worth looking at such calculations.

- Is such a modelled jet stable in its overall structure as well as stable with respect to Kelvin-Helmholtz instabilities?

- Does the jet form a cocoon?

- How is the beam collimated?

- Does the beam entrain material from the surrounding medium?

- Does the beam form quasi-periodic structures without getting totally disrupted?

- What is the role of the magnetic field and the relativistic particles that may well have the same energy density, or perhaps even exceed the energy density, of the thermal gas?

In all such calculations there is always the assumption, that we are dealing with a normal gas, consisting of protons, electrons, and heavier elements in "normal" abundances; we do not consider an electron-positron plasma, since there are no observations that demonstrate that such a plasma dominates in observed jets. The calculations show:

- Beams propagate well for beam Mach numbers $M_b \geq 6$.

- The hot spots can be understood as decelerating Mach disks.

- The backflow from the jet forms a cocoon for $T_b/T_m \geq 10$, where m refers to the surrounding medium, and T is the temperature.

- At extreme Mach numbers, the backflow itself can be supersonic.

- At low Mach numbers, lobes are formed.

- Low velocity cocoons and lobes are unstable to Kelvin-Helmholtz instabilities.

- Quasi-periodic structures of biconical shocks appear, but it is not clear whether they really correspond to the knots observed.

Further calculations by Norman et al. (1984a) demonstrate that regularly spaced biconical shocks and high pressure knots appear in all cases, which, in a plot of density ratio η_b (beam to surrounding medium) against beam Mach number, lie between a lower limiting line where cocoons form and an upper limiting line where reflection mode shocks appear ($\eta_b = (M_b/2)^{10/3}$ see Cohn 1983). This region covers density ratios both below and above unity, but stretches to higher and higher density ratios for higher and higher Mach numbers. The wavelength of the mode, when saturated, is insensitive to the Mach number and is, measured in spacing/width between 2.3 and 2.6. This appears to match the observations in a rather crude sense. Moreover, Smith et al. (1985) have demonstrated how a bright head with a sharp outer head can be produced along with wings and tails with arbitrarily complex subcomponent geometries. Norman et al. (1984b) have shown that the patterns of knots actually move at a fraction of order 0.3 (the simulations show a range of 0.14 to 0.66) of the jet flow speed, possibly corresponding to the apparent motion often observed in compact jets - there at near relativistic speeds. Kössl (1988) has shown that the role of magnetic fields and relativistic particles cannot yet be studied with present day computers, so that this essential part of the physics remains unclear.

2.2 Resume on Active Galactic Nuclei

This section serves to explain the general concepts (and personal biases) in what follows.

The definition of an active galactic nucleus (usually abbreviated to AGN) is a nonstellar energetic phenomenon in the centre of a galaxy. There appears to be a continuity of properties from nearly stellar luminosities $\sim 10^{39}$ erg s^{-1} to more than 10^{48} erg s^{-1} **assuming their radiation is emitted isotropically**. Many objects are known near 10^{45} erg s^{-1}, for which isotropic emission can reasonably be assumed. The variability time scales range from hours to longer than years. The geometric length scales from $\sim 10^3$ AU to over one Mpc and the photon energies from $\sim 3 \ 10^{-8}$ eV to $\sim 3 \ 10^{+8}$ eV. The dominant contribution to the luminosity is seen in two examples to be near 10 MeV (Bezler et al. 1984, Ballmoos et al. 1987) but this sample might increase rapidly as results from GRO become known. The peak is often suspected in the UV near 10 eV, and is often actually observed in the far-infrared near 10^{-2} eV. Many radio sources show a collinear structure on all scales. The compact structures often show knots that move at apparent speeds of up to $\sim 20c$ (Witzel et al.1988, Schalinski et al. 1988ab, Alberdi et al. 1988, Hummel et al. 1988, Krichbaum et al. 1990ab, Dhawan et al. 1990, Marcaide et al. 1990, Krichbaum and Witzel 1990, 1992, Krichbaum 1990, Schalinski et al. 1992ab) assuming $H_o = 50$ km s^{-1} Mpc^{-1}. There is a hint in the data that the motion may have a helical symmetry (Krichbaum 1990).

We shall now describe the current basic model. It is, however, not clear that there is a unique model at all, and the model we describe cannot account for all of the observations, but it has so far survived the test of time. The basic scenario consists of:

- Accretion onto a compact object, usually involving an accretion disk (invented by Lüst 1952 but reinvented several times, Shakura and Sunyaev 1973, see the review by Pringle 1981). This accretion disk is believed to be visible in the UV/X-ray range (e.g. Malkan 1983, Arnaud et al. 1985, Czerny and Elvis 1987, Kaastra and Barr 1989).

- A rotating black hole fed by accretion, the luminosity is derived from the gravitational field.

- A jet, perhaps always initially relativistic, with Lorentz factors of around 10.

- A disk-halo with clouds (perhaps stars and their winds) which we see in emission lines.

- A dusty disk out to at least about 300 pc, that may include a torus at its inner edge to provide the shadowing required by some interpretations of the data (Lawrence and Elvis 1982, Mushotzky 1982, Antonucci and Miller 1985, Krolik and Begelman 1988, Barthel 1989). There might be an analogous feature in our Galaxy i.e. an inner ring of dusty material (Genzel et al. 1986).

- A galaxy in a group of galaxies or in a young cluster.

2.3 Compact Jets

2.3.1 Apparent Superluminal Motion

Consider a signal which is sent out from the core of an AGN at time $t = 0$, and then the emission region moves with velocity βc along a straight line at an angle θ with respect to the line of sight. Let it do so for a time Δt and before it emits another signal. The light travel time difference is then $\Delta t - \Delta t\,\beta\,\cos\theta$, and the apparent distance on the sky traveled is $\Delta t\,\beta\,c\,\sin\theta$, so that the apparent velocity

$$v_\perp = c\,\frac{\beta\,\sin\theta}{1 - \beta\,\cos\theta}. \tag{38}$$

This apparent velocity $v_\perp$ has a maximum at $\sin\theta_\star = 1/\gamma$, where $\gamma = \frac{1}{\sqrt{1-\beta^2}}$. We also note that

$$v_\perp/c = 1 \quad\text{at}\quad \cos\theta_\star = \frac{1}{2} \pm \sqrt{\frac{5}{4} - \frac{1}{\beta}} \tag{39}$$

so that apparent superluminal motion is observable over a very large range of angles. Since, however, we often find fairly large values for the apparent speed, there must also be a strong selection effect which tends to bias us towards those sources with axes very close to the line of sight.

2.3.2 Lorentz Boosting

Isotropic emission of flux density S_o from a region which moves with relativistic speeds is boosted in the observer's frame to produce an anisotropy given by

$$S(\nu) = S_0\,(\nu/D)\,D^3 = S_o(\nu)\,D^{3-\alpha} \tag{40a}$$

where the Doppler factor

$$D = \frac{1}{\gamma\,(1 - \beta\cos\theta)}. \tag{40b}$$

This describes the Lorentz transformation of isotropically emitted radiation which is optically thin and arises from a single "blob". If we are dealing with a "jet" consisting of a series of "blobs" of finite lifetime then the number of "blobs" seen at any one time scales inversely with the Doppler factor thus leading to boosting by $D^{2-\alpha}$. For $\cos\theta = 1, ..., -1$, and $\gamma \gg 1$ we find $D = 2\,\gamma, ..., 1/2\gamma$. There are many papers exploring the selection effects that follow from these arguments (see the review given by Kellermann and Pauliny-Toth 1981 and the entire book devoted to such questions edited by Zensus and Pearson 1987). With the derivation of $v_\perp$ and the definition of D it is easily possible to visualize what ranges γ and θ may have for given limits or determinations of $v_\perp$ and D. This leads to the equation for a circle

$$(v_\perp/c)^2 + (D - \gamma)^2 = \gamma^2 - 1. \tag{41}$$

The resulting graph is shown in, e.g., Cohen and Unwin (1984). The circle centres on $v_\perp = 0$ and $D = \gamma$ and has the radius $\sqrt{\gamma^2 - 1}$.

   If one detects $\beta_\perp$ at a certain value, then one can derive a lower limit for the aspect angle from

$$\cos\theta \ \geq\ \frac{1}{1+\beta_\perp^2}\left(\frac{\beta_\perp^2\,\gamma}{\sqrt{\gamma^2-1}}-\sqrt{1-\frac{\beta_\perp^2}{\gamma^2-1}}\right). \tag{42}$$

If one finds a lower limit for D, no further limit for $v_\perp/c$ can be derived, since, for $\theta \to 0$, $D \to 2\,\gamma$ for $\gamma \gg 1$ and $v_\perp/c \to 0$.

2.3.3 Statistics

Consider a sample of sources that emit blobs with speed βc in random directions (Zensus and Pearson 1987), then the fraction of objects with apparent speed $> \beta_{\mathrm{app}}$ is given by

$$P(>\beta_{\mathrm{app}}) \ = \ \frac{1}{2}\left(\cos\theta_2-\cos\theta_1\right) \ = \ \frac{\sqrt{1-\beta_{\mathrm{app}}^2/(\gamma^2-1)}}{1+\beta_{\mathrm{app}}^2}. \tag{43}$$

When β is not known, then

$$P(>\beta_{\mathrm{app}}) \ < \ \frac{1}{1+\beta_{\mathrm{app}}^2}. \tag{44}$$

That means, for instance that in a randomly oriented sample one cannot have more than 5.9% with $\beta_{\mathrm{app}} > 4$ if we are dealing with one-sided jets, and twice that for two-sided jets. Such arguments can be used to derive limits on the Hubble constant, since the apparent speed is clearly inversely proportional to the Hubble constant; this line of argument suggests a Hubble constant near 100 km s^{-1} Mpc^{-1}.

2.3.4 Unified Models

Some implications of unified models have already been discussed above in Sect. 1.6.2. The reader is referred to the most important paper in this field, Orr and Browne (1982). Recent work has concerned the shading produced by a thick dusty torus near the nucleus, which obscures the broad emission line region from the view of the observer. This leads him/her to classify the object differently thus giving rise to selection effects. Barthel (1989) demonstrates that for a sample of radio galaxies and radio quasars with similar luminosity and redshift range as the 3CRR, such a model is expected to give rise to obscuration within 45 degrees from the symmetry plane. This means, that the majority of objects will not be recognized as having broad emission lines, even if they have such a region. That such conclusions are important, was already implied by the earlier work of Lawrence and Elvis (1982), Mushotzky (1982), and Antonucci and Miller (1985).

2.3.5 The Compton Catastrophe

Consider a homogeneous spherical blob at the synchrotron emission frequency (Kellermann and Pauliny-Toth 1969) where it just becomes optically thick to self-absorption, then

$$L_{\mathrm{IC}} \ \approx\ R^3\,u_{\mathrm{ph}}\,N_{\mathrm{rel}}\,\gamma^2 \ \approx\ \frac{u_{\mathrm{ph}}}{B^2/8\,\pi}\,L_{\mathrm{syn}} \tag{45}$$

with an obvious notation for the Inverse Compton luminosity, the radius of the region being considered, the energy density of the photon field, the "average" Lorentz factor of the relativistic electron population, the magnetic field, and the synchrotron luminosity. It follows, that

$$\frac{L_{IC}}{L_{syn}} = \frac{u_{ph}}{B^2/8\,\pi}.$$ (46)

The optical thickness condition requires $R\,\kappa_\nu = 1$ so that

$$R\,B^{\frac{p+2}{2}}\,\nu^{-\frac{p+4}{2}}\,N_{rel} = \text{constant}.$$ (47)

We also know that

$$L_{syn} = 4\,\pi\,D^2\,S_\nu \approx R^3\,N_{rel}\,B^{\frac{p+1}{2}}\,\nu^{-\frac{p-1}{2}}$$ (48)

from which it follows, that

$$N_{rel} \propto \frac{D^2}{R^3}\,S_\nu\,B^{-\frac{p+1}{2}}\,\nu^{\frac{p-1}{2}}$$ (49)

which implies using the optical thickness condition that

$$\left(\frac{D}{R}\right)^2\,B^{\frac{1}{2}}\,S_\nu\,\nu^{-\frac{5}{2}} \approx \text{constant}$$ (50a)

or

$$B^2 \approx \nu^{10}\,S_\nu^{-4}\,\theta^8$$ (50b)

which in turn leads to

$$\frac{u_{ph}}{B^2} \approx \frac{4\,\pi\,D^2\,S_\nu\,\nu}{4\,\pi\,R^2\,c\,B^2} \propto \left(\frac{S_\nu}{\theta^2\,\nu^2}\right)^5\,\nu \propto T_b^5\,\nu$$ (51)

so that finally

$$\frac{L_{IC}}{L_{syn}} \propto T_b^5\,\nu.$$ (52)

I leave it to the reader to work out the numerical factors. For the GHz frequency range we obtain $\frac{L_{IC}}{L_{syn}} \simeq 1$ for $T_b \simeq 5 \times 10^{11}$ K. This was all for first order Compton losses. Analogously we obtain for the second order

$$\frac{L_{IC}(2)}{L_{IC}(1)} = \frac{u_{ph}(IC,1)}{u_{ph}(syn)} = \frac{L_{IC}(1)}{4\,\pi\,R^2\,c}\,\frac{4\,\pi\,R^2\,c}{L_{syn}} = \frac{L_{IC}(1)}{L_{syn}} = \left(\frac{T_b}{5\times10^{11}\,K}\right)^5\,\frac{\nu}{GHz}$$ (53)

Hence we have for the second order Compton luminosity

$$L_{IC}(2) = L_{IC}(1)\left(\frac{T_b}{5\,10^{11}\,K}\right)^5\,\frac{\nu}{GHz}$$

$$= L_{syn}\left(\left(\frac{T_b}{5\,10^{11}\,K}\right)^5\,\frac{\nu}{GHz}\right)^2 = L_{syn}\,x^2$$ (54)

with x denoting the quantity inside the outer bracket. Hence summing up all Compton orders we get

$$L_{\mathrm{IC}}(\text{all Compton orders}) \ = \ L_{\mathrm{syn}} \left(\frac{T_b}{5 \times 10^{11} \ \mathrm{K}} \right)^5 \frac{\nu}{\mathrm{GHz}} \ (1 + x + x^2 + x^3 + ...). \quad (55)$$

This series diverges for $x \geq 1$, so as soon as

$$\left(\frac{T_b}{5 \times 10^{11} \ \mathrm{K}} \right)^5 \frac{\nu}{\mathrm{GHz}} \ \simeq \ 1 \qquad\qquad (56)$$

Compton losses increase dramatically. It follows that $L_{\mathrm{IC}}/L_{\mathrm{syn}} \simeq 1$ should be a limiting case. For "plausible" physical parameters we find that the first Compton order occurs in the keV photon energy for synchrotron photons near 100 GHz, hence $\gamma^2 \sim 2 \times 10^6$. An important condition for the simple Compton collision is however that $\gamma \, h \nu \ll m_e \, c^2 \simeq 0.5$ MeV. As $\gamma \simeq 10^3$ and $h\nu \leq 10^3$ eV, the second Compton order is substantially weakened due to the decreased cross-section (the quantum-mechanical correction due to Klein and Nishina) and the changed collision kinematics. That means that the Compton catastrophe is much milder than implied above. The result, however, remains that near $T_b \simeq 10^{12}$ K the Compton losses increase rapidly. We note that $L_{\mathrm{IC}} \simeq L_{\mathrm{syn}}$ means that the 2-point spectral index between corresponding spectral ranges should be close to -1, and this is confirmed observationally (Owen et al. 1981).

It is also possible to derive from the observed values for flux density, frequency, and radio component size, an expected X-ray flux density from the Inverse Compton process. In many cases this leads to expected X-ray flux densities far in excess of what is observed. Once again, relativistic boosting helps to explain this since the physical parameters in the emitter frame comoving with the emitter are different from the values in the observer's frame. It is possible to use this argument to derive a lower limit to the boosting factor (Marscher 1983, for an application see Biermann et al. 1987).

2.3.6 Variability

In a similar manner to the way one can explain apparent superluminal motion one can demonstrate that the length scales derived from time variability (as no signal can move faster than the speed of light) are underestimated by γD, where γ is again the Lorentz factor of the bulk relativistic motion, and D is the Doppler factor. This is particularly important when the detailed geometry of an emitting region is considered (Hughes et al. 1989ab, Heeschen et al. 1987, Quirrenbach et al. 1989ab, Witzel 1990, Wagner et al. 1990, Quirrenbach et al. 1991, Quian et al. 1991, Krichbaum et al. 1992) and most of the recent data point to apparent brightness temperatures $T_b \simeq 10^{19}$ K.

Taken together with what we have said earlier, this means that a) apparent superluminal motion, b) violation of the Compton limit, and c) extreme radio variability, can all be understood in terms of bulk relativistic motion with Lorentz factors near around 10. Recent work by A. Witzel's group suggests, however, that this picture is still missing a vital ingredient of physics, since within it the newly discovered correlated variability in the optical and radio regimes is totally unexplained. This clearly is a major test of our physical models for the variability of compact sources.

2.3.7 Standard Model

The most simple-minded and highly successful model was proposed by Blandford and Königl (1979) which we will now outline consider a conical jet with emissivity

$$\epsilon_\nu \propto N_{\rm rel}\, B^{3/2}\, \nu^{-1/2}, \tag{57a}$$

opacity

$$\kappa_\nu \propto N_{\rm rel}\, B^2\, \nu^{-3}, \tag{57b}$$

an electron energy distribution

$$N_{\rm rel}\, \gamma^{-2}, \tag{57c}$$

with

$$N_{\rm rel} \propto B^2 \propto r^{-2}, \tag{57d}$$

where r is the distance along the jet, or to within the factor of an angle θ, the lateral radius of the jet. At a given frequency ν there is a distance $r_\star$ where the emission becomes optically thick. Hence we have

$$S_\nu = \int_{r_\star}^{\infty} \pi \left(\frac{\theta}{2}\, r\right)^2 \epsilon_\nu\, dr \propto \int_{r_\star}^{\infty} r^2\, N_{\rm rel}\, B^{3/2}\, \nu^{-1/2}\, dr$$

$$\propto \int_{r_\star}^{\infty} r^{-3/2}\, \nu^{-1/2}\, dr \propto \left(r_\star\, \nu\right)^{-1/2}. \tag{58}$$

We also have

$$r_\star\, \kappa_\nu \propto r_\star\, N_{\rm rel}\, B^2\, \nu^{-3} \propto \left(r_\star\, \nu\right)^{-3} = {\rm constant}. \tag{59}$$

Hence we conclude, that $r_\star\, \nu = {\rm const}$ and so $S_\nu = {\rm const}$. We must also have

$$\theta = r_\star/D \sim 1/\nu, \tag{60}$$

where D is the distance, and so we find for the brightness temperature, that

$$T_b \propto \frac{S_\nu}{(\theta\, \nu)^2} \sim {\rm const}. \tag{61}$$

It follows that the emitted radio spectrum is flat, and that the brightness temperature is constant along the jet. At higher radio frequencies more compact sections of the jet become visible, but the "real" core is never seen in this model.

Since the observations show a clear correlation between flat radio spectra, compact structure, and bulk relativistic motion (Witzel et al. 1988), this standard model has served as a baseline to the present day. The new data (e.g. Krichbaum 1990), however, demonstrate that we have yet to consider moving as well as standing knots (shocks?), helical geometry, and yet perhaps even other physical concepts (see the above section on variability).

2.4 Suggested Reading List

1) Blandford and Königl (1979)
2) Kazanas (1989)
3) Norman and Scoville (1988)
4) Scoville and Norman (1988)
5) Kaastra and Barr (1989)
6) Camenzind (1990)

3 Shock Wave Acceleration in Jets and Hot Spots

3.1 The Basic Theory

In the following we introduce the arguments presented by Biermann and Strittmatter (1987), and in subsequent papers. Observations have shown convincingly from about 1976 (Rieke et al. 1976, Rieke et al. 1979, Rieke and Lebofsky 1980, Bregman et al. 1981, Rieke et al. 1982, Brodie et al. 1983, Röser and Meisenheimer 1986, 1987) that many extragalactic sources have a sharp cut-off $\sim 3 \times 10^{14}$ Hz in their non-thermal emission for both compact and extended sources. The first question that comes to mind is whether this might be due to selection effects. Conceivably this could be so for hot spots, where the data are too sparse to check for selection effects. On the other hand, Pic A might be a case where the cut-off appears to be at a higher frequency. From minimum energy arguments (see above) one can estimate that, at the cut-off frequency of the M87 jet, the loss time scale for relativistic electrons is $\sim$ 200 years, noticeably shorter than even the light travel time from the nucleus. This assumes, of course, an isotropic pitch angle distribution for the relativistic electron population and so we conclude, that local acceleration is required. Shockwaves in a plasma can accelerate particles in a process commonly referred to as 1st order Fermi acceleration, or diffusive shock acceleration (see, e.g., Drury 1983 for a comprehensive review). The process is based on repeated reflections of the energetic particles on the two sides of the shock, caused by irregularities in the magnetic field; since the two sides of a shock provide permanent compression with respect to each other, there is an energy gain. The shock density ratio r is related to the spectral index q of the particle number distribution in energy by

$$q = \frac{r+2}{r-1} \tag{62}$$

and the spectral index α of the resulting synchrotron emission from the electrons is then given by

$$\alpha = -\frac{q-1}{2}. \tag{63}$$

Assuming a standard equation of state with an adiabatic index of 5/3, the maximum shock density ratio is 4, so that the particle spectral index is 2, and the synchrotron emission spectral index is -0.5. For a relativistic equation of state the numbers are 4/3 for the adiabatic index, -1.5 for the particle distribution, and -0.25 for the synchrotron radiation spectral index. A possible shock configuration is a biconical shock in a jet flow, with or without a central Mach disk; if there is a Mach disk, then a slip contact discontinuity follows from the triple point of the three shocks (converging shock, Mach disk, diverging shock - all in rotational symmetry). We consider then the flow through one shock, and disregard for now any obliqueness (but see Jokipii 1987), i.e. assume that the normal to the shock is parallel to the flow. We will also assume that we have an isotropic phase-space distribution of the energetic particles, both protons (and heavier nuclei) and electrons. The mean free path for scattering is then given by

$$\lambda = r_g \frac{B^2/8\,\pi}{I(k)\,k}$$ (64)

where r_g is the gyration radius of a particle, and k is the resonant Alfven-wavenumber of the turbulent wavefield of the magnetic field, with $I(k)$ the energy in this field per wavenumber. The gyro-radius r_g is obviously given by

$$r_g = \frac{E}{e\,B},$$ (65)

where E is the energy of the particle.

Consider then a particle upstream and its Lorentz transformation as it moves down and back up. Upstream the particle has energy, momentum and pitch angle E_1, p_1, μ_1 respectively. Downstream the Lorentz transformation yields

$$E_2 = \gamma\,(E_1 + \beta c\,p_1\,\mu_1).$$ (66)

The probability of a transition through the shock per unit surface for an isotropic pitch angle distribution implies that the average value of the pitch angle $\langle \mu_1 \rangle = \frac{2}{3}$, and for highly relativistic particles we have $E_1 \cong p_1\,c$, so

$$E_2 = \gamma\,E_1\,(1 + \frac{2}{3}\,\beta).$$ (67)

Then the Lorentz transformation back upstream yields

$$E_3 = \gamma\,E_2\,(1 + \frac{2}{3}\,\beta)$$ (68)

since the minus sign occurs twice. After each cycle the energy increases by

$$\gamma^2\,(1 + \frac{2}{3}\,\beta)^2$$ (69)

which for small β is equal to $1 + \frac{4}{3}\,\beta$. The energy in the magnetic field per unit wavelength due to plasma-turbulence assuming no dissipation is usually found to be

$$I(k) = I_o/k_o\,(k/k_o)^{-x}, \text{ with } x = 5/3$$ (70)

i.e. it follows Kolmogorov's law (Kolmogorov 1941). This is seen in the case of solar wind turbulence (Matthaeus et al. 1982, Matthaeus and Goldstein 1982, Goldstein et al. 1984, Burlaga et al. 1987), and is supported by many other arguments regarding interstellar turbulence or new numerical experiments. We then define

$$b = \left(\int_{k_o}^{\infty} I(k)\,dk \right) / (B^2/8\,\pi).$$ (71)

The scattering coefficient can be written as

$$\kappa \cong \frac{1}{3} \frac{\lambda^2}{\tau_s} \, , \text{with } \tau_s \cong \frac{\lambda}{c} \tag{72}$$

which leads to an acceleration time (Drury 1983) of

$$\tau_{\text{acc}} = \frac{E}{\dot{E}} = \frac{80}{3 \, \pi} \left(\frac{c}{U_1^2} \right) \left(\frac{r_g}{b \, (x - 1)} \right) \left(\frac{r_{g,\text{max}}}{r_g} \right)^{x-1} \tag{73}$$

where U_1 is the shock speed. The concept is then, that protons (and heavier nuclei) reach high energies since they experience low losses, while electrons, with high losses, only reach moderate energies. Hence the protons are the first "messengers" far upstream to communicate to the flow that a shock is coming. Here they excite the turbulence, which then provides the wavefield for the electrons to scatter. The synchrotron losses for the protons and electrons, respectively, are given by

$$\tau_{p,\text{syn}} = \frac{6 \, \pi \, m_p^3 \, c}{\sigma_T \, m_e^2 \, \gamma_p \, B^2} \tag{74a}$$

$$\tau_{e,\text{syn}} = \frac{6 \, \pi \, m_e \, c}{\sigma_T \, \gamma_e \, B^2}. \tag{74b}$$

The energies of those particles which remain near the shock can increase until the acceleration time scale becomes equal to the loss time scale, $\tau_{\text{acc}} = \tau_{\text{syn}}$ and since the acceleration time scale increases with energy of the particle, and the loss time scale decreases with energy, this is bound to happen at a critical cut-off energy, which is, expressed as a Lorentz factor

$$\gamma_{p,\text{max}} = \left(\frac{27 \, \pi}{320} \, b \, (x - 1) \, \frac{e}{r_o^2 \, B} \right)^{1/2} \frac{U_1}{c} \frac{m_p}{m_e} \tag{75a}$$

$$\gamma_{e,\text{max}} = \left(\frac{27 \, \pi}{320} \, b \, (x - 1) \, \frac{e}{r_o^2 \, B} \right)^{1/2} \frac{U_1}{c} \left(\frac{m_e}{m_p} \right)^{\frac{2 \, (x-1)}{3-x}}. \tag{75b}$$

Here r_o is the classical electron radius. This implies a cut-off frequency for the synchrotron emission of

$$\nu_\star = \left(\frac{81 \, \pi}{5120} \, b \, (x - 1) \right) \left(\frac{U_1}{c} \right)^2 \left(\frac{m_e}{m_p} \right)^{\frac{4 \, (x-1)}{3-x}}. \tag{76}$$

With $x = 5/3$, we then find

$$\nu_\star = 3 \times 10^{14} \left(3 \, b \, \left(\frac{U_1}{c} \right)^2 \right) \text{ Hz.} \tag{77}$$

Expressed in terms of natural constants, the cut-off frequency is given by $\frac{c}{r_o} \left(\frac{m_e}{m_p} \right)^2$.

Given the definition of b, we would expect that the term $\left(3 \, b \, \left(\frac{U_1}{c} \right)^2 \right)$ is less than one, and so this cut-off frequency is an upper limit to any actual cut-off frequency, since obviously we normally do not know the level of the turbulence or the real shock speed.

3.2 Photon Interaction

Obviously it is of interest to consider photon losses, for which Stecker (1968) has provided the basic theoretical framework. The proton loss time scale for interactions with photons can be written as

$$\frac{1}{\tau_{p\gamma}} = \int_{\epsilon_{\mathrm{thr}}/2\,\gamma_p}^{\infty} d\epsilon\, n(\epsilon)\, \frac{c}{2\,\gamma_p^2\,\epsilon^2} \int_{\epsilon_{\mathrm{thr}}}^{2\,\gamma_p\,\epsilon} k_p(\epsilon')\, \sigma(\epsilon')\, \epsilon'\, d\epsilon' \tag{78}$$

with an obvious notation. Here the inner integral is done in the reference system of the proton. Cross-sections have been published by the Cambridge Bubble Chamber Group (1967). Observations suggest that in AGN environments the photon field (Bezler et al. 1984, Ballmoos et al. 1987) can be crudely described as

$$n(\epsilon) = \frac{N_o}{\epsilon_o}\, (\epsilon/\epsilon_o)^{-2} \text{ for } \epsilon_o < \epsilon < \epsilon^\star, \text{ and } = 0 \text{ otherwise.} \tag{79}$$

The different channels to consider are the following

$$\gamma + p \rightarrow p + \pi^o \rightarrow p + 2\,\gamma$$

$$\gamma + p \rightarrow n + \pi^+ \rightarrow p + e^- + \bar{\nu}_e + e^+ + \nu_e + \bar{\nu}_\mu + \nu_\mu$$

$$\gamma + p \rightarrow p + \pi^+ + \pi^- \rightarrow p + e^+ + \nu_e + 2\,\bar{\nu}_\mu + e^- + \bar{\nu}_e + 2\,\nu_\mu \tag{80}$$

and so on. We will consider below how we produce a large fraction of neutrinos in the resulting particle population. In this case the proton energy loss time scale can be expressed in a fashion similar to the preceding section:

$$\frac{1}{\tau_{p\gamma}} = \frac{1}{6\,\pi}\, \frac{B^2}{\ln(\epsilon^\star/\epsilon_o)}\, \frac{\bar{\sigma}_{\gamma p}}{m_p\, c}\, \gamma_p\, a$$

$$\text{with } a = \frac{N_o\, \epsilon_o\, \ln(\epsilon^\star/\epsilon_o)}{B^2/8\,\pi} \tag{81}$$

that is the ratio of the photon energy density to the magnetic field energy density. Hence the loss time scale can be written in the following form

$$\frac{1}{\tau_p} = \frac{1}{\tau_{p,\mathrm{syn}}} + \frac{1}{\tau_{p\gamma}} = \frac{1}{\tau_{p,\mathrm{syn}}}\, (1 + Aa)$$

$$\text{with } A = \frac{\bar{\sigma}_{\gamma p}}{\sigma_T}\, \frac{(m_p/m_e)^2}{\ln(\epsilon^\star/\epsilon_o)} \cong 200. \tag{82}$$

Analogously, we have for the electrons

$$\frac{1}{\tau_e} = \frac{1}{\tau_{e,\mathrm{syn}}}\, (1 + a) \tag{83}$$

and so we have finally for the upper limit to the cut-off frequency

$$\nu^\star = 3 \times 10^{14}\, \left(3\,b\, \left(\frac{U_1}{c}\right)^2\right)\, f(a) \text{ Hz}$$

$$\text{with } f(a) = \frac{(1 + Aa)^{1/2}}{(1 + a)^{3/2}}. \tag{84}$$

We note that the function $f(a)$ has the following limiting behaviour

$$f(a) \cong 1 \text{ for } Aa \ll 1$$

$$\cong 5.4 \text{ for } a \cong 0.5 \text{ its maximum}$$

$$\cong A^{1/2}/a \text{ for } Aa \gg 1. \tag{85}$$

Observations provide a number of consistency checks on this picture (discussed in Biermann and Strittmatter 1987 and subsequent papers):

1) Length scales: the gyroradii of the highest energy particles have to fit into the space provided, e.g. the scale of the hot spots, or the radius of the M87 jet. This is the case.

2) Time scales: sources should not vary faster than the acceleration and loss times (Sulentic et al. 1979, Warren-Smith et al. 1984) as is observed.

3) Secondary electron-positron production from p-p collisions should not dominate at infrared and optical frequencies. This was explored by Klemens (1987a, b), who demonstrated that these secondaries naturally produce synchrotron emission in X-rays and so could easily serve to explain the X-ray emission of the M87 jet.

4) The shape of the cut-off should be sufficiently sharp. This was explored first in Webb et al. (1984) and then in some detail by Fritz (1989), who showed that the cut-offs can be very sharp indeed.

5) There should be a correlation of the cut-off frequency with radiation intensity, which switches at $a \simeq 0.5$, from

$$\text{for } \leq 0.5 \ L \nearrow a \nearrow \nu_{cut} \nearrow$$

$$\text{for } \geq 0.5 \ L \nearrow a \nearrow \nu_{cut} \searrow \tag{86}$$

as is consistent with the data in a number of compact sources.

There are several additional consequences of this model:

a) All sources with such a cut-off near 3×10^{14}Hz in their emission spectra - if they can be understood with this model - have nearly relativistic flow. This includes certainly all of the flat spectrum sample chosen from the S5-survey (Witzel et al. 1988) and probably all edge-bright radio galaxies.

b) Because of momentum transfer across the walls of the jet, one might expect that the flow speed slows down along the jet, and likewise the shock velocities - obviously, if the obliqueness of the shocks and/or their pattern speed changes, then conditions might be different - reducing $\nu_\star$. There is some evidence to support this (Perez-Fournon et al. 1988).

c) The time scale for variability in jets and hot spots might be as short as years; in the M87 jet there is an observational claim of such variability.

d) Jets have an electron/proton plasma.

e) Hot spots are the sources of extremely high energy cosmic rays (see also Biermann 1991). A recent careful calculation (Rachen 1991) demonstrates that indeed the Fanaroff-Riley class II radio galaxies can account for the flux and spectrum, as well as the energy range of the ultra-high energy cosmic rays.

3.3 Highly Energetic Photons and Neutrinos

Let us consider the opacity with respect to pair creation over the lengthscale $r_{g,p,\max}$ and calculate the associated optical depth τ:

$$\tau = 5 \times 10^{-4} \, b^{1/2} \frac{U_1}{c} \frac{a}{(1 + Aa)^{1/2}} \frac{B^{1/2}}{\ln(\epsilon_*/\epsilon_o)} \frac{E}{m_e \, c^2}. \tag{87}$$

Note that for $\tau \geq 1$ the region is not transparent. As an example we will consider the parameters

$$a = 1, \; \epsilon_*/\epsilon_o = 10^9, \; b = 1, \; U_1/c = 1/\sqrt{3},$$

$$\Rightarrow \tau = 10^{-6} \, B^{1/2} \frac{E}{m_e \, c^2} \tag{88}$$

where B is measured in Gauss. We will consider three examples for the value of the magnetic field, and in each case indicate the maximum energy of the photons produced, the optical depth at that maximum energy, and the photon energy at which the optical depth is unity:

$$B = 10^{-4} : \pi^o = 6 \times 10^{21} \text{ eV} : \tau = 10^8 : \tau = 1 \text{ at } 5 \times 10^{13} \text{ eV}$$

$$B = 1 : \pi^o = 6 \times 10^{19} eV : \tau = 10^8 : \tau = 1 \text{ at } 5 \times 10^{11} \text{ eV}$$

$$B = 10^4 : \pi^o = 6 \times 10^{17} \text{ eV} : \tau = 10^8 : \tau = 1 \text{ at } 5 \times 10^9 \text{ eV}. \tag{89}$$

It follows that these processes can only be directly observed through neutrinos.

3.4 The Neutrino Luminosity

In order to obtain the neutrino luminosity, we consider a shock in a plasma flow, calculate the emissivity per volume and then integrate downstream along a streamline. The proton energy loss rate is

$$\frac{d\gamma_p}{dt} = -\frac{\sigma_T \, m_e^2 \, \gamma_p^2 \, B^2 \, (1 + Aa)}{6 \, \pi \, m_p^3 \, c}$$

with

$$A = \frac{\bar{\sigma}_{p\gamma}}{\sigma_T} \frac{(m_p/m_e)^2}{\ln(\epsilon_*/\epsilon_o)} \simeq 200. \tag{90}$$

This leads to a spatial dependence of the maximum value of the Lorentz factor

$$\gamma_{p,\max} = \frac{\gamma_p(0)}{1 + \gamma_p(0) \frac{\sigma_T \, m_e^2 \, \gamma_p^2 \, B^2 \, (1 + Aa)}{6 \, \pi \, m_p^3 \, c} \frac{x}{v}} \tag{91}$$

where $\gamma_p(0)$ is $\gamma_{p,\max}$ at $x = 0$ our starting point and v is the velocity. We then have to consider the distribution function of the protons, $K_p \, \gamma_p^{-2} \, d\gamma_p$, the energy of each proton, $\gamma_p \, m_p \, c^2$, the loss rate into photon interactions, $\frac{4}{3} N_o \, c \, \frac{\epsilon_o}{m_\pi c^2} \gamma_p \bar{\sigma}_{\gamma p}$, the fraction that goes into neutrinos, of order 3/4, then integrate over γ_p, and put

$$K_p \, m_p \, c^2 \, \ln \gamma_p(0) = a_p \frac{B^2}{8 \, \pi}, \tag{92}$$

to derive a neutrino luminosity of

$$L_\nu = \frac{3}{4} a_p \frac{B^2}{8\,\pi} \frac{m_p}{m_\pi} v \frac{\ln(1 + x^m/x_o^p)}{\ln(\gamma_p(0))} \frac{aA}{1 + Aa} \ \text{erg s}^{-1} \ \text{cm}^{-2} \tag{93}$$

where we have integrated along a cross-section of unity and used an obvious notation for x. Correspondingly the synchrotron luminosity of the relativistic electrons is

$$L_{\text{syn}} = a_e \frac{B^2}{8\,\pi} v \frac{\ln(1 + x^m/x_o^e)}{\ln(\gamma_e(0))} \frac{1}{1 + a} \tag{94}$$

The ratio then is

$$\frac{L_\nu}{L_{\text{syn}}} = 5 \frac{a_p}{a_e} \frac{\ln(1 + x^m/x_o^p)}{\ln(1 + x^m/x_o^e)} \frac{\ln(\gamma_e(0))}{\ln(\gamma_p(0))} \frac{(1 + a)aA}{1 + Aa} \tag{95}$$

As an example we will use $B = 1$ gauss, $v = 0.1\,c$, $\frac{U_1}{c} = \frac{1}{\sqrt{3}}$, $a = 1$, and $x_m = 3\,x_o^p$ which means $\gamma_e(0) = 2 \times 10^4$, $\gamma_p(0) = 4 \times 10^8$, $x_o^e = 5 \times 10^{13}$ cm, $x_o^p = 3 \times 10^{16}$ cm and so we finally obtain

$$\frac{L_\nu}{L_{\text{syn}}} = 0.92 \frac{a_p}{a_e} \simeq \frac{a_p}{a_e} \tag{96}$$

This ratio a_p/a_e is about 100 in our galaxy, in active galactic nuclei it might well be much lower, perhaps close to unity. One should also consider Inverse Compton, π^o photons, $\pi^\pm$, which yield pairs and hence synchrotron radiation, $\gamma\gamma$ pair creation, and further processes, all the way to hadronic cascades (see, e.g., Mannheim and Biermann 1989, Mannheim et al. 1991).

3.4.1 The Expected Neutrino Flux

Let us suppose the proton energy is E_p and the photon energy is ϵ in the observer's frame. In the proton's frame we have, for $\gamma = E_p/m_p\,c^2$, the photon energy $\epsilon' = \gamma\,\epsilon\,(1 - \beta\cos\theta)$. In the centre of mass system (cm) we then have $E_p^\star$ and $\epsilon^\star$ and it follows that

$$(E_p + \epsilon)^2 - (\epsilon\cos\theta + \beta E_p)^2 - \epsilon^2\sin^2\theta = (E_p^\star + \epsilon^\star)^2 =$$

$$s = (m_p\,c^2 + \epsilon')^2 - \epsilon'^2 = m_p^2\,c^4 + 2\,m_p\,c^2\,\epsilon'. \tag{97}$$

The Lorentz factor of the cm-system as opposed to the lab system is

$$\gamma_{\text{cm}} = E_p/(e_p^\star + E^\star). \tag{98}$$

The creation of a pion in the cm-system then means

$$E_\pi^\star = (s + m_\pi^2\,c^4 - m_p^2\,c^4)/2\sqrt{s}. \tag{99}$$

Since the direction is arbitrary in the cm-system, it follows that

$$\langle E_\pi \rangle = \gamma_{\text{cm}} E_\pi^\star = \frac{1}{2} E_p \left(1 - \frac{m_p^2\,c^4 - m_\pi^2\,c^4}{s}\right) \tag{100}$$

and in the lab-system

$$\langle E_\nu \rangle \simeq \frac{1}{4} \langle E_\pi \rangle = \frac{1}{4} \gamma_p \, m_\pi \, c^2 \, \frac{f + \frac{1}{2} m_\pi/m_p}{1 + 2 f \, m_\pi/m_p} \simeq \frac{1}{4} \gamma_p \, m_\pi \, c^2$$

$$\text{with } \epsilon' = f \, m_\pi \, c^2 \text{ and } f \geq 1 \qquad (101)$$

The spectrum of the neutrinos can be calculated similarly (Biermann 1989), and one finds typically a spectrum of E_ν^{-2} in numbers and E_ν^{-1} in neutrino flux density. As an example we consider the quasar 3C273, with the assumed parameters $a \simeq 1$, $a_p/a_e \simeq 1$ to obtain a flux of $4 \times 10^{-10} \, E_\nu^{-2} \, \mathrm{cm}^{-2} \, \mathrm{s}^{-1}$ where the unit of energy is in ergs ($\sim$ TeV). A complete sample of flat spectrum radio sources (northern sky, 5 GHz survey, down to 1 Jy) all added up yields a flux $3 \times 10^9 \, E_\nu^{-2} \, \mathrm{km}^{-2} \, \mathrm{yr}^{-1}$ The consequences from this and the previous section are then, that the neutrino luminosity could be high, and similar to the total non-thermal luminosity of a source and this leads to interesting further questions in high energy particle collision physics. Another new study of the neutrino emission from active galactic nuclei was made by Stecker et al. (1990).

3.5 UHE Cosmic Ray Sources

We will demonstrate in this section that the hot spots of Fanaroff-Riley II radio galaxies can account for particle energy, spectrum and flux of the cosmic rays above about 3×10^{18} eV. Using the radio galaxy counts by Peacock (1985) and Windhorst (1984) we can first calculate the input in synchrotron power averaged over all local space to obtain $1.1 \times 10^{-37} \, \mathrm{erg} \, \mathrm{cm}^{-3} \, \mathrm{s}^{-1}$. From Meisenheimer et al. (1989) and Rawlings and Sanders (1991) we can estimate the energy flow through the working surface - this involves various fudge factors which we assume all to average out to $(1 - 25) \, L_{\mathrm{syn}}$, which leads to a power input for cosmic ray acceleration of $3 \times 10^{-36} \, \mathrm{erg} \, \mathrm{cm}^{-3} \, \mathrm{s}^{-1}$. Hot spot spectra demonstrate that the relativistic particles have an energy spectrum $\propto E^{-2.00 \pm 0.14}$ (strictly speaking, they demonstrate this only for electrons, and we assume it for the protons). The confinement scale near the cut-off for the interaction with microwave background photons can be estimated as 3×10^{27} cm to yield an integral flux of $1.2 \times 10^{-18} \, \mathrm{cm}^{-2} \, \mathrm{s}^{-1} \, \mathrm{ster}^{-1}$ with a spectrum of $\sim E^{-1.3}$ and even flatter below 10^{18} eV. Careful calculation by Rachen (1991) taking into account all known cosmological evolution for radio galaxies refines this argument considerably, and yields the same basic result.

3.6 Relativistic Particles and the Broad Line Region

First we note that non-thermal radio emission, that is synchrotron emission from a population of relativistic electrons, appears to exist in most radio-weak quasars, in all Seyfert galaxies (e.g. Edelson and Malkan 1986), and, obviously, in radio-loud AGN. The spectrum is flat in a surprisingly large fraction of these sources, suggesting that a compact jet is present in many or possibly all cases. This suggests, without knowledge of their energetic importance, that there is a relativistic particle population in AGN. We conclude that all emission from radio-weak active galactic nuclei is consistent with an interpretation as thermal emission except radio and X-rays (for which we do not know). The ionization and heating of the broad line clouds can, at least in principle, be provided by the inner disk emission, the UV-X-ray bump (Arnaud et al. 1985). The heating of the far-infrared disk detected by Chini et al. (1989ab) and later by Sanders et al. (1989) is difficult to do with X-rays or the UV, although this is not yet entirely ruled out. Other galaxies, say M82 and our Galaxy - which are not known to contain an active nucleus - demonstrate over a large range of energy densities, that the energy density of relativistic particles and the other forms of energy in the interstellar medium are of the same order of magnitude. This leads then to the question: to what degree can relativistic particles (protons, nuclei, electrons, secondaries, neutrinos, et cetera) influence the energetics of the nuclear regions of AGNs? Here we concentrate on the effect of neutrinos: high energy protons and nuclei produce secondaries, and among these, neutrinos, in their collisions with photons and thermal matter. Since the cross- section of neutrinos with matter rises with energy linearly until about 3.5 TeV (Gaisser and Stanev 1985), we ask whether there are any objects that are optically thick to neutrinos of such energy and, in fact, stars, all along the main sequence, are optically thick. An optical thickness of one corresponds to a column density of about 3×10^{11} g cm^{-2}. Thomas (priv. comm.) has checked a sequence of stellar models from 1 to 30 solar masses along the main sequence and finds all of them optically thick both through the centre and also through a direction 20% off centre (measured in the star's radius). In MacDonald et al. (1991) we explored the consequences of this for stellar structure, using the UV-fluxes given by Gondhalekhar (1990) as a reference for energy fluxes, i.e. $10^{11\pm2}$ erg cm^{-2} s^{-1}. We calculated generalized zero-age main sequences for the star of mass 0.25, 0.5, 0.8, 1.0 M$_\odot$ for a range of neutrino fluxes. The results demonstrate that low mass stars expand due to the additional energy input, then switch over to a neutrino-supported structure with increasing neutrino flux, and at that moment move into the red giant region of the HR diagram. There, of course, their envelopes are more loosely bound than on the main sequence and so, a stellar wind can be generated by the relativistic nuclei population and hard X-rays. The stellar wind is dominated by such stellar mass loss. In the HR diagram the main sequence is peeled off from below, since low mass stars "go first". This then can be combined with the concept that ionization of stellar winds can provide the broad emission lines seen in AGN.

3.7 Suggested Reading

1) Biermann (1988)
2) Meisenheimer et al. (1989)
3) Biermann (1991)
4) Witzel and Quirrenbach (1990)
5) MacDonald et al. (1991)

4 Some Relevant Books

1) *Astrophysics of Active Galaxies and Quasistellar Objects*
ed. J.S. Miller, Univ. Science Books, 1985
2) *Astrophysical Radiation Hydrodynamics*
eds. K.H. Winkler and M. Norman, Reidel Publ. Comp., 1986
3) IAU Symposium 119 *Quasars*
eds. G. Swarup and V.K. Kapahi, Reidel Publ. Comp., 1986
4) *Superluminal Radio Sources*
eds. J.A. Zensus and T.J. Pearson, Cambridge Univ. Press, 1987

5) *Hot Spots in Extragalactic Radio Sources*
eds. K. Meisenheimer and H.-J. Röser, Springer Publ. Comp., 1989
6) IAU Symposium 134 *Active Galactic Nuclei*
ed. D.E. Osterbrock, J.S. Miller, Kluwer Publ. Comp., 1989
7) *Parsec Scale Radio Jets*
eds. J.A. Zensus and T.J. Pearson, Cambridge Univ. Press, 1990
8) *Beams and Jets in Astrophysics*
ed. P.A. Hughes, Cambridge Univ. Press, 1991
9) *Variability of Active Galaxies*
eds. W.J. Duschl, S.J. Wagner, M. Camenzind, Springer Publ. Comp. 1991

Acknowledgements: My students, collaborators and colleagues M. Diewald, H. Falcke, Dr. W. Krülls, S. Linden, K. Mannheim, M. Niemeyer, J. Rachen, Dr. R. Schaaf all have helped to put this manuscript together from my sparse lecture notes and to teach me the use of TEX; I thank them all for their generosity with their time and effort.

References

Alberdi, A., Marcaide, J.M., Elosegui, P., Schalinski, C.J., Witzel, A. (1988): "VLBI monitoring of the milliarcsecond structure of 4C39.25 at 2.8 and 3.6cm", in *Proc. 7th Workshop Meeting on European VLBI*, p. 154

Antonucci, R.R.J., Miller, J.S. (1985): "Spectropolarimetry and nature of NGC1068", *Astrophys. J.*, **297**, 621

Arnaud, K.A., Branduardi-Raymont, G., Culhane, J.L., Fabian, A.C., Hazard, C., McGlynn, T.A., Shafer, R.A., Tennant, A.F., Ward, M.J. (1985): "EXOSAT observations of a strong soft X-ray excess in MKN841", *Mon. Not. R. astr. Soc.*, **217**, 105

Ballmoos, P.V., Diehl, R.,, Schönfelder, V. (1987): "Centaurus A observations at MeV-gamma-ray energies", *Astrophys. J.*, **312**, 134

Barthel, P.D. (1989): "Is every quasar beamed?", *Astrophys. J.*, **336**, 606

Begelman, M.C., Rees, M.J., Blandford, R.D. (1979): "A twin-jet model for radio trails", *Nature*, **279**, 770

Bezler, M., Kendziorra, E., Staubert, R., Hasinger, G., Pietsch, W., Reppin, C., Trümper, J., Voges, W. (1984): "The high energy X-ray spectrum of 3C273", *Astron. Astrophys.*, **136**, 351

Biermann, P.L. (1989): "Lectures on photon and neutrino emission from shockwaves in active galactic nuclei", in *Proc. Erice Conference*, eds. M.M. Shapiro, J.F. Wefel, Kluwer Acad. Publ., p. 21

Biermann, P.L. (1991): "Highly luminous radio galaxies as sources of Cosmic Rays", in *Frontiers in Astrophysics*, eds. Silberberg, G. and Fazio, Cambridge Univ. Press, (in press)

Biermann, P.L., Kronberg, P.P., Schmutzler, T. (1989): "Extended X-ray emission from hot gas around the normal giant elliptical galaxy NGC5846", *Astron. Astrophys.*, **208**, 22

Biermann, P.L., Kühr, H., Snyder, W.A., Zensus, J.A. (1987): "The inverse Compton test for a large sample of compact radio sources", *Astron. Astrophys.*, **185**, 9

Biermann, P.L., Strittmatter, P.A. (1987): " Synchrotron emission from shockwaves in active galactic nuclei", *Astrophys. J.*, **322**, 643

Blandford, R.D., Königl, A. (1979): "Relativistic jets as compact radio sources", *Astrophys. J.*, **232**, 34

Bregman, J.N., Lebofsky, M.J., Aller, M.F., Rieke, G.H., Aller, H.D., Hodge, P.E., Glassgold, A.E., Huggins, P.J. (1981): "Multifrequency observations of the red QSO 1413+135", *Nature*, **293**, 714

Bridle, A. (1989): in *Parsec Scale Jets*, eds. J.A. Zensus and T.J. Pearson, Cambridge Univ. Press, 1990

Bridle, A.H., Perley, R.A. (1984): "Extragalactic Radio Jets", *Ann. Rev. Astron. Astrophys.*, **22**, 319

Brodie, J., Königl, A., Bowyer, S. (1983): "The discovery of optical emission knots in the inner jet of Cen A", *Astrophys. J.*, **273**, 154

Burlaga, L.F., Mish, W.H. (1987): *J. Geophys. Res. (A)*, **92**, 1261

Burns et al. (1983): "The inner radio structure of Cen A", *Astrophys. J.*, **273**, 128

Cambridge Bubble Chamber Group (1967): "Analysis of $\gamma - p$ reactions in a hydrogen bubble chamber to 6.0 BeV: cross sections and laboratory distributions", *Phys. Rev.*, **155**, 1477

Camenzind, M. (1986): "Hydromagnetic flows from rapidly rotating compact objects I. Cold relativistic flows from rapid rotators", *Astron. Astrophys.*, **162**, 32

Camenzind, M. (1987): "Hydromagnetic flows from rapidly rotating compact objects II. The relativistic axisymmetric jet equilibrium", *Astron. Astrophys.*, **183**, 341

Camenzind, M. (1990): "Magnetized disk-winds and the origin of bipolar outflows", *Rev. Mod. Astron.*, **3**, 234

Chini, R., Kreysa, E., Biermann, P.L. (1989a): "On the nature of radio quiet quasars", *Astron. Astrophys.*, **219**, 87

Chini, R., Biermann, P.L., Kreysa, E., Gemünd, H.-P. (1989b): "870 and 1300μm observations of radio quasars", *Astron. Astrophys. (Lett.)*, **221**, L3

Cohen, M.H. Unwin, S.C. (1984): "Superluminal effects and bulk relativistic motion", in IAU Symposium vol. 110, p. 95

Cohn, H. (1983): "The stability of a magnetically confined radio jet", *Astrophys. J.*, **269**, 500

Czerny, B., Elvis, M. (1987): "Constraints on quasar accretion disks from the opt/UV/soft X-ray big bump", *Astrophys. J.*, **321**, 305

Dhawan, V., Bartel, N., Rogers, A.E.E., Krichbaum, T.P., Witzel, A., Graham, D.A., Pauliny-Toth, I.I.K., Rönnäng, B.O., Hirabayashi, H., Inoue, M., Lawrence, C.R., Shapiro, I.I., Burke, B.F., Booth, R.S., Readhead, A.C.S., Morimoto, M., Johnston, K.J., Spencer, J.H., Marcaide, J.M. (1990): "Further Millimeter VLBI observations of 3C84 and other sources with 100 microarcsecond angular resolution", *Astrophys. J. (Lett.)*, **360**, L43

Drury, L.O'C (1983): "An introduction to the theory of diffusive shock acceleration of energetic particles in a tenuous plasma", *Rep. Progress in Physics*, **46**, 973

Edelson, R.A., Malkan, M. (1986): "Spectral energy distributions of AGN between 0.1 and 100 microns", *Astrophys. J.*, **308**, 59

Fanti, R. et al. (1988): "The radio structure of compact steep spectrum radio sources", in IAU Symposium 129, *The Impact of VLBI on Astrophysics and Geophysics*, eds. M.J. Reid, J.M. Moran, Kluwer Acad.Publ., p. 111

Fricke, K., Witzel, A. (1982): "Extragalactic Radio Sources", in Landolt-Börnstein, **VI**, 2c, eds. K. Schaifers, H.H. Voigt, Springer-Verlag, p. 315

Fritz, K.-D. (1989): "An explanation of abrupt infrared cut-offs of extragalactic radio sources by shock accelerated particles including self-generated turbulence", *Astrophys. J.*, **347**, 692

Gaisser, T.K. , Stanev, T. (1985): "Calculation of the neutrino flux from Cygnus X-3", *Phys. Rev. Lett.*, **54**, 2265

Genzel, R. et al. (1986): "158 micron mapping in Sagittarius A: rotation curve and mass distribution in the Galactic Centre", *Astrophys. J.*, **306**, 92

Goldstein, M.L., Burlaga, L.F., Matthaeus, W.H. (1984): "Power spectral signatures of interplanetary corotating and transient flows", *J. Geophys. Res. (A)*, **89**, 3747

Gondhalekhar, (1990): "Ultraviolet spectra of a large sample of quasars II: variability of Lyα and CIV emission lines", *Mon. Not. R. astr. Soc.*, **243**, 443

Heeschen, D.S. (1986): "Radio variability of extragalactic radio sources", *Astron. J.*, **94**, 1493

Heeschen, D.S., Krichbaum, Th., Schalinski, C.J., Witzel, A. (1987):"Rapid variability of extragalactic radio sources", *Astron. J.*, **94**, 1493

Hughes, P.A., Aller, H.D., Aller, M.F. (1989a): "Synchrotron emission from shocked relativistic jets I- the theory of radiowavelength variability and its relation to superluminal motion", *Astrophys. J.*, **341**, 54

Hughes, P.A., Aller, H.D., Aller, M.F. (1989b): "Synchrotron emission from shocked relativistic jets I- a model for the cm wave band quiescent and burst emission from BL Lacertae", *Astrophys. J.*, **341**, 68

Hummel, C.A., Schalinski, C.J., Krichbaum, T.P., Witzel, A., Johnston, K.J. (1988): "The quasar 0153+74", *Astron. Astrophys.*, **204**, 68

Jokipii, R. (1987): "Rate of energy gain and maximum energy in diffusive shock acceleration", *Astrophys. J.*, **313**, 842

Kaastra, J.S., Barr, P. (1989): "Soft and hard X-ray variability from the accretion disk of NGC5548", *Astron. Astrophys.*, **226**, 59

Kazanas, D. (1989): "On the nature of emission line clouds of quasars and active galactic nuclei", *Astrophys. J.*, **347**, 74

Kellermann, K.I. , Pauliny-Toth, I.I.K. (1969): "The spectra of opaque radio sources", *Astrophys. J. (Lett.)*, **155**, L71

Kellermann, K.I. , Pauliny-Toth, I.I.K. (1981): "Compact radio sources", *Ann. Rev. Astron. Astrophys.*, **19**, 373

Klemens, Y. (1987a): "Acceleration of secondary particles in active galactic nuclei", Ph.D. Thesis, Bonn

Klemens, Y. (1987b): in Proc. European Regional IAU Prague Meeting, p. 391

Kössl, D. (1988): "Magnetic effects in supersonic flows", Ph.D. Thesis München

Kolmogorov, A.N. (1941): C.R. Acad. URSS, **30**, 201

Kool, M. de, Begelman, M.C. (1989): "Production of self-absorbed synchrotron spectra steeper than $\nu^{5/2}$", *Nature*, **338**, 484

Krichbaum, T.P. (1990): "Images of parsec-scale jets from 7 mm VLBI", in *Parsec Scale Radio Jets*, eds. J.A. Zensus and T.J. Pearson, Cambridge Univ. Press, p. 83

Krichbaum, T.P., Hummel, C.A., Quirrenbach, A., Schalinski, C.J., Witzel, A., Johnston, K.J., Muxlow, T.W.B., Quian, S.J. (1990a): "The complex jet associated with the quasar 0836+71", *Astron. Astrophys.*, **230**, 271

Krichbaum, T.P., Booth, R.S., Kus, A.J., Rönnäng, B.O., Witzel, A., Graham, D.A., Pauliny-Toth, I.I.K., Quirrenbach, A., Hummel, C.A., Alberdi, A., Zensus, J.A., Johnston, K.J., Spencer, J.H., Rogers, A.E.E., Lawrence, C.R., Readhead, A.C.S., Hirabayashi, H., H., Inoue, M., Morimoto, M., Dhawan, V., Bartel, N., Shapiro, I.I., Burke, B.F., Marcaide, J.M. (1990b): "43 GHz-VLBI observations of 3C273 after a flux density outburst in 1988", *Astron. Astrophys.*, **237**, 3

Krichbaum, T.P., Quirrenbach, A., Witzel, A. (1992a): "Intra-day variability of compact extragalactic radio sources", in Proc. *Variability of Blazars*, eds. Valtaoja, E., Valtonen, M., Cambridge University Press, p. 331

Krichbaum, T.P., Witzel, A. (1990): "Astronomical results from recent 7 mm VLBI campaigns", in *Frontiers of VLBI*, eds. Hirabayashi, H. et al., Tokyo Univ.Press, p. 1

Krichbaum, T.P., Witzel, A. (1992b): "Structural variability of active galactic nuclei at 43 GHz", in Proc. *Variability of Blazars*, eds. Valtaoja, E., Valtonen, M., Cambridge University Press, p. 205

Krolik, J., Begelman, M.C. (1988): "Molecular tori in Seyfert galaxies: feeding the monster and hiding it", *Astrophys. J.*, **329**, 702

Kühr, H., Pauliny-Toth, I.I.K., Witzel, A., Schmidt, J. (1981a): "The 5-GHz strong source survey V. Survey of the area between declination 70 degrees and 90 degrees", *Astron. J.*, **86**, 854

Kühr, H., et al. (1981b): "A catalogue of extragalactic radio sources having flux densities greater than 1 Jy at 5 GHz", *Astron. Astrophys. Suppl. Ser.*, **45**, 367

Lawrence, A., Elvis, M. (1982): "Obscuration and the various kinds of Seyfert galaxies", *Astrophys. J.*, **256**, 410

Lüst, R. (1952): "The evolution of a gas mass rotating around a central body", *Zeitschrift für Nat. Forschung*, **7a**, 87

MacDonald, J., Stanev, T., Biermann, P.L. (1991): "Neutrino-heated stars and broad line emission from active galactic nuclei", *Astrophys. J.*, **378**, 30

Malkan, M.A. (1983): "UV excess of luminous quasars II: evidence for massive accretion disks", *Astrophys. J.*, **268**, 582

Mannheim, K., Biermann, P.L. (1989): "Photomeson production in quasars", *Astron. Astrophys.*, **221**, 211

Mannheim, K., Krülls, W., Biermann, P.L. (1991): "A novel mechanism for non-thermal X-ray emission", *Astron. Astrophys.*, **251**, 723

Marcaide, J.M., Alberdi, A., Elosegui, P., Marscher, A.P., Zhang, Y.F., Shaffer, D.B., Schalinski, C.J., Witzel, A., Jackson, N., Sandell, G. (1990): "Detection of a new component in the peculiar superluminal quasar 4C39.25", in Proc. *Parsec-Scale Radio Jets*, eds. Zensus, A., Pearson, T., Cambridge Univ. Press, p. 59

Marscher, A. (1983): "Accurate formula for the Self-Compton X-ray flux density from a uniform, spherical, compact radio source", *Astrophys. J.*, **264**, 296

Matthaeus, W.H., Goldstein, M.L. (1982): "Measurements of the rugged invariants of magneto-hydrodynamic turbulence in the solar wind", *J. Geophys. Res. (A)*, **87**, 6011

Matthaeus, W.H., Goldstein, M.L., Smith, C. (1982): "Evaluation of magnetic helicity in homogeneous turbulence", *Phys. Rev. Lett.*, **48**, 1256

Meisenheimer, K., Röser, H.-J., Hiltner, P.R., Yates, M.G., Longair, M.S., Chini, R., Perley, R.A. (1989): "The synchrotron spectra of radio hot spots", *Astron. Astrophys.*, **219**, 63

Miley, G.K. (1973): "Brightness and polarization distribution of head-tail galaxies at 1415 MHz", *Astron. Astrophys.*, **26**, 413

Miley, G. (1980): "The structure of extended radio sources", *Ann. Rev. Astron. Astrophys.*, **18**, 165

Miley, G.K., Wellington, K.J., van der Laan, H. (1975): "The structure of the radio galaxy NGC1265", *Astron. Astrophys.*, **38**, 381

Mushotzky, R. (1982): "The X-ray spectrum and time variability of narrow emission line galaxies", *Astrophys. J.*, **256**, 92

Norman, C., Scoville, N. (1988): "The evolution of starburst galaxies to active galaxies", *Astrophys. J.*, **332**, 124

Norman, M.L., Smarr, L., Winkler, K.-H. (1984a): "Fluid dynamical mechanisms for knots in astrophysical jets", in *Numerical Astrophysics: A Festschrift for J.R. Wilson*, ed. J. Centrella

Norman, M.L., Smarr, L., Winkler, K.-H., Smith, M.D. (1982): "Structure and dynamics of supersonic jets", *Astron. Astrophys.*, **113**, 285

Norman, M.L., Winkler, K.-H., Smarr, L.L. (1984b): "Knot production and disruption via nonlinear Kelvin-Helmholtz pinch instabilities", in *Energy Transport in Extragalactic Radio Sources*, eds. A. Bridle and J. Eilek, p. 150

Orr, M.J.L., Browne, I.W.A. (1982): "Relativistic beaming and radio statistics", *Mon. Not. R. astr. Soc.*, **200**, 1067

Owen, F.N., Helfand, D.J., Spangler, S.R. (1981): "The correlation of X-ray emission with strong millimeter activity in extragalactic radio sources", *Astrophys. J. (Lett.)*, **250**, L55

Peacock, J.A. (1985): "The high redshift evolution of radio galaxies and quasars", *Mon. Not. R. astr. Soc.*, **217**, 601

Perez-Fournon, I., Colina, L., Gonzalez-Serrano, J.I., Biermann, P.L. (1988): "CCD photometry of the M87 jet", *Astrophys. J. (Lett.)*, **329**, L81

Perley, R.A. (1989): in *Hot Spots and Jets in Extragalactic Radio Sources*, eds. K. Meisenheimer and H.-J. Röser, Springer-Verlag

Perley, R.A., Willis, A.G., Scott, J.S. (1979): "The structure of the radio jets in 3C449", *Nature*, **281**, 437

Pringle, J.E. (1981): *Ann. Rev. Astron. Astrophys.*, **19**, 137

Quian, S.J., Quirrenbach, A., Witzel, A., Krichbaum, T.P., Hummel, C.A., Zensus, J.A. (1991): "A model for the rapid radio variability in the quasar 0917+624", *Astron. Astrophys.*, **241**, 15

Quirrenbach, A., Witzel, A., Krichbaum, Th., Hummel, C.A., Alberdi, A., Schalinski, C. (1989a): "Rapid variability of extragalactic radio sources", *Nature*, **337**, 442

Quirrenbach, A., Witzel, A., Quian, S.J., Krichbaum, T., Hummel, C.A., Alberdi, A. (1989b): "Rapid radio polarization variability in the quasar 0917+624", *Astron. Astrophys. (Lett.)*, **226**, L1

Quirrenbach, A., Witzel, A., Wagner, S., Sanchez-Pons, F., Krichbaum, T.P., Wegner, R., Anton, K., Erkens, U., Haehnelt, M., Zensus, J.A., Johnston, K.J. (1991): "Correlated radio and optical variability in the BL Lac object 0716+714", *Astrophys. J.*, **372**, 71

Rachen, J. (1991): M.Sc. Thesis, Bonn

Rawlings, S., Saunders, R. (1991): "Evidence for a common central engine mechanism in all extragalactic radio sources", *Nature*, **349**, 138

Rieke, G.H., Grasdalen, G.L., Kinman, T.D.,, Hintzen, P., Wills, B.J., Wills, D. (1976): "Photometric and spectroscopic observations of the BL Lacertae object AO 0235+164", *Nature*, **260**, 754

Rieke, G.H., Lebofsky, M.J. (1980): "Identification of infrared sources in 'empty' fields", in IAU Symp. 92 *Objects at High Redshifts*, p. 263

Rieke, G.H., Lebofsky, M.J., Kinman, T.D. (1979): "A possibly new type of QSO identified through infrared measurements", *Astrophys. J. (Lett.)*, **232**, L151

Rieke, G.H., Lebofsky, M.J., Wisniewski, W.Z. (1982): "Abrupt cut-offs in the optical-infrared spectra of nonthermal sources", *Astrophys. J.*, **263**, 73

Röser, H.-J., Meisenheimer, K. (1986): "CCD photometry of the jet in 3C273", *Astron. Astrophys.*, **154**, 15

Röser, H.-J., Meisenheimer, K. (1987): "A bright optical counterpart of the western hot spot in Pictor A", *Astrophys. J.*, **314**, 70

Sanders, R.H. (1983): "The reconfinement of jets", *Astrophys. J.*, **266**, 73

Sanders et al. (1989): "Continuous energy distributions of quasars: shapes and origins", *Astrophys. J.*, **347**, 29

Schalinski, C.J., Witzel, A., Alef, W., Campbell, J., Alberdi, A. (1988a): "Radio Astronomical analysis of IRIS-experiments", in *Proc. 7th Workshop Meeting on European VLBI*, p. 121

Schalinski, C.J., Witzel, A., Krichbaum, Th.P., Hummel, C.A., Biermann, P.L:, Johnston, K.J., Simon, R.S. (1988b): "Bulk relativistic motion in a complete sample of flat spectrum radio sources", in Proc. *The Impact of VLBI on Astrophysics and Geophysics*, eds. M.J. Reid and J.M. Moran, Kluwer Publ., p. 71

Schalinski, C.J., Witzel, A., Hummel, C.A., Krichbaum, T.P., Quirrenbach, A., Johnston, K.J. (1992a): "Monitoring of the milliarcsecond structure of S5-1928+738: Apparent superluminal motion along a fixed path?", in *Variability of Blazars*, eds. E. Valtaoja and M. Valtonen, M., Cambridge Univ. Press, p. 221

Schalinski, C.J., Witzel, A., Krichbaum, T.P., Hummel, C.A., Quirrenbach, A., Johnston, K.J. (1992b): "Structural variability of blazars from the complete S5-VLBI-sample", in *Variability of Blazars*, eds. E. Valtaoja and M. Valtonen, M., Cambridge Univ. Press, p. 225

Schlickeiser, R., Biermann, P.L., Crusius-Wätzel, A. (1991): "On a nonthermal origin of steep far-infrared turnovers in radio quiet galactic nuclei", *Astron. Astrophys.*, **247**, 283

Scoville, N., Norman, C. (1988): "Broad emission lines from the mass-loss envelopes of giant stars in active galactic nuclei", *Astrophys. J.*, **332**, 163

Shakura, N.I., Sunyaev, R.A. (1973): "Black holes in binary systems. observational appearance", *Astron. Astrophys.*, **24**, 337

Smith, M., Norman,M.L., Winkler, K.-H., Smarr, L. (1985): "Hot spots in radio galaxies: a comparison with hydrodynamic simulations", *Mon. Not. R. astr. Soc.*, **214**, 67

Stecker, F.W. (1968): "Effect of photo-meson production by the universal radiation field on high energy cosmic ray", *Phys. Rev. Lett.*, **21**, 1016

Stecker, F.W., Done, C., Salamon, M.H., Sommers, P. (1990): "High energy neutrinos from active galactic nuclei", preprint

Stocke, J.T., Rieke, G.H., Lebofsky, M.J. (1981): "New observational constraints on the M87 jet", *Nature*, **294**, 319

Sulentic, J.W., Arp, H., Lorre, J. (1979): "Some properties of the knots in the M87 jet", *Astrophys. J.*, **233**, 44

van der Laan, H. (1966): "Model for variable extragalactic radio sources", *Nature*, **211**, 1131

Wagner, S., Sanchez-Pons, F., Quirrenbach, A., Witzel, A. (1990): "Simultaneous optical and radio monitoring of rapid variability in quasars and BL Lac objects", *Astron. Astrophys. (Lett.)*, **235**, L1

Warren-Smith, R.F., King, D.J., Scarrott, S.M. (1984): "Polarimetry and photometry of M87: Is the jet fading?", *Mon. Not. R. astr. Soc.*, **210**, 415

Webb, G.M., Drury, L.O'C, Biermann, P.L. (1984): "Diffusive shock acceleration of energetic electrons subject to synchrotron losses", *Astron. Astrophys.*, **137**, 185

Wills, B.J., Browne, I.W.A. (1986): "Relativistic beaming and quasar emission lines", *Astrophys. J.*, **302**, 56

Windhorst, R. (1984): Ph.D. Thesis Leiden

Witzel, A. (1987): in *Superluminal Radio Sources*, eds. J.A. Zensus and T.J. Pearson, p. 83

Witzel, A. (1990): "Intra-day variability of extragalactic radio sources", in *Parsec-Scale Radio Jets*, eds. Zensus, A., Pearson, T., Cambridge Univ. Press, p. 206

Witzel, A., Quirrenbach, A. (1990): in *Propagation Effects in Space VLBI*

Witzel, A., Schalinski, C.J., Johnston, K.J., Biermann, P.L., Krichbaum, T.P., Hummel, C.A., Eckart, A. (1988): "The occurrence of bulk relativistic motion in compact radio sources", *Astron. Astrophys.*, **206**, 245

Zdziarski, A. et al. (1987): "Effects of electron-positron pair opacity for spherical accretion onto black holes", *Astrophys. J. (Lett.)*, **315**, L113

Zensus, A., Pearson, T. (1987): in *Superluminal Radio Sources*, eds. J.A. Zensus and T.J. Pearson, Cambridge Univ. Press, p. 28

Methods in Astronomical Image Processing with Special Applications to the Reduction of CCD Data

Steven Jörsäter

Stockholm Observatory, S-133 36 Saltsjöbaden, Sweden

1 A Brief Introductory Note

Image processing is today an established branch of computer science. In astronomy, however, the meaning of "image processing" is rather different from that in other disciplines. In these lectures I have aimed at describing the reasons for the special nature of astronomical image processing, its basic features and tools, and examples of typical basic applications in astronomy. I do not have room to discuss the many advanced mathematical methods employed in current astronomical image processing - they have much more in common with main-stream image processing than the material covered here.

2 History of Astronomical Imaging

Since ancient times, astronomy has been the science of mapping the heavens. For many centuries the naked eye was the only available instrument. This was also true after the introduction of the telescope when the eye took the role of an astronomical instrument in the modern sense - i.e. a detector device that is connected to a telescope. Gradually substantial data bases consisting of stellar charts, maps of the of moon etc. were built up. A revolution came about in the latter half of the last century when the photographic technique had advanced enough to be of use in astronomy. The main importance was, of course, that the photographic plate *integrates* the light shone upon it and thus enables it to reach much deeper than the eye which can only integrate for a fraction of a second. A very important side effect was that data could now conveniently be stored and compared to other data obtained at other times and with different instruments. Furthermore, various photographic techniques could be employed in making copies of the originals thereby enhancing different features in the images. Astronomical image processing was born and the capability of the photographic technique was, and still is, impressive and remains in extensive use today, see e.g. the examples in Laustsen, Madsen and West (1987).

3 Astronomical Image Data

3.1 Images in Various Formats

Photographic techniques can produce results which are very pleasing to the eye and can also help to enhance faint features. These capabilities have been, and still are, of great scientific value. Astronomy has, however, gradually become a more physical science and the desire to accurately measure colours, fluxes, positions etc. in the images is great. The advent of the digital computer in the fifties and sixties started a new revolution in image manipulation. Digital images can easily be added, filtered, deprojected etc. and, if the original data are well calibrated, it is easy to measure physical quantities directly from the images. The digital revolution has been rather slow in this area, however, and photographic techniques are still being employed in wide-field applications. The reason for this is that an image consists of vast amounts of data and the early primitive computers available before the 1970s were not capable of manipulating the millions of picture elements needed to construct a high quality image. Only now has image processing come of age and become commercially heavily exploited. The requirements on computers for speed and storage are, even with today's supermachines, at the very limit of the technology.

3.2 Digitized Image Data

There are two basic kinds of modern digital astronomical data. One kind is the rapidly dominating direct digital data where the detector electronics directly delivers digital data. This will be discussed more in the following section. The other kind is analogue data recorded on photographic plates which has been digitized afterwards using a photometer to scan the plates. This is a rather cumbersome procedure since it involves two difficult measurements, first the telescope observations and then the scanning measurement. In addition, photographic plates are rather difficult to calibrate. The reasons that this technique has been used, and is still being used, are twofold. The first reason is that digital detectors have only recently become generally available so that plates were in many cases the only alternative. The second reason is that the photographic plate is still unsurpassed in terms of wide- angle viewing. The largest CCDs available today (which are very expensive) contain 16 million pixels (picture elements) whereas a Schmidt plate may easily contain 200 million pixels. For wide- angle astronomy, such as that pursued with Schmidt telescopes, photographic plates remain the only practical option although this will probably change in the near future. After digitization and careful calibration these data may be used as direct digital image data.

3.3 Digital Image Data

The detectors that produce direct image digital data have essentially revolutionized astronomy. In this class I include detectors that do not store (except momentarily) data in analogue form. In the optical it is the CCDs (Charge Couple Devices), that have dominated. These detectors have a number of wonderful properties. Their sensitivity is very high, having close to 100 % quantum efficiency (i.e. every incoming photon is detected). They furthermore have excellent linearity, geometric stability and they are physically robust and have large tolerance to overexposure (see e.g. Mclean 1989, *ibid* Chap. 4). Calibrating CCDs, however, is a rather complex procedure which will be discussed in detail below. The interesting thing about the CCD, from a philosophical point of view, is that the incoming data is truly digital (as it is in the form of photons). It is then converted to a charge which after being counted in an analogue way is then redigitized. Many kinds of detectors produce digital data, such as IR arrays and various kinds of photon counting detectors. Radio data, in particular radio interferometric data, are rather different in nature but the data are also stored in digital form and may be included in this class.

4 Philosophy of Astronomical Image Processing

4.1 Properties of Digital Astronomical Images

We will here attempt to define the kinds of images that the rest of these lectures are about. We define images as 1 to 3-D digital data coming more or less straight from an astronomical detector in a general sense. Typical of modern instruments is that they produce large quantities of data, up to tens of GBytes from one campaign. What do we want from these images? This will vary. Often, however, only small quantities of end data are required such as "What is the magnitude of ..." or "Is it statistically certain that spirals are less common in clusters...". Thus the term "data reduction" may be very apt - from 10 GBytes of data we may only require a simple yes or no!

4.2 Human Image Processing

The eye and brain together constitute a very powerful image processing unit. The eye is superb at detecting special features such as straight lines in images. It has sophisticated automatic on- line colour correction facilities. Large-scale patterns are almost instantaneously detected using the enormous parallel capacity of the brain. Computer algorithms tend to operate on images on a pixel-by-pixel basis. This makes it difficult to detect global patterns. It is therefore often frustrating to try to design algorithms that outperform the eye for large scale patterns. When doing image processing and creating images for display the user must bear in mind that he or she is producing something which is to be decoded by another system, in this case the human visual system. Quite often we try to rely on effects that fool the eye such as 3-D graphics. It is important to realize how the human visual system works in order to produce high-quality graphic output.

4.3 Astronomical vs. Computer Science Image Processing

The central purpose of astronomical image processing is to process and visualize images obtained from telescopes. This makes it rather different in flavour compared to other kinds of image processing. The main reason for this is that astronomers normally want *quantitative* data from their images such as colour indices or magnitudes. In computer science, however, one is mainly concerned with *qualitative* image processing. Very often the applications are related to attempts to mimic human vision. Thus complex pattern recognition and edge detectors are important areas in the intelligent interpretation of images required, for example, in robots. These areas of research are, as mentioned above, different from typical data reduction type image processing in astronomy and it is important for the student to keep this in mind when confronted with the subject from the computer science view. This approach is of considerable interest, however, when trying to interpret more complex and/or theoretical data which depend on many variables (multivariate data).

5 Basic Tools of Astronomical Image Processing

5.1 Display Applications

The most fundamental task for astronomical image processing is to be able to present results so that they are easily interpreted by the human eye. Modern display hardware with pseudo-colour/true-colour abilities offer many possibilities in addition to more old-fashioned ones such as plotting and contouring.

5.2 Calibration of Intensity Scales

This is one of the most important tasks for astronomical image processing. Essentially all instruments yield data that must be calibrated before they can be readily interpreted in physical units. Linear detectors such as CCDs offer the most straightforward situation where simple adding and scaling is sufficient. The most difficult part is often to find the background level that is present in the image. An important way to proceed is to form the mean, median and various dispersion moments of the image. CCD data is particularly difficult since it often contains a few pixels with very wild numbers. The so-called kappa*sigma clipping is a way to suppress the influence of a few outliers. Iterations are performed where the mean and the standard deviation is calculated only over pixels in the range $-\kappa * \sigma_i < M_i(I) < \kappa * \sigma_i$ where κ is a number in the range 2 to 4 and σ_i and $M_i(I)$ are the respective standard deviation and mean, evaluated in each iteration. This iteration procedure typically never converges in a true sense but it approaches a good value after a few iteration cycles. Interpreting the histogram of the pixel values is clearly an important tool. The actual rescaling is, in the linear case, easy and is done relative to calibration source measurements.

5.3 Calibration of Length Scales

Various astronomical applications require that the number of pixels along an axis in a detector be calibrated in terms of a physical quantity. The most common applications are astrometric calibration and wavelength calibration in spectroscopy. The astrometric calibration tries to relate the pixel position to the actual position in angular measures. Various optical and geometrical distortions must be corrected for. In spectroscopy, typically wavelength is a function of position along one of the axes in the detector but more complex situations arise e.g. in echelle spectrographs where the dispersion runs along oblique (semi-)parallel tracks in the image. The final stage in the calibration process often involves image re-shaping.

5.4 Image Re-shaping

A very frequent application is that of image re-shaping. A common situation is that two images of the same field have been obtained with different filters, telescopes etc. and need to be aligned. This process almost always involves rebinning of the pixels which is a complex task and has to be used with caution. A rebinning operation always leads to degraded resolution, the problem is to minimize that. The rotation of images is another important image re-shaping application.

5.5 Feature Enhancement

Faint or otherwise disguised features often need to be enhanced. A common method of boosting extended features is to smooth images. Display methods are frequently used whereby contours are placed at a critical level relative to the feature looked for.

5.6 Noise Suppression

This application is closely related to the former since feature enhancement very frequently means contrasting a faint object relative to a noisy background. The most common way to achieve noise suppression is smoothing, employing e.g. median filters, but other types of filtering are also important.

5.7 Noise and Error Analysis

The noise in an image is not only bad. It is an important diagnostic tool which can be used to check the "sanity" of the image and the detector with which it was obtained. This is an important application of astronomical image processing. In particular, it is of fundamental importance to find and remove bad pixels, in e.g. CCD imaging, which always occur. These pixels frequently contain wild numbers and may corrupt the whole analysis if one is not cautious. Never trust a straight mean or standard deviation in a CCD image where bad-pixel removal has not taken place.

6 Image Processing Packages: Design of AIPS and MIDAS

I will now discuss two major astronomical image processing packages, AIPS and MI-DAS. Although there are many packages now in operation these two packages are fairly representative of what is available and both have advantages as well as problems. Both systems have been developed by astronomical institutes and are available essentially free of charge to the community.

6.1 AIPS

AIPS or Astronomical Image Processing System has been developed by the National Radio Astronomy Observatory (NRAO) in the USA (see NRAO 1990). This is one of the oldest packages available, its design began as far back as 1978. It is also one of the most successful astronomical packages, a major reason being the support given by the NRAO, not the least regarding bug fixing. Due to the primitive state of computers and especially software standards at that time, it was decided to make it operating system independent (i.e. it contains its own operating system). Today it is mainly run under UNIX and to a limited extent, under VMS. It is available also for a few other operating systems but these are of very marginal importance today. The main emphasis of AIPS is to process radio aperture synthesis data. Amongst AIPS present features are:

- Powerful commands
- Global keywords with the possibility to save parameters at any time
- On-line help, history and log facilities
- Intermediate-level control language
- For its time, an impressive and uniform user interface
- Device-independent interfaces
- Support of multiple graphics standards (today mostly X-windows and Postscript)

AIPS communicates with the outside world through a command parser language called POPS (People Oriented Parsing Service) which is also used by other program packages. This parser language represented the state of the art at the time of its creation and is still useful. Within POPS, AIPS supports:

- Tasks - powerful, often multipurpose stand-alone modules
- Verbs - which are basic subcommands within the AIPS program module itself
- Pseudo-Verbs - operating system type commands such as editing procedures
- Adverbs - keywords with fixed names but often having a different function in different tasks/verbs
- Procedures - files which contain valid AIPS commands

The AIPS tasks are the major building blocks of AIPS. They are often complex, may have dozens of different functions, and may take many hours of CPU time to execute. They have a multiparameter input in a user interface which uses the (essentially) fixed POPS keywords. An effort has been made to let adverbs represent the same things in different tasks, such as INTAPE with obvious meaning. Other adverbs, such as IN2SEQ (a file sequence parameter) is more difficult to standardize and has rather different meaning in different tasks. The rather complex nature of the tasks make them very well suited for accomplishing fixed jobs such as reducing standard VLA data but provide little flexibility for unusual applications.

The main data structures within AIPS are POPS adverbs, MAPS which are up to 7 dimensional images and UVDATA which contain lists of UV-plane data and associated information related to aperture synthesis interferometry. Various predefined special files are also utilized such as graphic files, CC (Clean Components) files and history files.

It has been said that if you want to see the advantages with global keywords, then look at AIPS but a similar statement could be made about the disadvantages. AIPS has, at run time, only fixed keywords.

The programming interface to AIPS is rather cumbersome. A reason for this is the AIPS goal of having an independent operating system environment which means that customized routines must be employed to do essentially everything, resulting in very deep nesting of subroutines. The standard language is FORTRAN 77.

There is now a major effort in creating a new AIPS called AIPS++. This package, which should provide the same functionality as the present AIPS plus a lot more, is from a programmer's point of view redesigned from scratch. More powerful user interface and control facilities will be provided and the whole package will become more modular to correct the major drawback of the present AIPS. The new package is expected to replace AIPS by 1993.

6.2 MIDAS

MIDAS stands for Munich Image and Data Analysis System and has been designed and developed by the European Southern Observatory (ESO) in Munich, Germany (see ESO 1991). MIDAS was originally developed for VAX VMS in 1982. The porting to UNIX took place much later and was not complete until 1988. MIDAS is a general purpose image processing system with applications mainly in optical astronomy and related to ESO instruments in particular. MIDAS command procedures have taken much of their syntax from VMS command procedures and have powerful control ability. The major constituents of MIDAS are:

- The Monitor - the command parser which also supports primitive commands. Rather powerful symbolic manipulation of images and tables is allowed.
- The standard interfaces which provide an easy interface between user code and MIDAS data structures
- The application programs and command procedures which number many hundreds. Each of these typically performs rather limited tasks such as calculating the Fourier transform of an image.

The data structures used are:

- Images (.BDF files) which may have up to 4 dimensions. Only a small minority of the applications support more than 2-D operations, however.
- Tables which may be manipulated in a powerful way using spreadsheet program type commands.
- Keywords - MIDAS variables which may be, real, integer or characters and are dynamically creatable
- Descriptors which are just like Keywords but are tagged to an image or a table.

7 Reduction of CCD Data

The standard procedure for initial processing of CCD data contains the following steps, relevant to both imaging and spectroscopy (see e.g. Lauer 1989):

- Bias subtraction
- Clipping

- Preflash subtraction
- Dark subtraction
- Flat fielding
- Sky subtraction
- Extinction
- Deconvolution methods on correction
- Rebinning/Combining

The various steps will now be described and some discussion of CCD properties is included. This presumably represents the most common application of Astronomical Image Processing today.

7.1 Bias Subtraction

Bias, in the context of CCDs, is the name for the signal added before digitization to avoid having the noise create negative values at readout for a cell. A typical value is 200 ADU (Analogue-Digital Units) and full well is typically 16384 ADU. The bias is in principle an absolutely smooth and constant value which is added to all cells and ideally it should only be one number to subtract. In practice it is affected by noise, the level varies somewhat due to the temperature and other effects on the electronics, and it may also vary with the position on the chip for various reasons. In order to get a hold on the real level of bias one normally *overscans* the CCD (see Mclean 1989, *ibid* p. 127), i.e. one reads out more columns or rows than there are in the CCD. The signal in the remaining area should reflect the bias and normally does so very well. The first step, when trying to determine the bias of the CCD, is to inspect several frames to try to determine if there is a spatial variation. If there is none, we are only looking for one number to subtract. If there is one, we need to form averages of a number of bias frames (say 10) in order not to increase noise and subtract the average image from the images to be reduced. The level may change, however, so we must determine the mean of the overclocked areas in the image. If there is no structure in the bias frame we may simply determine this number and subtract it from the image. If we need to subtract the bias image we must determine the bias drift by determining the mean of the overscanned columns in the image and the bias mean frame, scale the latter so it matches the bias level of the former and subtract it. Care must be taken when forming the means so that they are not corrupted by cosmic rays which cannot directly hit the overscanned data (they do not correspond to real pixels anyway!) but charge transfer problems may still affect nearby overclocked areas. Means should be formed using kappa*sigma clipping described above. Bias subtraction is to be applied to all images, science frames as well as dark frames and flat fields.

7.2 Clipping

After having subtracted the bias we are now ready to determine what areas contain signal. This may be smaller than the physical area of the chip since e.g. a spectrograph slit may only cover a part of the image. After careful inspection of the data frames we can now clip the frames and remove overscanned columns and rows as well as areas that do not contain data.

7.3 Preflash Subtraction

In some cases one must preflash the CCD in order to improve charge transfer which is poor in some chips at low light levels. This introduces photon noise, of course, and is avoided if possible. Should you have preflashed data, you need to form a mean of a set of images of the preflash (bias subtracted!) and subtract it from the science and flat field frames.

7.4 Dark Subtraction

We must now form a dark frame. Some CCDs have very low dark current and dark subtraction may not be required. If it is required you need to form a dark image since dark currents, unlike bias, are *never* uniform over the image. During a run you will typically have acquired several series of darks with different exposure times. Let's assume you have 5 darks with 10 minutes exposure time and 5 with 30 minutes. The first thing you want to do is to form means of each group of 5 (using kappa*sigma clipping!) and compare the individual frames with the means to ensure that differences are not systematical in nature. The second thing to do is to scale the two means according to exposure time (to, say, 1 minute of exposure time) to check that the dark current is, indeed, linear with time to first order, as it should be. You then form a mean of the two (or more) means and scale it according to the exposure time of the science frame and subtract it. Flat fields seldom need to be dark subtracted since their exposure times are normally short.

7.5 Flat Fielding

In CCD astronomy the quality of the measurements are frequently directly correlated to the care taken when doing the flat fielding.

Why are flat fields needed? The detector characteristics of CCDs are often described as being close to ideal. One feature is, however, that each of the pixels tends to operate as a near perfect linear detector but with a somewhat different characteristic response coefficient from its neighbours. In order to be able to treat the image as one unit the various pixels need to be normalized relative to one another. This is accomplished by taking a picture of an (ideally) perfectly uniformly lit surface such as a small portion of the dawn sky. Since CCDs are linear this flat field image can be used right away (after, of course, bias subtraction and, where applicable, preflash and dark frame subtraction). The science frame is simply divided by this frame.

What are the main problems with flat fields? The flat fields must represent the conditions of the science frame observation as closely as possible. Colour temperature is here an important factor since the various pixels tend to have different responses in different colours. A flat field surface should therefore ideally have exactly the same colour as the object (including background). This is in practice impossible and it is thus a very important error source. Obtaining flat fields from the dawn or dusk night sky is sufficient for many applications but clearly not sufficient when photometric accuracy better than 1% is needed. A method frequently used is to separate the high spatial frequency variation, which is assumed to be less critical and hence determinable from an observation of e.g. the dawn sky, from the low spatial frequency variation which requires much less signal for determination and can be determined from a low S/N observation of the night sky. Flat fielding of long slit spectra requires a special technique - here the flat fielding is normally done by observing a continuum source, thereby compensating for the varying response along the slit. The dispersion direction flatness is corrected simultaneously with the calibration - a known standard object is observed and the data is corrected accordingly.

Figure 1 presents a set of CCD images showing the effect of flat-fielding. The images have been obtained with a Texas Instrument 800-by-800 pixel CCD and a focal reducer on the Danish 1.5 m telescope at La Silla, Chile. First is shown the raw image (a) of a galaxy and an uneven background caused by flat-fielding effects. The lookup table (colour coding) used is such that bright areas correspond to dark in the image and vice versa. The very faintest areas are depicted as black again increasing the contrast of the extreme areas. A flat field (b) is then shown in the same filter illustrating the uneven CCD response. Finally is shown the resulting flattened image (c) where the raw image (a) has been divided by the flat field (b). The dramatic improvement in background flatness is obvious. A few other image errors are evident in the flattened image - note the two bright "hot" columns and the optical distortion of the stars towards the edges caused by the optics in the focal-reducing lens system.

7.6 Sky Subtraction

The background sky level is substantial and must be removed in most astronomical exposures, images as well as spectra. In images the sky background is typically, like the bias, described by one number. If your object is small and you have plenty of clean background the problem is relatively straightforward, you simply determine the background by inspecting the histogram, or you just take the median, or form a kappa*sigma clipped mean. The only essential remaining problem here is that the sky background is not well defined, it consists of airglow which is fairly uniform but also of the light from a myriad of stars and galaxies. If it is critical to your application you must thus study the spatial statistics of background variation. This is much easier if you have several deep background frames exposed around your target observation. Beware, however, that the background level varies rapidly with time since airglow does. This is also true if your object is large, forcing you to obtain sky level from separate exposures. You cannot obtain an object frame of say 15 minutes duration and then obtain a sky frame of the same duration nearby, and still expect the background level to be reasonably accurate (it could be wrong by a factor of 20%). Instead, you should make several consecutive overlapping frames, so that you can monitor the change in airglow as a function of time between the frames and compensate for it. Long slit spectra constitute again a different problem. Here the sky background varies dramatically in the dispersion direction since the airglow itself contains many lines. Since the line features are very sharp you need to wavelength-calibrate and rebin your spectra before you do the subtraction. The main reason for this is that the slit direction is normally not exactly perpendicular to the dispersion direction because of optical distortions. As with direct images, the best case is when you have portions of blank sky along your slit. You can then take the mean in the slit direction over the blank sky area and subtract that from the whole spectrum.

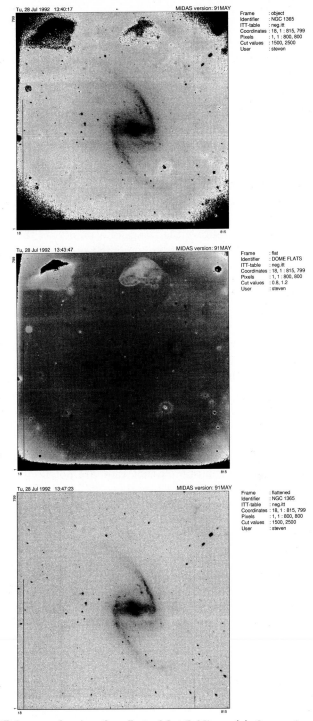

Fig. 1. MIDAS: A set of CCD images showing the effect of flat-fielding - (a) the raw image, (b) a flat field, (c) the flattened image

7.7 Extinction Correction

This step normally involves multiplication by the atmospheric extinction factor which must be found either from observations during the night or, less accurately, by using some standard extinction curve for the sight in question. It typically involves only one number (or a function of wavelength in spectra) and has little to do with image processing but is included for completeness.

7.8 Deconvolution Methods

Now is the time to improve the image of a point, i.e. the point spread function (PSF). If various observations are to be combined or compared they should preferably have the same PSF. This is certainly not always necessary but different PSFs may give very strange results for compact objects when e.g. a continuum image is subtracted from a narrow band image. Since the atmospheric seeing tends to vary from night to night we need to correct the data if we have a number of different exposures. The easiest and safest thing to do is to convolve the data with a suitable function (typically a Gaussian) so that all images get the same resolution as the image with the poorest resolution. This is often not desirable, however, and then deconvolution methods must be used to improve the worst data. Many different methods are in use such as CLEAN and MEM (Högbom 1974, Cornwell and Braun 1989), both of which are in heavy use in radio astronomy but may perfectly well be employed in CCD reductions. These, as well as the Richardson-Lucy method and other methods (see e.g. Adorf et al. 1991) are frequently used to deconvolve Hubble Space Telescope data in order to remove the effects of the primary mirror spherical aberration. The common feature of all deconvolution schemes, aimed at achieving super-resolution, is that they are potentially dangerous in the sense that uncritical use invariably leads to spurious data and that they require high S/N data as well as good spatial sampling in order to work. If the latter two requirements are fulfilled (which they seldom are in astronomy) the resolution may be improved by a small factor and thus enable the common resolution to be somewhat better than the worst resolution in the data set.

Figure 2 presents a set of VLA radio aperture synthesis images showing the effect of the deconvolution algorithm CLEAN. The raw aperture synthesis image (a) of a 21-cm HI-line channel of a galaxy (basically the Fourier transform of the interferometric images) is shown first. Due to incomplete coverage of the aperture plane the sources are surrounded by strong sidelobes. The dirty beam (b) or the point spread function is then shown. In this case the deviation from a Gaussian shape is substantial. The grey coding span is from -2% (white) to 8% (black) of the central peak value. Finally is seen the resulting CLEAN image (c). The contrast enhancement is strong. The grainy appearance is due to noise.

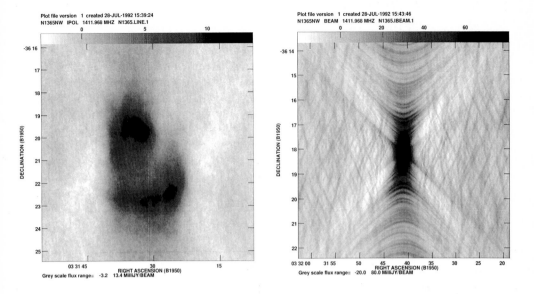

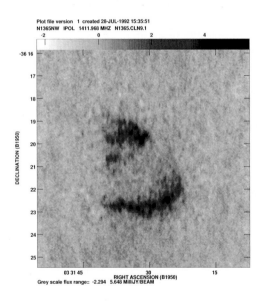

Fig. 2. AIPS: A set of VLA radio aperture synthesis images showing the effect of the deconvolution algorithm CLEAN - (a) the raw aperture synthesis image, (b) the dirty beam, (c) the CLEAN image

7.9 Rebinning/Combining

The final stage of the standard reduction path is to assemble and combine the images so that the actual astronomical data may be extracted. This situation can be of many different kinds. One may have many images of the same object in the same filter but taken with different telescope positions so that the data need to be resampled to ensure pixels coincide before the images may be added. This addition may be sophisticated and take the detailed S/N characteristics of each image into account in order to optimize the S/N for the resulting image. Other scenarios involve combining optical and radio data, data obtained in different filters and/or with different telescopes etc. The common denominator of all this is the need to rebin (resample) the data onto a common pixel grid. Information is always lost in the rebinning process and care has to be taken to minimize the effects. In most applications rather simple-minded rebinning using linear interpolation is sufficient and this is the method normally used. The data is now ready for display and qualitative as well as quantitative analysis.

8 Summary and Prospects for the Future

Astronomical Image Processing is the tool used today by astronomers to inspect and manipulate their data and has essentially replaced darkroom work and magnifying glasses. In addition to this, it has put many new advanced image processing techniques at the astronomer's disposal. The computer hardware has up till now only been marginally adequate for the steadily more complex tasks it has been put to solve. In the near future we are likely to get vastly more powerful hardware but also more exacting applications, so the race may go on. To name a few areas of probable importance we can list:

- More on-line processing
- Expert type systems that guide the astronomers in their reduction efforts
- Automatic observing programmes which will yield huge data quantities
- Yet more advanced filtering - optical inferometry is an example
- Animation and movies consisting of many thousands of images - also from simulations
- Real-time applications - this is already in operation in solar astronomy
- 3-D modeling of galaxies and nebulae

References

Adorf, H.-M., Walsh, J.R., Hook and Hook, R.N. (1991): "Restoration Experiments at the ST-ECF" in "The Restoration of HST images and Spectra", Space Telescope Science Institute, Eds. R.L. White and R.J. Allen, p. 121

Cornwell, T. and Braun, R. (1989): "Deconvolution in Synthesis Imaging in Astronomy", in A.S.P. Conf. Series 6, p. 167

ESO (1991): " MIDAS Users Guide", Image Processing Group at the European Southern Observatory

Högbom, J. (1974): *Astron. Astrophys. Suppl. Ser.* **15**, 417

Lauer, T. (1989): "The Reduction of WF/PC Camera Images", *Publ. Astr. Soc. Pac.* **101**, 445

Laustsen, S., Madsen, C. and West, R.M. (1987): "Exploring the Southern Sky", The European Southern Observatory

Mclean, I.S. (1989): "Electronic and Computer-Aided Astronomy - from Eyes to Electronic Sensors", Ellis Hoorwood Ltd

NRAO (1990): " AIPS Cookbook", National Radio Astronomy Observatory

Multivariate Methods for Data Analysis

Fionn Murtagh

Space Telescope – European Coordinating Facility[1], European Southern Observatory, Karl-Schwarzschild-Straße 2, D-8046 Garching bei München.

1 Introduction

In observational astronomy, multivariate data analysis may well be an activity which is downstream of image processing. These methods can also be of importance for the analysis of ever-growing databases.

We will describe a number of techniques – principal components analysis (PCA), cluster analysis, and graph-theoretic methods. A short bibliography following each section gives examples of applications.

Some general remarks on analyzing data follow.

The data array to be analysed crosses *objects* with *variables* or *parameters*. The former are usually taken as the rows of the array, and the latter as the columns. The object-parameter dependencies can take many forms. *Quantitative* data is real valued data, positive or negative, defined relative to some zero point. Another form of data is *qualitative* or *categorical*, i.e. the object-parameter relation falls into one of a number of categories; in its most simple form, this is a yes/no dependency indicating the presence or absence of the parameter. The coding of a categorical variable or parameter may take the general form of values "a", "b", etc. for the different categories, or "1", "2", etc. (where in the latter case, the values have "qualitative" significance only). A final form of data to be mentioned here is *ordinal* data where a rank characterises the object on the parameter.

In the great majority of multivariate data analysis methods, the notion of distance (or similarity) is central. In clustering, objects are clustered on the basis of their mutual similarity, for instance, and in principal components analysis the points are considered as vectors in a metric space (i.e. a space with a distance defined).

Some of the problems which arise in deciding on a suitable distance are as follows. If the data to be analysed is all of one type, a suitable distance can be chosen without undue difficulty. If the values on different coordinates are quite different – as is more often the case – some *scaling* of the data will be required before using a distance. Equally, we may view the use of a distance on scaled data as the definition of another, new distance. More troublesome is the case of data which is of mixed type (e.g. quantitative,

---

[1] Affiliated to Astrophysics Div., Space Science Dept., European Space Agency.

categorical, ordinal values). Here, it may be possible to define a distance which will allow all coordinates to be simultaneously considered. It may be recommendable, though, to redefine certain coordinate variables (e.g. to consider ordinal values as quantitative or real variables). As a general rule, it is usually best to attempt to keep "like with like". A final problem area relates to missing coordinate values: as far as possible care should be taken in the initial data collection phase to ensure that all values are present.

Proximity between any pair of items will be defined by *distance, dissimilarity* or *similarity*. *Distance* is simply a more restrictive *dissimilarity,* – it satisfies certain axioms. Both *distances* and *dissimilarities* measure identical items by a zero value, and by increasingly large (positive) values as the proximity of the items decreases. *Similarities* are mathematical functions which treat pairs of items from the other perspective: large values indicate large proximity, while small (positive) or zero values indicate little or no proximity.

The most commonly used distance for quantitative (or continuous) data is the *Euclidean* distance. If $\mathbf{a} = \{a_j : j = 1, 2, ..., m\}$ and $\mathbf{b} = \{b_j : j = 1, 2, ..., m\}$ are two real-valued vectors then the unweighted squared Euclidean distance is given by

$$d^2(a, b) = \sum_j (a_j - b_j)^2 = (\mathbf{a} - \mathbf{b})'(\mathbf{a} - \mathbf{b})$$

where $\mathbf{a}$ and $\mathbf{b}$ are taken as column vectors, and $'$ denotes transpose,

$$= \|\mathbf{a}\|^2 + \|\mathbf{b}\|^2 - 2\mathbf{a}'\mathbf{b} \tag{1}$$

where $\|.\|$ is the norm, or distance from the origin.

When *binary* data is being studied (i.e. categorical data with presence/absence values for each object, on each parameter) mutual possession of a property contributes 0 to this distance, mutual non-possession also contributes 0, and presence/absence contributes 1.

With a suitable coding, the Euclidean, Hamming or other distances may be used in a wide variety of circumstances. Consider in Fig. 1 part of a set of records from a fictionalized catalogue. Part of two records, x and y, are shown. In order to carry out a comparison of such records, we can as a preliminary recode each variable, as shown. The Hamming and the squared Euclidean distances are both then

$$d(x, y) = 0 + 0 + 0 + 0 + 1 + 1 + 1 = 3 \tag{2}$$

in this case. User choice is required in defining such a disjunctive form of coding. In the case of quantitative variables this coding may be especially useful when one is in the presence of widely varying values: specifying a set of categories may make the distances less vulnerable to undesirable fluctuations.

The possible preponderance of certain variables in, for example, a Euclidean distance leads to the need for a *scaling* of the variables, i.e. for their *centring* (zero mean), *normalization* (unit variance), or *standardization* (zero mean and unit standard deviation). If a_{ij} is the jth coordinate of vector $\mathbf{a}_i$ (i.e. the ij^{th} table entry), $\bar{a}_j$ is the mean value on coordinate (variable) j, and σ_j is the standard deviation of variable j, then we standardize a_{ij} by transforming it to

$$(a_{ij} - \bar{a}_j)/\sigma_j$$

Table 1. Two records (x and y) with three variables (Component, time to failure, maintenance carried out) showing disjunctive coding

Record x:	S1, 18.2, X
Record y:	S1, 6.7, —

	Component				Time to failure		Maintenance?
	S1	S2	S3	—	≤ 10	> 10	Yes
x	1	0	0	0	0	1	1
y	1	0	0	0	1	0	0

where

$$\bar{a}_j = \sum_{i=1}^{n} a_{ij}/n$$

$$\sigma_j^2 = \sum_{i=1}^{n} (a_{ij} - \bar{a}_j)^2/n \tag{3}$$

and n is the number of rows in the given table.

Standardization, defined in this way, is widely used, but nothing prevents some alternative scaling being decided on: for instance we could divide each value in the data matrix by the row mean, which has the subsequent effect of giving a zero distance to row vectors each of whose elements are a constant times the other. We may regard the resultant distance as a weighted Euclidean distance of the following form:

$$d^2(a, b) = \sum_{j} (w_1 a_j - w_2 b_j)^2. \tag{4}$$

In the following, we will initially regard the input data as being quantitative (and return later to qualitative and mixed, quantitative/qualitative data). In the next section, we will assume that the data is sufficiently homogeneous so as not to require rescaling (but we will return to this aspect in Sect. 2.2.4 below).

2 Principal Components Analysis

2.1 The Problem

The $n \times m$ data array which is to be analysed may be viewed immediately as a set of n row-vectors, or alternatively as a set of m column-vectors. PCA seeks the best, followed by successively less good, summarizations of this data. Cluster analysis, as will be seen, seeks groupings of the objects or attributes. By focussing attention on particular groupings, Cluster analysis can furnish a more economic presentation of the data. PCA has this same objective but a different summarization of the data is aimed at.

In Fig. 2a, three points are located in $\mathbb{R}^2$. We can investigate this data by means of the coordinates on the axes, taken separately. We might note, for instance, that on axis 1 the points are fairly regularly laid out (with coordinates 1, 2 and 3), whereas on axis 2 it appears that the points with projections 4 and 5 are somewhat separated from the point with projection 2. In three-dimensional space a model could be constructed; however in higher-dimensional spaces this is impossible.

Given, for example, the array of 4 objects by 5 attributes,

$$\begin{pmatrix} 7 & 3 & 4 & 1 & 6 \\ 3 & 4 & 7 & 2 & 0 \\ 1 & 7 & 3 & -1 & 4 \\ 2 & 0 & -6 & 4 & 1 \end{pmatrix}$$

the projections of the 4 objects onto the plane constituted by axes 1 and 3 is simply

$$\begin{pmatrix} 7 & 4 \\ 3 & 7 \\ 1 & 3 \\ 2 & -6 \end{pmatrix}.$$

Thus far, the projection of points onto axes or planes is a trivial operation. PCA, however, first obtains *better* axes. Consider Fig. 2b where a new axis has been drawn as nearly as possible through all points. It is clear that if this axis went precisely through all points, then a second axis would be redundant in defining the locations of the points; i.e. the cloud of three points would be seen to be one-dimensional.

$$Array \ for \ Figure \ 2 \ = \begin{pmatrix} 1 & 2 \\ 2 & 4 \\ 3 & 5 \end{pmatrix}$$

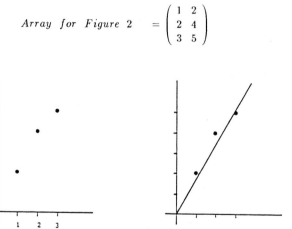

Fig. 2. Points and their projections onto axes

PCA seeks the axis which the cloud of points is closest to (usually the Euclidean distance defines *closeness*). This criterion can be shown to be mathematically identical to another criterion: that the projections of points on the axis sought for be as elongated as possible. This second criterion is that the *variance* of the projections be as great as possible.

If, in general, the points under examination are m-dimensional, it will be very rare in practice to find that they approximately lie on a one-dimensional surface (i.e. a line). A second best-fitting axis, orthogonal to the first already found, will together constitute a best-fitting plane. Then a third best-fitting axis, orthogonal to the two already obtained, will together constitute a best-fitting three-dimensional subspace.

Let us take a few simple examples in two-dimensional space (Fig. 3).

Consider the case where the points are *centred* (i.e. the origin is located at the centre of gravity). We will seek the best-fitting axis, and then the next best-fitting axis. Figure 3a consists of just two points, which if centred must lie on a one-dimensional axis. In Fig. 3b, the points are arranged at the vertices of a triangle. The vertical axis, here, accounts for the greatest variance, and the symmetry of the problem necessitates the positioning of this axis as shown. In the examples of Fig. 3, the positive and negative orientations of the axes are arbitrary since they are not integral to our objective in PCA.

Central to the results of a PCA are the coordinates of the points (i.e. their projections) on the derived axes. These axes are listed in decreasing order of importance, or best-fit. Planar representations can also be output: the projections of points on the plane formed by the first and second new axes; then the plane formed by the first and third new axes; and so on, in accordance with user-request.

It is not always easy to remedy the difficulty of being unable to visualize high-dimensional spaces. Care must be taken when examining projections, since these may give a misleading view of interrelations among the points.

2.2 Mathematical Description

2.2.1 Introduction

The mathematical description of PCA which follows is important because other techniques (e.g. Correspondence Analysis or Discriminant Analysis) may be viewed as variants on it. It is one of the most straightforward geometric techniques, and is widely employed. These facts make PCA the best geometric technique to start with, and the most important to understand.

The following section will present the basic PCA technique. It answers the following questions:

1. How is PCA formulated as a geometric problem?
2. Having defined certain geometric criteria, how is PCA formulated as an optimization problem?
3. Finally, how is PCA related to the eigen-decomposition of a matrix?

Section 2.2.3 relates the use of PCA on the n (row) points in space $\mathbb{R}^m$ and the PCA of the m (column) points in space $\mathbb{R}^n$. Secondly, arising out of this mathematical relationship, the use of PCA as a data reduction technique is described. This section, then, answers the following questions:

1. How is the PCA of an $n \times m$ matrix related to the PCA of the transposed $m \times n$ matrix?
2. How may the new axes derived – the principal components – be said to be linear combinations of the original axes?

In practice the variables on which the set of objects are characterised may often have differing means (consider the case of n students and m examination papers: the latter may be marked out of different totals, and furthermore it is to be expected that different

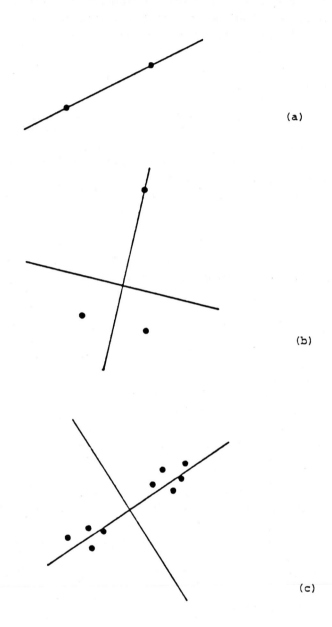

Fig. 3. Some examples of PCA of centred clouds of points

examiners will set different standards of rigour). To circumvent this and similar difficulties, PCA is usually carried out on a transformed data matrix. Section 2.2.4 describes this transformation.

As an aid to the memory in the mathematics of the following sections, it is often convenient to note the dimensions of the vectors and matrices in use; it is vital of course that consistency in dimensions be maintained (e.g. if $\mathbf{u}$ is a vector in $\mathbb{R}^m$ of dimensions $m \times 1$, then pre-multiplication by the $n \times m$ matrix X will yield a vector $X\mathbf{u}$ of dimensions $n \times 1$).

2.2.2 The Basic Method

Consider a set of n objects measured on each of m attributes or variables. The $n \times m$ matrix of values will be denoted by $X = \{x_{ij}\}$ where i is a member of the set of objects and j a member of the attribute set. The objects may be regarded as row vectors in $\mathbb{R}^m$ and the attributes as column vectors in $\mathbb{R}^n$.

In $\mathbb{R}^m$, the space of objects, PCA searches for the best-fitting set of orthogonal axes to replace the initially-given set of m axes in this space. An analogous procedure is simultaneously carried out for the dual space, $\mathbb{R}^n$. First, the axis which best fits the objects/points in $\mathbb{R}^m$ is determined. If $\mathbf{u}$ is this vector, and is of unit length, then the product $X\mathbf{u}$ of $n \times m$ matrix by $m \times 1$ vector gives the projections of the n objects onto this axis.

The criterion of goodness of fit of this axis to the cloud of points will be defined as the squared deviation of the points from the axis. Minimizing the sum of distances between points and axis is equivalent to maximizing the sum of squared projections onto the axis (see Fig. 4), i.e. to maximizing the variance (or *spread*) of the points when projected onto this axis.

The squared projections of points on the new axis, for all points, is

$$(X\mathbf{u})'(X\mathbf{u}). \tag{5}$$

Such a quadratic form would increase indefinitely if $\mathbf{u}$ were arbitrarily large, so $\mathbf{u}$ is chosen – arbitrarily but reasonably – to be of unit length, i.e. $\mathbf{u}'\mathbf{u} = 1$. We seek a maximum of the quadratic form $\mathbf{u}'S\mathbf{u}$ (where $S = X'X$) subject to the constraint that $\mathbf{u}'\mathbf{u} = 1$. This is done by setting the derivative of the Lagrangian equal to zero. Differentiation of

$$\mathbf{u}'S\mathbf{u} - \lambda(\mathbf{u}'\mathbf{u} - 1) \tag{6}$$

where λ is a Lagrange multiplier gives

$$2S\mathbf{u} - 2\lambda\mathbf{u}. \tag{7}$$

The optimal value of $\mathbf{u}$ (let us call it $\mathbf{u}_1$) is the solution of

$$S\mathbf{u} = \lambda\mathbf{u}. \tag{8}$$

The solution of this equation is well-known: $\mathbf{u}$ is the eigenvector associated with the eigenvalue λ of matrix S. Therefore the eigenvector of $X'X$, $\mathbf{u}_1$, is the axis sought, and the corresponding largest eigenvalue, λ_1, is a figure of merit for the axis, – it indicates the amount of variance explained by the axis (see next section).

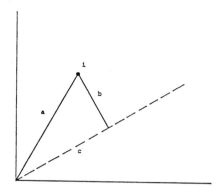

For each point i,

  b = distance of i from new axis,

  c = projection of vector i onto new axis,

  a = distance of point from origin.

By Pythagoras, $a^2 = b^2 + c^2$; since a is constant, the choice of new axis which minimizes b simultaneously maximizes c (both of which are summed over all points, i).

Fig. 4. Projection onto an axis

The second axis is to be orthogonal to the first, i.e. $\mathbf{u'u}_1 = 0$, and satisfies the equation

$$\mathbf{u'}X'X\mathbf{u} - \lambda_2(\mathbf{u'u} - 1) - \mu_2(\mathbf{u'u}_1) \qquad (9)$$

where λ_2 and μ_2 are Lagrange multipliers. Differentiating gives

$$2S\mathbf{u} - 2\lambda_2\mathbf{u} - \mu_2\mathbf{u}_1. \qquad (10)$$

This term is set equal to zero. Multiplying across by $\mathbf{u'_1}$ implies that μ_2 must equal 0. Therefore the optimal value of $\mathbf{u}$, $\mathbf{u}_2$, arises as another solution of $S\mathbf{u} = \lambda\mathbf{u}$. Thus λ_2 and $\mathbf{u}_2$ are the second largest eigenvalue and associated eigenvector of S.

The eigenvectors of $S = X'X$, arranged in decreasing order of corresponding eigenvalues, give the line of best fit to the cloud of points, the plane of best fit, the three-dimensional hyperplane of best fit, and so on for higher-dimensional subspaces of best fit. $X'X$ is referred to as the *sums of squares and cross products* matrix.

It has been assumed that the eigenvalues decrease in value: equal eigenvalues are possible, and indicate that equally privileged directions of elongation have been found. In practice, the set of equal eigenvalues may be arbitrarily ordered in any convenient fashion. Zero eigenvalues indicate that the space is actually of dimensionality less than expected (the points might, for instance, lie on a plane in three-dimensional space).

2.2.3 Dual Spaces and Data Reduction

In the dual space of attributes, $\mathbb{R}^n$, a PCA may equally well be carried out. For the line of best fit, $\mathbf{v}$, the following is maximized:

$$(X'\mathbf{v})'(X'\mathbf{v})$$

subject to

$$\mathbf{v}'\mathbf{v} = 1. \tag{11}$$

In $\mathbb{R}^m$ we arrived at

$$X'X\mathbf{u}_1 = \lambda_1\mathbf{u}_1. \tag{12}$$

By similar reasoning, in $\mathbb{R}^n$, we have

$$XX'\mathbf{v}_1 = \mu_1\mathbf{v}_1. \tag{13}$$

Pre-multiplying the first of these relationships by X yields

$$(XX')(X\mathbf{u}_1) = \lambda_1(X\mathbf{u}_1) \tag{14}$$

and so $\lambda_1 = \mu_1$ (because we have now arrived at two eigenvalue equations which are identical in form). We must be a little more attentive to detail before drawing a conclusion on the relationship between the eigenvectors in the two spaces: in order that these be of unit length, it may be verified that

$$\mathbf{v}_1 = \frac{1}{\sqrt{\lambda_1}}X\mathbf{u}_1 \tag{15}$$

satisfies the foregoing equations. (The eigenvalue is necessarily positive, since if zero there are no associated eigenvectors.) Similarly,

$$\mathbf{v}_k = \frac{1}{\sqrt{\lambda_k}}X\mathbf{u}_k \tag{16}$$

and

$$\mathbf{u}_k = \frac{1}{\sqrt{\lambda_k}}X'\mathbf{v}_k \tag{17}$$

for the kth largest eigenvalues and eigenvectors (or principal axes). Thus successive eigenvalues in both spaces are the same, and there is a simple linear transformation which maps the optimal axes in one space into those in the other.

In some software packages, the eigenvectors are rescaled so that $\sqrt{\lambda}\mathbf{u}$ and $\sqrt{\lambda}\mathbf{v}$ are used instead of $\mathbf{u}$ and $\mathbf{v}$. In this case, the *factor* $\sqrt{\lambda}\mathbf{u}$ gives the new, rescaled projections of the points in the space $\mathbb{R}^n$ (i.e. $\sqrt{\lambda}\mathbf{u} = X'\mathbf{v}$).

The coordinates of the new axes can be written in terms of the old coordinate system. Since

$$\mathbf{u} = \frac{1}{\sqrt{\lambda}}X'\mathbf{v} \tag{18}$$

each coordinate of the new vector $\mathbf{u}$ is defined as a linear combination of the initially-given vectors:

$$u_j = \sum_{i=1}^{n} \frac{1}{\sqrt{\lambda}} v_i x_{ij} = \sum_{i=1}^{n} c_i x_{ij} \tag{19}$$

(where $i \leq j \leq m$ and x_{ij} is the $(i,j)^{th}$ element of matrix X). Thus the jth coordinate of the new vector is a *synthetic* value formed from the jth coordinates of the given vectors (i.e. x_{ij} for all $1 \leq i \leq n$).

Since PCA in $\mathbb{R}^n$ and in $\mathbb{R}^m$ lead respectively to the finding of n and of m eigenvalues, and since in addition it has been seen that these eigenvalues are identical, it follows that the number of *non-zero eigenvalues* obtained in either space is less than or equal to $\min(n, m)$.

It has been seen that the eigenvectors associated with the p largest eigenvalues yield the best-fitting p-dimensional subspace of $\mathbb{R}^m$. A measure of the approximation is the percentage of variance explained by the subspace

$$\sum_{k \leq p} \lambda_k / \sum_{k=1}^{n} \lambda_k \tag{20}$$

expressed as a percentage.

2.2.4 Practical Aspects

Since the variables or attributes under analysis are often very different (some will "shout louder" than others), it is usual to standardize each variable in the following way. If r_{ij} are the original measurements, then the matrix X of (i, j)-value

$$x_{ij} = \frac{r_{ij} - \bar{r}_j}{s_j \sqrt{n}} \tag{21}$$

where

$$\bar{r}_j = \frac{1}{n} \sum_{i=1}^{n} r_{ij} \tag{22}$$

and

$$s_j^2 = \frac{1}{n} \sum_{i=1}^{n} (r_{ij} - \bar{r}_j)^2 \tag{23}$$

is input to the PCA. The matrix to be diagonalized, $X'X$, is then of $(j, k)^{th}$ term:

$$\rho_{jk} = \sum_{i=1}^{n} x_{ij} x_{ik} = \frac{1}{n} \sum_{i=1}^{n} (r_{ij} - \bar{r}_j)(r_{ik} - \bar{r}_k) / s_j s_k \tag{24}$$

which is the correlation coefficient between variables j and k.

Using the definitions of x_{ij} and s_j above, the distance between variables j and k is

$$d^2(j, k) = \sum_{i=1}^{n} (x_{ij} - x_{ik})^2 = \sum_{i=1}^{n} x_{ij}^2 + \sum_{i=1}^{n} x_{ik}^2 - 2 \sum_{i=1}^{n} x_{ij} x_{ik} \tag{25}$$

and, substituting, the first two terms both yield 1, giving

$$d^2(j, k) = 2(1 - \rho_{jk}).$$ (26)

Thus the distance between variables is directly proportional to the correlation between them.

The distance between row vectors is

$$d^2(i, h) = \sum_j (x_{ij} - x_{hj})^2 = \sum_j (\frac{r_{ij} - r_{hj}}{\sqrt{n}s_j})^2 = (\mathbf{r}_i - \mathbf{r}_h)' M (\mathbf{r}_i - \mathbf{r}_h)$$ (27)

where $\mathbf{r}_i$ and $\mathbf{r}_h$ are column vectors (of dimensions $m \times 1$) and M is the $m \times m$ diagonal matrix of jth element $1/ns_j^2$. Note that the row points are now centred but the column points are not: therefore the latter may well appear in one quadrant on output listings.

Analysis of the matrix of $(j, k)^{th}$ term ρ_{jk} as defined above is PCA on a *correlation* matrix. Inherent in this, as has been seen, is that the row vectors are centred and reduced.

If, instead, centring was acceptable but not the rescaling of the variance, we would be analysing the matrix of $(j, k)^{th}$ term

$$c_{jk} = \frac{1}{n} \sum_{i=1}^{n} (r_{ij} - \bar{r}_j)(r_{ik} - \bar{r}_k).$$ (28)

In this case we have PCA of the *variance-covariance* matrix.

The following should be noted.

1. We may speak about carrying out a PCA on the correlation matrix, on a variance-covariance matrix, or on the "sums of squares and cross-products" matrix. These relate to intermediate matrices which are most often determined by the PCA program. The user should however note the effect of transformations on his/her data (standardization, centring) which are inherent in these different options.

2. Rarely is it recommendable to carry out a PCA on the "sums of squares and cross-products" matrix; instead some transformation of the original data is usually necessary. In the absence of justification for treating the data otherwise, the most recommendable strategy is to use the option of PCA on a correlation matrix.

3. All the quantities defined above (standard deviation, variances and covariances, etc.) have been in *population* rather than in *sample* terms. That is to say, the data under examination is taken as all we can immediately study, rather than being representative of a greater underlying distribution of points. Not all packages and program libraries share this viewpoint, and hence discrepancies may be detected between results obtained in practice from different sources.

2.3 Short Bibliography

PCA has been a fairly widely used technique in astronomy. The following list indicates a few of the types of problems to which PCA can be applied. As general introductions to the technique, the reader is recommended to refer to Murtagh and Heck (1987), Chatfield and Collins (1980) and Kendall (1980).

Bijaoui, A. (1974): "Application astronomique de la compression de l'information", *Astron. Astrophys.*, **30**, 199–202

Brosche, P. (1973): "The manifold of galaxies: Galaxies with known dynamical properties", *Astron. Astrophys.*, **23**, 259–268

Brosche, P., Lentes, F.T. (1984): "The manifold of globular clusters", *Astron. Astrophys.*, **139**, 474–476

V. Bujarrabal, V., Guibert, J., Balkowski, C. (1981): "Multidimensional statistical analysis of normal galaxies", *Astron. Astrophys.*, **104**, 1–9

Buser, R. (1978): "A systematic investigation of multicolor photometric systems. I. The UBV, RGU and *uvby* systems", *Astron. Astrophys.*, **62**, 411–424

Chatfield, C., Collins, A.J. (1980): *Introduction to Multivariate Analysis*, Chapman and Hall

Christian, C.A. (1982): "Identification of field stars contaminating the colour-magnitude diagram of the open cluster Be 21", *Astrophys. J. Suppl.*, **49**, 555–592

Christian, C.A., Janes, K.A. (1977): "Multivariate analysis of spectrophotometry". *Publ. Astr. Soc. Pac.*, **89**, 415–423

Deeming, T.J. (1964): "Stellar spectral classification. I. Application of component analysis", *Mon. Not. R. astr. Soc.*, **127**, 493–516

Efstathiou, G., Fall, S.M. (1984): "Multivariate analysis of elliptical galaxies", *Mon. Not. R. astr. Soc.*, **206**, 453–464

Faber, S.M. (1973): "Variations in spectral energy distributions and absorption line strengths among elliptical galaxies", *Astrophys. J.*, **179**, 731–754

Fracassini, M., Pasinetti, L.E., Antonello, E., Raffaelli, G. (1981): "Multivariate analysis of some ultrashort period Cepheids (USPC)", *Astron. Astrophys.*, **99**, 397–399

Galeotti, P. (1981): "A statistical analysis of metallicity in spiral galaxies", *Astrophys. Space Sci.*, **75**, 511–519

Heck, A. (1976): "An application of multivariate statistical analysis to a photometric catalogue", *Astron. Astrophys.*, **47**, 129–135

Heck, A., Egret, D., Nobelis, Ph., Turlot, J.C. (1986): "Statistical confirmation of the UV spectral classification system based on IUE low-dispersion spectra", *Astrophys. Space Sci.*, **120**, 223–237

Kendall, M. (1980): *Multivariate Analysis*, Griffin, London

Kerridge, S.J., Upgren, A.R. (1973): "The application of multivariate analysis to parallax solutions. II. Magnitudes and colours of comparison stars", *Astron. J.*, **78**, 632–638

Koorneef, J. (1978): "On the anomaly of the far UV extinction in the 30 Doradus region", *Astron. Astrophys.*, **64**, 179–193

Massa, D. (1980): "Vector space methods of photometric analysis. III. The two components of ultraviolet reddening", *Astron. J.*, **85**, 1651–1662

Murtagh, F., Heck, A. (1987): *Multivariate Data Analysis*, Kluwer, Dordrecht

Nicolet, B. (1981): "Geneva photometric boxes. I. A topological approach of photometry and tests.", *Astron. Astrophys.*, **97**, 85–93

Okamura, S., Kodaira, K., Watanabe, M. (1984): "Digital surface photometry of galaxies toward a quantitative classification. III. A mean concentration index as a parameter representing the luminosity distribution", *Astrophys. J.*, **280**, 7–14

Pelat, D. (1975): "A study of H I absorption using Karhunen-Loève series", *Astron. Astrophys.*, **40**, 285–290

Upgren, A.R., Kerridge, S.J. (1971): "The application of multivariate analysis to parallax solutions. I. Choice of reference frames", *Astron. J.*, **76**, 655–664

Vader, J.P. (1986): "Multivariate analysis of elliptical galaxies in different environments", *Astrophys. J.*, **306**, 390–400

Whitney, C.A. (1983): "Principal components analysis of spectral data. I. Methodology for spectral classification", *Astron. Astrophys. Suppl. Ser.*, **51**, 443–461

Whitmore, B.C. (1984): "An objective classification system for spiral galaxies. I. The two dominant dimensions", *Astrophys. J.*, **278**, 61–80

3 Cluster Analysis

3.1 The Problem

Automatic classification algorithms are used in widely different fields in order to provide a description or a reduction of data. A clustering algorithm is used to determine the inherent or natural groupings in the data, or provide a convenient summarization of the data into groups. The term "classification" can be employed in different senses: here, we will be solely concerned with un-supervised clustering, with no prior knowledge on the part of the analyst regarding group memberships.

As is the case with principal components analysis, and with most other multivariate techniques, the objects to be classified have numerical measurements on a set of variables or attributes. Hence, the analysis is carried out on the rows of an array or matrix. If we have not a matrix of numerical values, to begin with, then it may be necessary to skilfully construct such a matrix. The objects, or rows of the matrix, can be viewed as vectors in a multidimensional space (the dimensionality of this space being the number of variables or columns).

Needless to say, a clustering analysis can just as easily be implemented on the columns of a data matrix. There is usually no direct relationship between the clustering of the rows and clustering of the columns, as was the case for the dual spaces in principal components analysis.

Motivation for clustering includes the following:

1. Analysis of data: here, the given data is to be analyzed in order to reveal its fundamental features. The significant interrelationships present in the data are sought. This is the multivariate statistical use of clustering, and the validity problem (i.e. the validation of clusters of data items produced by an algorithm) is widely seen as a major current difficulty.
2. User convenience: a synoptic classification is to be obtained which will present a useful decomposition of the data. This may be a first step towards subsequent analysis where the emphasis is on heuristics for summarizing information. Appraisal of clustering algorithms used for this purpose can include algorithmic efficiency and ease of use.

3.2 Mathematical Description

Most published work in cluster analysis involves the use of either of two classes of clustering algorithm: hierarchical or non-hierarchical (often partitioning) algorithms. Hierarchical algorithms in particular have been dominant in the literature and so we concentrate on these methods.

3.2.1 Hierarchical Methods

The single linkage hierarchical clustering approach outputs a set of clusters (to use graph theoretic terminology, a set of maximal connected subgraphs) at each level – or for each threshold value which produces a new partition. The following algorithm, in its general structure, is relevant for a wide range of hierarchical clustering methods which vary only in the update formula used in Step 2. These methods may, for example, define a criterion of compactness in Step 2 to be used instead of the connectivity criterion used here. Such hierarchical methods will be studied in Sect. 3.2.2, but the single linkage method with which we begin is one of the oldest and most widely used methods (its usage is usually traced to the early 1950s). An example is shown in Fig. 5 – note that the dissimilarity coefficient is assumed to be symmetric, and so the clustering algorithm is implemented on half the dissimilarity matrix.

Single linkage hierarchical clustering

Input: An $n(n-1)/2$ set of dissimilarities.

Step 1: Determine the smallest dissimilarity, d_{ik}.

Step 2: Agglomerate objects i and k: i.e. replace them with a new object, $i \cup k$; update dissimilarities such that, for all objects $j \neq i, k$:

$$d_{i \cup k, j} = \min \{d_{ij}, d_{kj}\}.$$

Delete dissimilarities d_{ij} and d_{kj}, for all j, as these are no longer used.

Step 3: While at least two objects remain, return to Step 1.

Equal dissimilarities may be treated in an arbitrary order. There are precisely $n - 1$ agglomerations in Step 2 (allowing for arbitrary choices in Step 1 if there are identical dissimilarities). It may be convenient to index the clusters found in Step 2 by $n+1, n+2,$ $\ldots, 2n - 1$, or an alternative practice is to index cluster $i \cup k$ by the lower of the indices of i and k.

The title *single linkage* arises since, in Step 2, the interconnecting dissimilarity between two clusters ($i \cup k$ and j) or components is defined as the least interconnecting dissimilarity between a member of one and a member of the other. Other hierarchical clustering methods are characterized by other functions of the interconnecting linkage dissimilarities.

Compared to other hierarchic clustering techniques, the single linkage method can give rise to a notable disadvantage for summarizing interrelationships. This is known as *chaining*. An example is to consider four subject-areas, which it will be supposed are characterized by certain attributes: computer science, statistics, probability, and measure theory. It is quite conceivable that "computer science" is connected to "statistics" at some threshold value, "statistics" to "probability", and "probability" to "measure theory", thereby giving rise to the fact that "computer science" and "measure theory" find themselves, undesirably, in the same cluster. This is due to the intermediaries "statistics" and "probability".

3.2.2 Agglomerative Algorithms

In the last section, a general agglomerative algorithm was discussed. A wide range of these algorithms have been proposed at one time or another. Hierarchic agglomerative algorithms may be conveniently broken down into two groups of methods. The first group is that of linkage methods – the single, complete, weighted and unweighted average linkage methods. These are methods for which a graph representation can be used.

The second group of hierarchic clustering methods are methods which allow the cluster centres to be specified (as an average or a weighted average of the member vectors of the cluster). These methods include the centroid, median and minimum variance methods.

	1	2	3	4	5
1	0	4	9	5	8
2	4	0	6	3	6
3	9	6	0	6	3
4	5	3	6	0	5
5	8	6	3	5	0

	1	2U4	3	5
1	0	4	9	8
2U4	4	0	6	5
3	9	6	0	3
5	8	5	3	0

Agglomerate 2 and 4 at
dissimilarity 3.

Agglomerate 3 and 5 at
dissimilarity 3.

	1	2U4	3U5
1	0	4	8
2U4	4	0	5
3U5	8	5	0

	1U2U4	3U5
1U2U4	0	5
3U5	5	0

Agglomerate 1 and 2U4 at
dissimilarity 4.

Finally agglomerate 1U2U4 and
3U5 at dissimilarity 5.

Resulting
dendrogram:

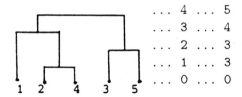

... 4 ... 5
... 3 ... 4
... 2 ... 3
... 1 ... 3
... 0 ... 0

Ranks   Criterion
or   values
levels. (or linkage
         weights).

Fig. 5. Construction of a dendrogram by the single linkage method

The latter may be specified either in terms of dissimilarities, alone, or alternatively in terms of cluster centre coordinates and dissimilarities. A very convenient formulation, in dissimilarity terms, which embraces all the hierarchical methods mentioned so far, is the *Lance-Williams dissimilarity update formula*. If points (objects) i and j are agglomerated into cluster $i \cup j$, then we must simply specify the new dissimilarity between the cluster and all other points (objects or clusters). The formula is:

$$d(i \cup j, k) = \alpha_i d(i, k) + \alpha_j d(j, k) + \beta d(i, j) + \gamma \mid d(i, k) - d(j, k) \mid \qquad (29)$$

where α_i, α_j, β, and γ define the agglomerative criterion. Values of these are listed in the second column of Table 2.

In the case of the single link method, using $\alpha_i = \alpha_j = \frac{1}{2}$, $\beta = 0$, and $\gamma = -\frac{1}{2}$ gives us

$$d(i \cup j, k) = \frac{1}{2} d(i, k) + \frac{1}{2} d(j, k) - \frac{1}{2} \mid d(i, k) - d(j, k) \mid \qquad (30)$$

which, it may be verified by taking a few simple examples of three points, i, j, and k, can be rewritten as

$$d(i \cup j, k) = \min \{ d(i, k), d(j, k) \}. \qquad (31)$$

This was exactly the update formula used in the agglomerative algorithm given in the previous section. Using other update formulas, as given in Column 2 of Table 2, allows the other agglomerative methods to be implemented in a very similar way to the implementation of the single link method.

In the case of the methods which use cluster centres, we have the centre coordinates (in Column 3 of Table 2) and dissimilarities as defined between cluster centres (Column 4 of Table 2). The Euclidean distance must be used, initially, for equivalence between the two approaches. In the case of the *median method*, for instance, we have the following (cf. Table 2).

Let **a** and **b** be two points (i.e. m-dimensional vectors: these are objects or cluster centres) which have been agglomerated, and let **c** be another point. From the Lance-Williams dissimilarity update formula, using squared Euclidean distances, we have:

$$d^2(a \cup b, c) = \frac{d^2(a, c)}{2} + \frac{d^2(b, c)}{2} - \frac{d^2(a, b)}{4} = \frac{\|\mathbf{a} - \mathbf{c}\|^2}{2} + \frac{\|\mathbf{b} - \mathbf{c}\|^2}{2} - \frac{\|\mathbf{a} - \mathbf{b}\|^2}{4}. \qquad (32)$$

The new cluster centre is $(\mathbf{a} + \mathbf{b})/2$, so that its distance to point **c** is

$$\|\mathbf{c} - \frac{\mathbf{a} + \mathbf{b}}{2}\|^2. \qquad (33)$$

That these two expressions are identical is readily verified. The correspondence between these two perspectives on the one agglomerative criterion is similarly proved for the centroid and minimum variance methods.

The single linkage algorithm discussed in the last section, duly modified for the use of the Lance-Williams dissimilarity update formula, is applicable for all agglomerative strategies. The update formula listed in Table 2 is used in Step 2 of the algorithm.

The following properties make the minimum variance agglomerative strategy particularly suitable for synoptic clustering:

Table 2. Specifications of seven hierarchical clustering methods

Hierarchical clustering methods (and aliases).	Lance and Williams dissimilarity update formula.	Coordinates of centre of cluster, which agglomerates clusters i and j.	Dissimilarity between cluster centres g_i and g_j.
Single link (nearest neighbour).	$\alpha_i = 0.5$ $\beta = 0$ $\gamma = -0.5$ (More simply: $min\{d_{ik}, d_{jk}\}$)		
Complete link (diameter).	$\alpha_i = 0.5$ $\beta = 0$ $\gamma = 0.5$ (More simply: $max\{d_{ik}, d_{jk}\}$)		
Group average (average link, UPGMA).	$\alpha_i = \frac{|i|}{|i|+|j|}$ $\beta = 0$ $\gamma = 0$		
McQuitty's method (WPGMA).	$\alpha_i = 0.5$ $\beta = 0$ $\gamma = 0$		
Median method (Gower's, WPGMC).	$\alpha_i = 0.5$ $\beta = -0.25$ $\gamma = 0$	$g = \frac{g_i + g_j}{2}$	$\|g_i - g_j\|^2$
Centroid (UPGMC).	$\alpha_i = \frac{|i|}{|i|+|j|}$ $\beta = -\frac{|i||j|}{(|i|+|j|)^2}$ $\gamma = 0$	$g = \frac{|i|g_i + |j|g_j}{|i|+|j|}$	$\|g_i - g_j\|^2$
Ward's method (minimum variance, error sum of squares.	$\alpha_i = \frac{|i|+|k|}{|i|+|j|+|k|}$ $\beta = -\frac{|k|}{|i|+|j|+|k|}$ $\gamma = 0$	$g = \frac{|i|g_i + |j|g_j}{|i|+|j|}$	$\frac{|i||j|}{|i|+|j|}\|g_i - g_j\|^2$

Notes: $|i|$ is the number of objects in cluster i; g_i is a vector in m-space (m is the set of attributes), — either an intial point or a cluster centre; $\|.\|$ is the norm in the Euclidean metric; the names UPGMA, etc. are due to Sneath and Sokal (1973); finally, the Lance and Williams recurrence formula is:

$$d_{i\cup j.k} = \alpha_i d_{ik} + \alpha_j d_{jk} + \beta d_{ij} + \gamma \mid d_{ik} - d_{jk} \mid .$$

1. Since it is based on a decomposition of the variance of the points, this method shares the same mathematical framework as principal components analysis.
2. The two properties of cluster homogeneity and cluster separability are incorporated in the cluster criterion. For summarizing data, it is unlikely that more suitable criteria could be devised.
3. As in the case of other geometric strategies, the minimum variance method defines a cluster centre of gravity. This mean set of cluster members' coordinate values is the most useful summary of the cluster.

Further details on this method may be found in Murtagh and Heck (1987).

3.2.3 Partitioning Methods

We will conclude this section with a short look at non-hierarchical clustering methods.

A large number of assignment algorithms have been proposed. The single-pass approach usually achieves computational efficiency at the expense of precision, and there are many iterative approaches for improving on crudely-derived partitions.

The following is an example of a single-pass algorithm. The general principle followed is: make one pass through the data, assigning each object to the first cluster which is close enough, and making a new cluster for objects that are not close enough to any existing cluster.

Single-pass overlapping cluster algorithm

Input: n objects, threshold t, dissimilarity on objects.

Step 1: Read object 1, and insert object 1 in membership list of cluster 1. Let representative of cluster 1 be given by object 1. Set i to 2.

Step 2: Read i^{th} object. If $diss(\ i^{th}$ object, cluster $j\ ) \leq t$, for any cluster j, then include the ith object in the membership list of cluster j, and update the cluster representative vector to take account of this new member. If $diss(\ i^{th}$ object, cluster $j) > t$, for all clusters j, then create a new cluster, placing the ith object in its membership list, and letting the representative of this cluster be defined by the ith object.

Step 3: Let i to $i + 1$. If $i \leq n$, go to Step 2.

The cluster representative vector used is usually the mean vector of the cluster's members; in the case of binary data, this representative might then be thresholded to have 0 and 1 coordinate values only. In Step 2, it is clear that overlapping clusters are possible in the above algorithm. In the worst case, if threshold t is chosen too low, all n objects will constitute clusters and the number of comparisons to be carried out will be $O(n^2)$. The dependence of the algorithm on the given sequence of objects is an additional disadvantage of this algorithm. However, its advantages are that it is conceptually very simple, and for a suitable choice of threshold will probably not require large processing time. In practice, it can be run for a number of different values of t.

As a non-hierarchic strategy, it is hardly surprising that the variance criterion has always been popular (for some of the same reasons as were seen in Sect. 3.2.2 for the hierarchical approach based on this criterion). We may, for instance, minimize the following sum of squared deviations which is often considered as tantamount to the within-class variance

$$V_{opt} = \min_P \sum_{p \in P} \sum_{i \in p} ||\mathbf{i} - \mathbf{p}||^2 \tag{34}$$

where the partition P consists of classes p of centre $\mathbf{p}$, and we desire the minimum of this criterion over all possible partitions, P. To avoid a nontrivial outcome (e.g. each class being singleton, giving zero totalled within class variance), the number of classes (k) must be set.

A hierarchical clustering, using the minimum variance criterion, then provides a solution, – not necessarily optimal – at level $n - k$ when n objects are being processed. An alternative approach uses iterative refinement, as follows.

Iterative optimization algorithm for the variance criterion
Step 1: Arbitrarily define a set of k cluster centres.
Step 2: Assign each object to the cluster to which it is closest (using the Euclidean distance, $d^2(i, p) = ||\mathbf{i} - \mathbf{p}||^2$).
Step 3: Redefine cluster centres on the basis of the current cluster memberships.
Step 4: If the totalled within class variances is better than at the previous iteration, then return to Step 2.

We have omitted in Step 4 a test for convergence (the number of iterations should not exceed e.g. 25). Cycling is also possible between solution states. This algorithm could be employed on the results (at level $n - k$) of a hierarchic clustering in order to improve the partition found. It is however a suboptimal algorithm – the minimal distance strategy used by this algorithm is clearly a *sufficient* but not a *necessary* condition for an optimal partition.

The initial cluster centres may be chosen arbitrarily (for instance, by averaging a small number of object-vectors); or they may be chosen from prior knowledge of the data. In the latter case we may for example "manually" choose a small set of stars and galaxies, determine their (parameter-space) centres of gravity, and use these as the basis for the classification of objects derived from a digitized image.

Yet another approach to optimizing the same minimum variance criterion is the exchange method.

Exchange method for the minimum variance criterion
Step 1: Arbitrarily choose an initial partition.
Step 2: For each $i \in p$, see if the criterion is bettered by relocating i in another class q. If this is the case, we choose class q such that the criterion V is least; if it is not the case, we proceed to the next i.
Step 3: If the maximum possible number of iterations has not been reached, and if at least one relocation took place in Step 2, return again to Step 2.

Two remarks may be made: we will normally require that an object not be removed from a singleton class; and the change in variance brought about by relocating object i from class p to class q can be shown to be

$$\frac{|p|}{|p| - 1}||\mathbf{i} - \mathbf{p}||^2 - \frac{|q|}{|q| - 1}||\mathbf{i} - \mathbf{q}||^2 \tag{35}$$

Hence, if this expression is positive, i ought to be relocated from class p (to class q if another, better, class is not found).

Späth (1985) offers a thorough treatment of iterative optimization algorithms of the sort described.

In terminating, it is necessary to make some suggestion as to when these partitioning algorithms should be used in preference to hierarchical algorithms. We have seen that the number of classes must be specified, as also the requirement that each class be non-empty. A difficulty with iterative algorithms, in general, is the requirement for parameters to be set in advance (Anderberg 1973 describes a version of the ISODATA iterative clustering method which requires 7 pre-set parameters). As a broad generalization, it may thus be asserted that iterative algorithms ought to be considered when the problem is clearly defined in terms of numbers and other characteristics of clusters; but hierarchical routines often offer a more general-purpose and user-friendly option.

3.3 Short Bibliography

For a general introduction to this topic the reader is referred to Murtagh and Heck (1987), Gordon (1981) and Everitt (1980).

Anderberg, M.R. (1973): *Cluster Analysis for Applications*, Academic Press, New York

Barrow, J.D., Bhavsar, S.P., Sonoda, D.H. (1985): "Minimal spanning trees, filaments and galaxy clustering", *Mon. Not. R. astr. Soc.*, **216**, 17–35

Bianchi, R., Butler, J.C., Coradini, A., Gavrishin, A.I. (1980): "A classification of lunar rock and glass samples using the G-mode central method", *The Moon and the Planets*, **22**, 305–322

Butchins, S.A. (1982): "Automatic image classification", *Astron. Astrophys.*, **109**, 360–365

Coradini, A., Fulchignoni, M., Gavrishin, A.I. (1976): "Classification of lunar rocks and glasses by a new statistical technique", *The Moon*, **16**, 175–190

Carusi, A., Massaro, E.: (1978): "Statistics and mapping of asteroid concentrations in the proper elements' space", *Astron. Astrophys. Suppl. Ser.*, **34**, 81–90

Cowley, C.R., Henry, R.: (1979): "Numerical taxonomy of Ap and Am stars", *Astrophys. J.*, **233**, 633–643

Davies, J.K., Eaton, N., Green, S.F., McCheyne, R.S., Meadows, A.J. (1882): "The classification of asteroids", *Vist. Astron.*, **26**, 243–251, 1982.

Everitt, B. (1980): *Cluster Analysis*, Heinemann Educational Books, London

Giovannelli, F., Coradini, A., Lasota, J.P., Polimene, M.L. (1981): "Classification of cosmic sources: a statistical approach", *Astron. Astrophys.*, **95**, 138–142

Gordon, A.D. (1981): *Classification*, Chapman and Hall, London

Heck, A., Albert, A., Defays, D., Mersch, G. (1977): "Detection of errors in spectral classification by cluster analysis", *Astron. Astrophys.*, **61**, 563–566

Heck, A., Egret, D., Nobelis, Ph., Turlot, J.C. (1986): "Statistical confirmation of the UV spectral classification system based on IUE low-dispersion stellar spectra", *Astrophys. Space Sci.*, **120**, 223–237

Huchra, J.P., Geller, M.J. (1982): "Groups of galaxies. I. Nearby groups", *Astrophys. J.*, **257**, 423–437

Jarvis, J.F., Tyson, J.A. (1981): "FOCAS: faint object classification and analysis system", *Astron. J.*, **86**, 476–495

Jasniewicz, G. (1984): "The Böhm-Vitense gap in the Geneva photometric system", *Astron. Astrophys.*, **141**, 116–126

Materne, J. (1978): "The structure of nearby clusters of galaxies. Hierarchical clustering and an application to the Leo region", *Astron. Astrophys.*, **63**, 401–409

Mennessier, M.O. (1985): "A classification of miras from their visual and near-infrared light curves: an attempt to correlate them with their evolution", *Astron. Astrophys.*, **144**, 463–470

Moles, M., del Olmo, A., Perea, J. (1985): "Taxonomical analysis of superclusters. I. The Hercules and Perseus superclusters", *Mon. Not. R. astr. Soc.*, **213**, 365–380

Murtagh, F., Heck, A. (1987): *Multivariate Data Analysis*, Kluwer, Dordrecht

Paturel, G. (1979): "Etude de la région de l'amas Virgo par taxonomie", *Astron. Astrophys.*, **71**, 106–114

Perea, J., Moles, M., del Olmo, A. (1986): "Taxonomical analysis of the Cancer cluster of galaxies", *Mon. Not. R. astr. Soc.*, **222**, 49–53

Pirenne, B., Ponz, D., Dekker, H. (1985): "Automatic analysis of interferograms", *The Messenger*, No. 42, 2–3

Späth, H. (1985): *Cluster Dissection and Analysis: Theory, Fortran Programs, Examples*, Ellis Horwood, Chicester

4 Graph-Theoretic Methods

4.1 The Problem

Since a graph representation captures the notion of relative separation, it provides a powerful tool for clustering-type problems. A number of such methods will be looked at, together with their applicability for processing point patterns. The latter do not need to be in the plane, and methods described below generalize easily to higher dimensions.

4.2 Mathematical Description

4.2.1 Terminology

A graph G is a set (V, E) of vertices and edges. The set of edges is often weighted, and one may therefore have

$$E : V \times V \longrightarrow \mathbb{R}^+ \qquad (36)$$

i.e. E is the set of all pairs (i, j) such that $i, j \in V$, and the associated weight $w(i, j)$ is a positive real. If the graph is *directed*, then nodes and arcs replace vertices and edges.

A *planar graph* is one which can be represented in the plane, without cross-overs of edges. A planar graph can be shown to have less than or equal to $3n - 6$ edges, when n vertices are in question. This provides a bound on such structures as the Delauney triangulation, to be looked at below.

A *path* in a graph is a sequence of distinct edges of the form $(x_1, x_2), (x_2, x_3), \ldots$. A *circuit* is a path whose final vertex is the same as its first. A *Hamiltonian* path, or circuit, is a path or circuit which traverses every vertex once and once only (save for the last vertex in the circuit case). The Hamiltonian path or circuit problem is not easily solvable. The decision as to whether or not a given graph has a Hamiltonian circuit belongs to the difficult class of problems which are known as NP-complete. A polynomial-time algorithm for this problem is therefore not thought to exist. The *travelling salesperson problem* (TSP) is an alternative term for the Hamiltonian circuit problem. A salesperson must visit each and every customer, and not double back on any occasion. The graph of this problem may not be *complete*: there may be less than $n(n-1)/2$ vertices, where $n = |V|$, the cardinality of the vertex set. There is a relationship between the TSP and the minimal spanning tree (MST), which is looked at in the next section. In fact, as discussed by Garey and Johnson (1979, p. 131), the MST can be used in a clever way to provide an approximate TSP where the total length of the latter is no worse than twice the total weight of the MST.

4.2.2 Minimal Spanning Tree

A *tree* is a graph which does not contain a circuit: it is therefore a "skeleton" graph. A *spanning tree* is one which includes every vertex. A *minimal spanning tree* (MST) has minimal totalled edge weights, when compared to any other possible spanning tree. The MST has proven to be a useful data structure for picking out plausible clusters of points (vertices) or for detecting outlier points.

Figure 6 shows aspects of the construction of a MST on five vertices. Read the set of boxes on the left hand-hand side in a top-to-bottom direction, noticing that edges are drawn if the associated weights are less than a threshold, which is being raised from box to box. The vertices are considered as five points in the usual Euclidean plane. Thus the edge weights are identical in this case to Euclidean distances between appropriate points. The closely related single linkage hierarchical clustering is also shown. The thresholded graphs define *components*, i.e. connected subsets of the set of vertices. The MST can always be determined in $O(n^2)$ time. In fact, for various types of data, and using various speed-up tactics, the MST may be determined in an even more favourable order-of-magnitude time. A number of references for efficient algorithms for the MST are given in Murtagh (1985).

The MST is used for detecting separated clusters of points, or even clusters of different densities by studying the $n-1$ edge weights in the MST. Note that although Fig. 6 showed a 2-dimensional example, the MST can easily be constructed in 3-dimensional space or any arbitrary m-dimensional one. If we find some very large edge weight in the MST, then probably by cutting this edge of the MST, we are left with two clearly separated clusters. Zahn (1981) defines an *inconsistent edge* as one whose weight is significantly greater than the average of nearby edge weights at either extremity. Nearby edges may be defined, e.g., as all edges within path-length 2 of the end-vertices of the examined edge.

An operation on the MST to avoid the effects of anomalous and otherwise noisy points is *pruning*. "Hairs" are removed from the MST by removing edges with the following property: vertices of degree 1 connected to vertices of degree 3 or greater.

If we find that some of the MST's edges are of small weight, and some of large weight, using e.g. a histogram of edge weights, then this signals clusters of differing densities.

Some examples of point patterns where the MST can perform well are shown in Fig. 7.

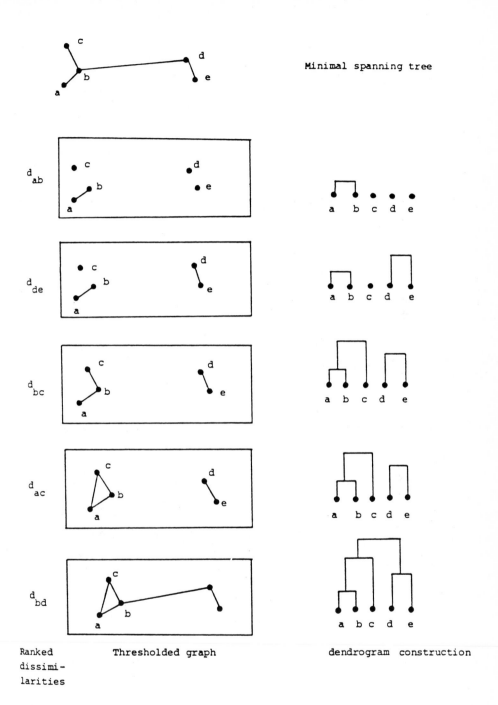

Minimal spanning tree

d_{ab}

d_{de}

d_{bc}

d_{ac}

d_{bd}

Ranked
dissimi-
larities

Thresholded graph

dendrogram construction

Fig. 6. The minimal spanning tree and single link hierarchy

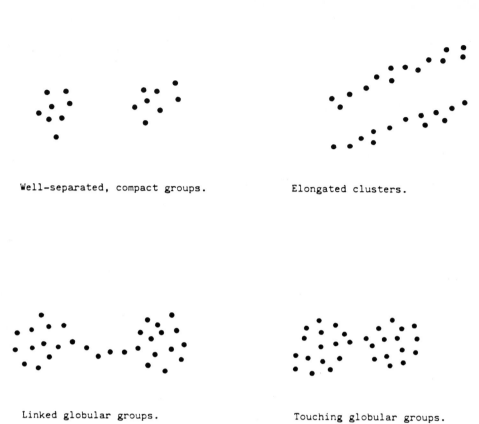

Well-separated, compact groups.

Elongated clusters.

Linked globular groups.

Touching globular groups.

Concentric groups.

Groups characterised by differing densities.

Fig. 7. Point patterns which the MST can aid in analyzing

4.2.3 Other Graph Methods for Analysing Point Patterns

The MST is a robust and flexible method which does not require excessive computation time or storage space. Other data structures have also been proposed for the analysis of point patterns.

The *Relative Neighborhood Graph* (RNG) is defined as follows. Points i and j are relative neighbors if

$$d(i, j) \leq \max\{d(i, k), d(j, k)\} \tag{37}$$

for all $k = 1, \ldots, n, k \neq i, j$. Points i and j are therefore at least as close to one another as they are to any other point. This implies that relative neighbors are such that their *lune* (see Fig. 8) is empty. The RNG is the graph for which edges are defined between relative neighbors. The RNG results in a greater number of edges than does the MST.

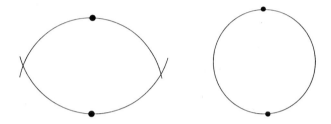

Fig. 8. A pair of relative neighbors, indicating their *lune*; and a vertex-pair in a Gabriel Graph indicating their *disk*

Another graph structure is the *Gabriel Graph* (GG) or least-squares adjacency graph. An edge in graph GG exists whenever

$$d^2(i, j) \leq d^2(i, k) + d^2(j, k) \tag{38}$$

for $k \neq i, j$. The region of exclusion, here, is defined by a disk (Fig. 8).

If DT is the Delauney triangulation (to be looked at below), we have the following result:

$$\text{MST} \subseteq \text{RNG} \subseteq \text{GG} \subseteq \text{DT} \tag{39}$$

i.e. an edge in the MST is necessarily an edge in the RNG of the same graph, but the reverse does not need to be valid; and such a relationship holds for the other graph structures also.

The *Voronoi diagram* or *Dirichlet tesselation* is defined as follows. Consider the convex polygonal region

$$\{x \mid d(x, i) < d(x, j)\} \quad \text{for all points } i \neq j \tag{40}$$

where i and j belong to a finite set of n points in the plane or in a space of higher dimension. Such polygons may be bounded or may extend to infinity (Fig. 9). They are constructed from the intersection of the perpendicular bisectors of the line joining point i with point j. The polygon could be described as the "territory" of point i: it is the part of the plane nearer to i than to any other point j.

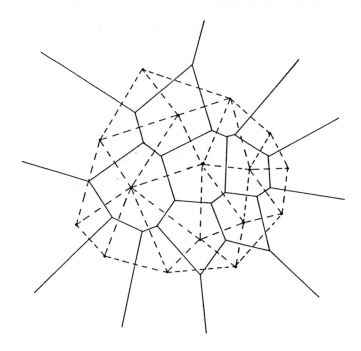

Fig. 9. The complete lines show the Voronoi diagram of 16 points; the hatched lines show the Delauney triangulation

A Voronoi diagram on n points has at most $2n - 5$ vertices and $3n - 6$ edges. The *dual* of the Voronoi diagram is known as the *Delauney triangulation* (Fig. 9).

Interest in the Voronoi diagram has arisen from the possible meaning which can be attributed to the polygons surrounding each point. In ecology, animal territorial rights might prevail. The Voronoi diagram has also been used as a model encompassing information relating to pancakes, filaments, voids and clusters of galaxies. It has been used, in an interesting fashion, as a basic framework within which the basic laws of galaxy evolution take place.

The Voronoi diagram has been generalized in various ways: to metrics other than the Euclidean one; to the Voronoi diagram of circles and lines rather than points; k-order Voronoi diagrams are based on k-nearest neighbors; weighted Voronoi diagrams define tiles in terms of weighted distances, i.e. the regions of influence defined by the polygons are not due to the same "expansion factors" in all tiles (Aurenhammer and Edelsbrunner 1984); and finally the problem of Voronoi diagrams in 3-dimensional space ("Voronoi foam") and higher dimensions has been tackled. These areas are briefly surveyed in Lee and Preparata (1984). For computational issues in 2-dimensional space, see Preparata and Shamos (1985).

4.3 Short Bibliography

As a general reference to the topics that follow, the reader is referred to:
Tucker, A. (1980): *Applied Combinatorics*, Wiley, New York

Minimal spanning tree:

Barrow, J.D., Bhavsar, S.P., Sonoda, D.H. (1985): "Minimal spanning trees, filaments and galaxy clustering", *Mon. Not. R. astr. Soc.*, **216**, 17–35
Bhavsar, S.P., Ling, E.N. (1988): "Are the filaments real?", *Astrophys. J. (Lett.)*, **331**, L63–L68
Garey, M.R., Johnson, D.S. (1979): *Computers and Intractability. A Guide to the Theory of NP-Completeness*, W.H. Freeman, San Francisco
Murtagh, F. (1985): *Multidimensional Clustering Algorithms*, COMPSTAT Lectures Volume 4, Physica-Verlag, Vienna and Würzburg
Zahn, C.T. (1971): "Graph-theoretical methods for detecting and describing Gestalt clusters", *IEEE Transactions on Computers*, **C-20**, 68–86

Voronoi diagrams:

Ahuja, N. (1982): "Dot pattern processing using Voronoi neighborhoods", *IEEE Transactions on Pattern Analysis and Machine Intelligence*, **PAMI-4**, 336–343
Aurenhammer, F., Edelsbrunner, H. (1984): "An optimal algorithm for constructing the weighted Voronoi diagram in the plane", *Pattern Recognition*, **17**, 251–257
Bhaskar, S.K., Rosenfeld, A., Wu, A. (1989): "Models for neighbor dependency in planar point patterns", *Pattern Recognition*, **22**, 533–559
Icke, V., van de Weygaert, R. (1987): "Fragmenting the universe. I. Statistics of two-dimensional Voronoi foams", *Astron. Astrophys.*, **184**, 16–32
Icke, V., van de Weygaert, R. (1990): " The Galaxy Distribution as a Voronoi Foam", *Quart. J. R. astr. Soc.*, **32**, 85–112
Lee, D.T., Preparata, R.P. (1984): "Computational geometry – a survey", *IEEE Transactions on Computers*, **C-33**, 1072–1101
Preparata, R.P. Shamos, M.I.(1985): *Computational Geometry*, Springer-Verlag, New York

Other graph-theoretic structures:

Matula, D.W., Sokal, R.R. (1980): "Properties of Gabriel Graphs relevant to geographic variation research and the clustering of points in the plane", *Geographical Analysis*, **12**, 205–222
Urquhart, R. (1982): "Graph theoretical clustering based on limited neighborhood sets", *Pattern Recognition*, **15**, 173–187 ("Erratum" (1982): *Pattern Recognition*, **15**, 427)

Printing: Druckerei Zechner, Speyer
Binding: Buchbinderei Kränkl, Heppenheim

I. **Appenzeller,** University of Heidelberg, **H.J. Habing,** Leiden; **P. Léna,** University of Paris (Eds.)

Evolution of Galaxies
Astronomical Observations

Proceedings of the Astrophysics School I, Organized by the European Astrophysics Doctoral Network at Les Houches, France, 5–16 September 1988

1989. X, 391 pp. (Lecture Notes in Physics. Managing Ed.: W. Beiglböck. Eds.: H. Araki, E. Brézin, J. Ehlers, U. Frisch, K. Hepp, R.L. Jaffe, R. Kippenhahn, H.A. Weidenmüller, J. Wess, J. Zittartz. Vol. 333) Hardcover DM 76,–   ISBN 3–540–51315–9

These eight lectures have been written up in a clear and pedagogical style in order to serve as an introduction for students to fields of modern astrophysical and astronomical research where textbooks are otherwise not available. The first four lectures cover topics in galactic astronomy (formation, structure and evolution of galaxies) and the remaining four are devoted to observational methods and astronomical instrumentation. The lecturers in the European Astrophysics Doctoral Network rank among the most highly respected specialists, and their lectures are carefully edited and updated before publication.

H.M. Maitzen; J. van Paradijs, University of Amsterdam (Eds.)

Galactic High–Energy Astrophysics
High–Accuracy Timing and
Positional Astronomy

EADN Astrophysics School IV, Organized by the European Astrophysics Doctoral Network at Graz, Austria 1991

C.B. de Loore, University of Brussels (Ed.)

Late Stages of Stellar Evolution Computational Methods in Astrophysical Hydrodynamics

Proceedings of the Astrophysics School II, Organized by the European Astrophysics Doctoral Network at Ponte de Lima, Portugal, 11–23 September 1989

1991. VIII, 390 pp. (Lecture Notes in Physics. Managing Ed.: W. Beiglböck. Eds.: H. Araki, E. Brézin, J. Ehlers, U. Frisch, K. Hepp, R.L. Jaffe, R. Kippenhahn, H.A. Weidenmüller, J. Wess, J. Zittartz. Vol. 373) Hardcover DM 73,–   ISBN 3–540–53620–5

This collection of 7 lectures is intended as a textbook for graduate students who want to learn about modern developments in astronomy and astrophysics. The first part surveys various aspects of the late stages of stellar evolution, including observation and theory. B.C. de Loore's long article on stellar structure is followed by reviews on supernovae, circumstellar envelopes, and the evolution of binaries. The second part deals with the important problem of modelling stellar evolution based on computational hydrodynamics.

S. Beckwith; T. Ray, Dublin Institute for Advanced Studies (Eds.)

Star Formation and Techniques in Infrared and mm–Wave Astronomy

EADN Astrophysics School V, Organized by the European Astrophysics Doctoral Network at Berlin, Germany, 21 September – 2 October 1992

Lecture Notes in Physics

For information about Vols. 1–374
please contact your bookseller or Springer-Verlag

New Series m: Monographs